Brief Candle
in the Dark

ALSO BY RICHARD DAWKINS

The Selfish Gene

The Extended Phenotype

The Blind Watchmaker

River Out of Eden

Climbing Mount Improbable

Unweaving the Rainbow

A Devil's Chaplain

The Ancestor's Tale

The God Delusion

The Greatest Show on Earth

The Magic of Reality (*with Dave McKean*)

An Appetite for Wonder

www.richarddawkins.net/bcd

Brief Candle in the Dark

My Life in Science

Richard Dawkins

HARPER LUXE

An Imprint of HarperCollins*Publishers*

HarperCollins books may be purchased for educational, business, or sales promotional use. For information, please e-mail the Special Markets Department at SPsales@harpercollins.com.

FIRST HARPERLUXE EDITION

HarperLuxe™ is a trademark of HarperCollins Publishers

Library of Congress Cataloging-in-Publication Data has been applied for.

ISBN 978-0-06-241699-5

15 16 17 18 19   ID/RRD   10 9 8 7 6 5 4 3 2 1

For Lalla

'Out, out, brief candle: Life's but a walking shadow, a poor player, that struts and frets his hour upon the stage and then is heard no more . . .'
William Shakespeare, *Macbeth*, Act V, scene v

'Science as a candle in the dark'
Carl Sagan, subtitle to *The Demon-haunted World* (1995)

'Better to light a candle than curse the darkness'
Anon.

Contents

Brief Candle
in the Dark

Flashback at a feast

What am I doing here in New College Hall, about to read my poem to a hundred dinner guests? How did I get here–a subjective 25-year-old, objectively bewildered to find himself celebrating his seventieth circuit of the sun? Looking around the long candlelit table with its polished silver and sparkling wineglasses, reflecting flashes of wit and sparkling sentences, I indulge my mind in a series of quick-firing flashbacks.

Back to childhood in colonial Africa amid big, lazy butterflies; the peppery taste of nasturtium leaves stolen from the lost Lilongwe garden; the taste of mango, more than sweet, spiced by a whiff of turpentine and sulphur; boarding school in the pine-scented Vumba mountains of Zimbabwe, and then, back 'home' in England, beneath the heavenward spires of

Salisbury and Oundle; undergraduate days damsel-dreaming among Oxford's punts and spires, and the dawning of an interest in science and the deep philosophical questions which only science can answer; early forays into research and teaching at Oxford and Berkeley; the return to Oxford as an eager young lecturer; more research (mostly collaborating with my first wife Marian, whom I can see at the table here in New College), and then my first book, *The Selfish Gene*. Those swift memories take me to the age of thirty-five, halfway to today's landmark birthday. They milestone the years covered by my first book of memoirs, *An Appetite for Wonder*.

My thirty-fifth birthday recalled to me an article by the humorist Alan Coren about his own. Coren was mock-depressed by the thought that he had reached half-time and it was now downhill all the way. I didn't feel the same, perhaps because I was just putting the finishing touches to my first, rather youthful book and was looking forward to publication and its aftermath.

One aspect of that aftermath was being pitchforked by the unexpectedly high sales of the book into the company of those who are regularly asked, by journalists scratching around for column inches, to list their ideal gathering of dinner-party guests. In the days when I used to respond to that kind of request I would invite some great scientists as a matter of course, but also writers and creative spirits of all kinds. Indeed, any of those lists would probably have included at least fifteen of those actually attending my birthday dinner today, among them novelists, playwrights, television personalities, musicians, comedians, historians, publishers, actors and multinational business tycoons.

Thirty-five years ago, I think to myself as I spot familiar faces around the table, such a mix of literary and artistic guests at a scientist's birthday dinner would have seemed unlikely. Has the *Zeitgeist* changed since C. P. Snow lamented the chasm between scientific and literary culture? What has happened in the years covered by my dinnertime musing? My reverie jumps me into the middle of the period and I conjure the giant, unforgettable figure of Douglas Adams, sadly absent from the feast. In 1996, when I was fifty-five and he ten years younger, I had a televised conversation with him for a Channel Four documentary called *Break the Science Barrier*. The programme's purpose was precisely to show that science needed to burst into the wider culture, and my interview with Douglas was the high spot. Here is part of what he said:

I think the role of the novel has changed a little bit. In the nineteenth century, the novel was where you went to get your serious reflections and question-ings about life. You'd go to Tolstoy and Dos-toyevsky. Nowadays, of course, you know the scientists actually tell much more about such issues than you would ever get from novelists. So I think that for the real solid red meat of what I read I go to science books, and read some novels for light relief.

Could this be a part of what has changed? Have nov-elists, journalists and others of the kind C. P. Snow would have planted firmly in his first culture increas-ingly come to embrace the second? Might Douglas, if he were still living, now go back to the novel and un-cover some of what he had moved to science to find, twenty-five years after he read English at Cambridge: Ian McEwan, say, or A. S. Byatt? Or other novelists who love science, such as Philip Pullman or Martin Amis, William Boyd or Barbara Kingsolver? Then there are highly successful science-inspired plays, in the tradition of Tom Stoppard and Michael Frayn. Could this starry dinner party, put together for me by my wife Lalla Ward, an artist and actress herself well read in science, be some kind of symbol of a change in the culture—as well as a personal landmark in my life? Are we witnessing a constructive merger between sci-entific and literary cultures, perhaps the 'third' culture towards which my literary agent John Brockman has been working behind the scenes as he nurses his online *salon* and grows his glittering list of science authors? Or the merging of cultures that I aspired to in my own *Unweaving the Rainbow* where, under the influence of Lalla, I tried to reach out to the world of literature and bridge the gap from science? *Où sont les C. P. Snows d'antan?*

Two anecdotes tell (and if you don't like digressive anecdotes you might find you're reading the wrong book). One of my guests at this New College dinner, the explorer and adventurer Redmond O'Hanlon, author of grotesquely funny travel books such as *Into the Heart of Borneo* and *In Trouble Again*, would throw literary parties and dinners with his wife Belinda to which, it seemed, the whole of literary London was invited. Novelists and critics, journalists and editors, poets and publishers, literary agents and literary lions descended on their remote corner of the Oxfordshire countryside, the house crammed full of stuffed snakes, shrunken heads, leathery corpses and leatherbound books, exotic curios of anthropology and—one suspects—anthropophagy. These evenings were always notable for the company and, when it included Salman Rushdie, for the bodyguards keeping their own company upstairs.

On one of these occasions, Lalla and I happened to have staying with us Nathan Myhrvold, Chief Technology Officer at Microsoft and one of the most inventive geeks of Silicon Valley. Nathan is a mathematical physicist by training. After his PhD at Princeton he worked in Cambridge with Stephen Hawking during the period when Stephen was still just capable of talking, though not intelligibly to any but his close associates, who acted as interpreters for the benefit of the rest of the world. Nathan became one of these highly qualified physicist amanuenses. True to his promise, he is now one of high tech's most innovative thinkers. When Redmond and Belinda invited us we told them we had a house guest and they, hospitable as ever, told us to bring him along.

Nathan is too polite to monopolize a conversation. His neighbours at the long table presumably asked him what he did, and the conversation grew into a discussion of string theory and other arcana of modern physics. And the literary illuminati were held spellbound. They began as usual, no doubt, exchanging aphoristic wit with their neighbours. But inexorably a wave of curious interest in science swept from Nathan's end down the long table, and the evening became a kind of informal seminar on the weirdness of modern physics. When a seminar includes intelligences of the calibre of this bunch of dinner companions, interesting things happen. Lalla and I basked in reflected glory as sponsors of the unexpected guest at this archetypal 'third culture' evening. And afterwards Redmond telephoned and told Lalla that never, in all his years of throwing such parties, had he seen his eminent literary guests reduced to such stunned silence.

The second anecdote is almost a mirror image. The playwright and novelist Michael Frayn came, with his wife the distinguished writer Claire Tomalin, to stay with Lalla and me when his remarkable play *Copenhagen* was being performed at the Oxford Playhouse. The play is about the relationship between two giants of modern physics, Niels Bohr and Werner Heisenberg, and about a riddle in the history of science: why did Heisenberg visit Bohr in Copenhagen in 1941, and what was Heisenberg's role in the war (see also page 283)? After the performance, Michael was conducted to an upper room in the Playhouse, where he was quizzed by the assembled physicists of Oxford. It was a privilege to hear this aristocrat of literary and philosophical learning field questions from the cream of Oxford scientists, including several fellows of the Royal Society. Once again, an evening for third culture warriors to treasure: an evening to surprise–and delight–the C. P. Snow of thirty years earlier.

I dare to hope that my books, starting with *The Selfish Gene* in 1976, are among those that changed the cultural landscape, along with the works of Stephen

Hawking, Peter Atkins, Carl Sagan, Edward O. Wilson, Steve Jones, Stephen Jay Gould, Steven Pinker, Richard Fortey, Lawrence Krauss, Daniel Kahneman, Helena Cronin, Daniel Dennett, Brian Greene, the two M. Ridleys (Mark and Matt), the two Sean Carrolls (physicist and biologist), Victor Stenger and others, and the critical and journalistic buzz that they engendered. I'm not talking about science journalists explaining science to the lay public, although that's a good thing as well. I'm talking about books by professional scientists, aimed at professionals in their own and other disciplines, but written in language such that the general reading public can look over their shoulders and join in. I'd like to think that I might have been one of those who helped to kick-start this 'third culture'.

Unlike *An Appetite for Wonder*, this second volume of my autobiography is not simply chronological, not even a single flashback from my seventieth birthday. Rather, it is a series of flashbacks divided into themes, punctuated by digression and anecdote. Since we are dispensing with rigid chronology, the order of themes is somewhat arbitrary. I said in the first volume that 'insofar as anything was the making of me, Oxford was': so why not start with my return to those glowing limestone walls?

Oh, the things that are done by a don

From 1970 to 1990 I was University Lecturer in Animal Behaviour in the Zoology Department at Oxford, promoted to Reader from 1990 to 1995. The lecturing duties were not especially onerous, at least by American standards. As well as lecturing on animal behaviour, I was one of those who inaugurated a new option in evolution (evolution had always, naturally, been a core feature of the course, but the new option allowed students the opportunity to take deeper advantage of Oxford's long-time expertise in the subject). In addition to students reading zoology or biological sciences, I lectured to those reading human science and psychology, both of which were honour schools with an examination paper in animal behaviour.

I also gave an annual service course for zoology students on computer programming. Incidentally, this showed up an astonishing *variance* in student ability– a far wider gap between the strongest and the weakest than I ever noticed in the rest of their course. The weakest never really got it, despite my best efforts, and despite the fact that they had no trouble with the non-computational parts of the course. The strongest? Well, Kate Lessells turned up late for the practical class having missed all the sessions in the first half of the term. I protested: 'You've never touched a computer before, and you've missed four weeks. How can you expect to do today's practical exercise?'

'What did you say in the lectures?' was the imperturbable reply of this steady-eyed, tomboyish young woman.

I was baffled: 'You really mean you want me to condense four weeks of lectures into five minutes?'

She nodded, still imperturbable, and with what might have been an ironic half-smile.

'Right,' I said, and I'm not sure whether I thought I was issuing a challenge to her or a challenge to myself: 'You asked for this.' I did condense four hours of lectures into five minutes. She simply nodded after every sentence without taking a note and without speaking a word. Then this formidably clever young woman sat down at the console, completed the exercise and left the room. That, at least, is my memory of what occurred. I may be exaggerating a little, but nothing in Kate's subsequent career leads me to think so.

Apart from lecturing and running practical classes in the Department of Zoology, my other teaching duties consisted of tutorials; these I conducted in New College (it was new in 1379; today one of Oxford's oldest), where I became a fellow in 1970. Most lecturers and professors at Oxford and Cambridge are also fellows of one of the thirty or forty semi-independent colleges or halls that make up these two federal universities. My salary was paid partly by the University of Oxford (where my duties consisted mostly of lecturing and researching in the Department of Zoology) and partly by New College, for which I had to do a minimum six hours' tutoring per week–often of students from other colleges, by exchange arrangement with their own tutors, which was the normal practice in the biological sciences, though not so much in other subjects. Tutorials were usually one-on-one when I started teaching, although paired tutorials gradually became more common. As an undergraduate myself, I had loved the tutorial system, and had far preferred one-on-one tutorials, in which I would read my essay aloud to the tutor, who would either take notes and then discuss it, or interrupt the reading to comment. Nowadays, Oxford tutors are more likely to take two or even three students at a time in the same hour, and essays are typically not read aloud but handed in and read ahead of time.

Through my early years at New College, our pupils were all male. In 1974, those of us in the fellowship who wanted to admit women narrowly failed to secure the necessary two-thirds majority. Some of the opposition was frankly misogynistic. The most deplorable examples have now, thankfully, long receded into the past, so there's no need for me to repeat their awful arguments. I was pleased to be able to deploy statistics in a college meeting to annul some of their more egregious claims about the academic abilities of women.

In fact, we won the first vote in 1974, to change our statutes and make the admission of women *possible*. But–a typical parliamentary manoeuvre–the price of victory was a concession: we agreed to hold a separate, subsequent vote the following term on the *actual* admission of women undergraduates. We presumed that the second vote would go our way too, but it didn't. Whether those opponents who bargained the concession had cleverly foreseen the absence of a crucial voter on sabbatical leave in America, I don't know. But the upshot was that New College was unexpectedly not among the first five male colleges to admit women, although we were among the first to change our statutes to allow us to do so (and the very first, long before my time, formally to debate it). We didn't achieve the final step until 1979, along with the majority of Oxford's other colleges. Though we couldn't admit women students in 1974, our change of statutes did allow us to elect women fellows. Unfortunately the first woman we elected, though a distinguished scholar in her field, showed signs of being pretty misogynistic herself: she wasn't fond of female students or female junior colleagues (as I learned from one of them who became a close friend of mine). We were more fortunate in subsequent elections, and New College is now a flourishing mixed-sex community, with all the benefits that that brings.

New recruits

One of my most grievous responsibilities was to admit young biologists to New College in the first place. What was hard was the obligation to turn down so many good and keen candidates because the competition was so severe. Every November, from all around Britain and far beyond, crowds of eager young people descend on Oxford for their interviews, many of them noticeably shivering with cold in their thin, unaccustomed suits. The colleges put them up in student rooms, which have been vacated by all the incumbent undergraduates except a few volunteers who stay behind as 'shepherds' to look after them, show them around, and try to make sure that cold is all they shiver from.

In addition to interviewing candidates, until it was abolished I also had to read their answers to the Oxford entrance examination, and contribute to setting the questions for that quirkily idiosyncratic paper ('Why do animals have heads?' 'Why has the cow got four legs and the milking stool three?' Neither of those two questions was mine, by the way.) In both the entrance exam and the interviews, we were not after factual knowledge for its own sake. Exactly what we were testing is harder to define: intelligence, yes, but not just IQ-type intelligence. I suppose it was something like 'ability to reason constructively in the particular ways required by the subject', in my case biology: lateral thinking, biological intuition, maybe 'teachability'–even an attempt to assess: 'Would it be a rewarding experience to teach this person? And is this the kind of person who would benefit from an Oxbridge education, especially from our unique tutorial system?'

Here's a digression whose relevance will become apparent. In 1998, I was invited to present the trophy at the final of *University Challenge*. This is a general knowledge quiz show on BBC television, in which representatives of universities (the colleges of Oxford and Cambridge are treated as separate entities for this purpose) compete in a complicated form of knockout competition. The standard of general knowledge displayed can be astonishingly high–the popular *Who Wants to Be a Millionaire?* is very low-grade by comparison, and presumably gains its appeal from the element of gambling for large rewards. In my speech in Manchester, presenting the trophy to the 1998 winners of *University Challenge* (Magdalen College, Oxford, who defeated Birkbeck, London, in the final), I said (according to a quote in Wikipedia, which agrees with my memory):

> I'm conducting a campaign at Oxford with my colleagues to make them abolish the A-Level [the nationwide school-leaving exam, which tests specialist knowledge] as the criterion for admitting students, and substitute *University Challenge*. I'm very serious about this; the kind of mind you need to win *University Challenge*–it's not the knowledge that matters–but it's the retentive mind to pick things up wherever you are that you need at university too.

I told a story of an undergraduate studying history at Oxford, who was unable to find Africa on a map of the world. When I said to a colleague that she should never have been admitted to our (or any) university, he protested that maybe she had missed the relevant geography lesson at school. That is precisely not the point. If you need a geography lesson in order to know where Africa is—if, by age seventeen, you have somehow failed to imbibe such knowledge by osmosis or simple curiosity—you surely don't have the sort of mind that would benefit from a university education. That is an extreme illustration of why I suggested a *University Challenge*-style general knowledge test as part of our entrance procedure: not general knowledge for its own sake, but general knowledge as a litmus test for a teachable mind.

My suggestion (tongue somewhat but not wholly in cheek) has yet to be taken seriously. But Oxford did (still does) make an effort to test more than just factual knowledge narrowly relevant to the proposed subject of study. A typical question that I might ask in an interview (derived from Peter Medawar) was:

> The artist El Greco had a reputation for painting his figures especially long and thin. It's been suggested that the reason for this is that he had a defect

in his vision, which made everything appear stretched in the vertical direction. Do you think that is a plausible theory?

Some students immediately got it, and I would rate them highly: 'No, it's a bad theory because when he looked at his own paintings they would be even *more* stretched.' Some didn't get it at first, but I was able to coax them through a line of reasoning until they did see the point. Some of them were then clearly intrigued, perhaps annoyed with themselves for not seeing the point straight away, and I'd rate them fairly highly too, as teachable. Some might put up a fight, and I'd give them credit for that too: 'Perhaps El Greco's vision was defective only when he looked at more distant objects like the model, not when he looked at close objects like his canvas.' Others just didn't get the point at all, even when I tried to lead them towards it, and I marked them down as less likely to benefit from an Oxford education.

I'm going to expand a little on the kinds of questions that Oxford tutors ask at interview. This is partly because I think the art of interviewing for university entrance is interesting in its own right. But also, if I betray some inside tips it might actually help prospective students hoping to get into one of those (now rather few) universities that still bother to interview.

Here's a similar conundrum to the 'El Greco question' which I would sometimes use:

Why do mirrors reverse from left to right but don't turn the image upside down? And is it a problem in psychology, physics, philosophy or what?

Again, I was mostly testing the teachability of students, their ability to be led through a chain of reasoning even if they didn't immediately solve the puzzle. Actually, that particular puzzle is surprisingly difficult. It helps if you reframe it and think not about a mirror but about a glass door, say a door in a hotel that has LOBBY written on it. When you see it from the other side, it says YᗺᗺO⅃ not ⅃OᗺᗺY. It is easier to explain why this is so for the glass door than for the mirror. Generalizing to the mirror is then straightforward physics: a good example of the value of reframing a problem to make it tractable.

Or I might remind them that the image on our retina is upside down, yet we see the world the right way up. 'Talk me through an explanation for that.' Another favourite question to test their biological intuition began: 'How many grandparents do you have?' Four. 'And how many great-grandparents?' Eight. 'And how many great-great-grandparents?' Sixteen. 'Right then, how many ancestors do you think you had two thousand years ago, in the time of Christ?' The brighter ones tumbled to the fact that you can't go on doubling up indefinitely, because the number of ancestors rapidly overtakes the billions of people in the world now, let alone the comparatively small number that were alive in the time of Christ. That proved to be a good line of reasoning to coax them to the conclusion that we are all cousins, with numerous shared ancestors who lived not so long ago. Another way to put the question might be: 'How far back in time do you think you'd have to go before you hit an ancestor that you and I share?' I treasure the answer given me by one young woman from rural Wales. She looked me up and down with an implacable gaze and then slowly delivered her verdict: 'Back to the apes.'

I'm afraid she didn't get in (although not for that reason). Nor did the young man from a public school[1] who lolled back in his chair (the image of him putting his feet on the desk has got to be a false memory born of the impression he projected) and drawled his reply to one of my best sallies: 'That's a pretty damned silly question isn't it?' I must say I was tempted by him, but the competition was too stiff so I recommended him to a pugnacious colleague in another college, who accepted

---

[1] The strange English term for a private school.

him. That young man later went out to do field research in Africa and was said to have stared down a charging elephant.

A colleague in philosophy was fond of this interview question, and I agree that it's a good one: 'How do you know you aren't dreaming at this moment?' Another colleague liked this one:

> A monk [not sure why it had to be a monk, just to give colour, I suppose] set off at dawn to walk up a long, winding path from the bottom to the top of a mountain. It took him all day. When he reached the summit he spent the night in a mountain hut. Then, at the same time next morning, he walked all the way down the same path again to the bottom of the mountain. Can we be sure that there is some particular spot along the path, such that the monk passed that spot at exactly the same time on both days?

The answer is yes, but not everybody is able to see, or explain, why. The trick, again, is to reframe it. Imagine that, as the monk sets off to walk up the mountain, another monk simultaneously sets off on the reverse journey along the same path, from the top to the bottom. Obviously the two monks have to meet at some particular point along the path at some particular time during the day. I was amused by this conundrum, but I don't think I ever used it in an interview because, once you've got the point, it doesn't, unlike the El Greco question (or the mirror question or the inverted retinal image question or indeed the dreaming question) lead on to anything further. But once again, it illustrates the power of reframing: I suppose it's an aspect of 'lateral thinking'.

A question that I never used, but that might be good for testing the sort of mathematical intuition that biologists need (as opposed to mathematical skills like algebraic manipulation or arithmetic calculation, which also do no harm), is this. Why do so many influences–gravity, light, radio waves, sound etc.–obey an inverse square law? As you move away from a source, the strength of the influence diminishes steeply as the square of the distance. Why should this be? One way to express the answer intuitively is this. The influence, whatever it is, radiates outwards in all directions, plastering itself over the inner surface of an expanding sphere. The larger the area of the expanding surface, the more 'thinly spread' the influence is. The area of a sphere (as we remember from Euclidean geometry and could prove if we set our minds to it but wouldn't bother in the interview) is proportional to the square of the radius. Hence the inverse square law. That's mathematical intuition without (necessarily) mathematical manipulation, and it's a valuable quality to look for in a biology student.

The interview might proceed to a further, less mathematical but still interesting, discussion of possible biological applications, which would help in judging the teachability of a student. Female silkworm moths attract a mate by emitting a chemical, a 'pheromone'. Males can detect it at an astonishingly great distance. Would we expect the inverse square law to apply here? On the face of it, perhaps yes, but the student might point out that the pheromone will be blown in a particular direction by the wind. How does that affect matters? The student might also point out that, even when there's no wind, the pheromone would not diffuse outwards in an expanding sphere, if only because half of the sphere would be stopped by the ground and most of the other half would be too high. Which might prompt the tutor to divulge the following intriguing fact, almost certainly unknown to the student.

Because of an interaction between pressure and temperature gradients, sound travels through the sea further (and more slowly) at some depths than at others. There's a layer called the SOFAR Channel or Sound Channel in which, because sound is reflected back into the layer from its margins, it propagates in something more like an expanding ring than an expanding sphere. The distinguished whale expert and conservationist Roger Payne reckons that when whales with a very loud song position themselves in the Sound Channel, their songs could theoretically be heard right across the Atlantic (itself a captivating idea to inspire the student being interviewed). Would the inverse square law apply to these whale songs? If the sound were 'plastered' over the inside of an expanding ring, the student might reason that the area of 'plastering' would be proportional to something closer to the radius than the square of the radius (circumference of circle directly proportional to radius). But it obviously wouldn't be a perfectly flat circle. A legitimate answer to the question, and one that I'd applaud, would be: 'This is getting too complicated for my intuition. Let's phone a physicist.'

Like, I think, most college tutors, I developed a loyalty to many of the candidates I interviewed. I was unwillingly obliged to reject far more than half of them, and it often hurt me to do so. I would make strenuous efforts to see them placed in other Oxford colleges, urging on colleagues the virtues of 'my' candidates. I used to resent it when another college accepted from its own list of applicants a candidate whom I thought clearly less qualified than one whom we, at

New College, had had to reject through sheer force of numbers. But I suppose my colleagues had the same loyalties to 'their' candidates. The Oxford system of allowing all the colleges to admit candidates separately has little to be said in its favour and much against. My guess is that the sheer complexity of the system puts not a few candidates off applying to Oxford at all. And that's a better reason for being put off than the absurd misperception that Oxford is 'posh' or 'snobbish' (as it admittedly used to be, but is no longer–quite the reverse).

For most of my adult life I have looked younger than I am (a point we'll come back to in the chapter on television), and this led to an amusing incident during one interview season. Exhausted and parched after a whole day of interviewing entrance candidates, I repaired to the King's Arms pub, just outside New College. I was standing at the bar waiting for my beer when a tall young man came striding over, put his arm sympathetically around my shoulder and said: 'Well, how did you get on then?' I recognized him as one of the candidates I had just interviewed. He must have remembered my face, too, as one that he'd seen during the day, and thought I was one of his rivals. Andrew Pomiankowski won his place at New College, got an exceptionally high first-class degree, went on to do a

DPhil (Oxford-and Sussex-speak for PhD) with John Maynard Smith at Sussex University, and is now a professor of evolutionary genetics at University College, London. He was just one of many very clever pupils whom it was my privilege to teach.

Here's another story of an outstanding pupil well suited to the tutorial system. When I was tutoring in my room in New College, I often overran the hour and my next student would wait outside. I hadn't realized that my voice was audible through the door until, on one occasion, I was holding forth on some topic or other when the door suddenly burst open and the next student came charging in shouting indignantly: 'No no no, I really cannot agree with that.' All credit to Simon Baron-Cohen. I'm sure he was right and I was wrong. He is now a professor at Cambridge, famous for his pioneering work on autism (if a little less famous than his cousin Sacha Baron-Cohen, the amusingly scandalizing actor).

My graduate student, star pupil and indeed later mentor Alan Grafen–of whom much more in later chapters–was not one of my own New College undergraduates, but his friend and collaborator Mark Ridley was. Less mathematical than Alan, Mark is a prodigiously knowledgeable scholar, historian of biology, synthesizer, critical thinker, learned reader and elegantly stylish writer. He went on to write many important books, including what was to become one of the two leading textbooks of evolution: the kind that American campus bookstores seem to order by the cubic yard, restocking with new editions at regular intervals. Alan and Mark have worked together on several occasions, including one field project on albatrosses, camping on a Galápagos island with a very clever young woman from Germany, Catie Rechten. Alan later told me that, when he and Mark were flying out to Galápagos together, he became aware of a strange, low murmuring sound coming from the next seat. On investigation, it turned out to be Mark, reciting Latin poetry to himself. Yes, that's Mark, and no doubt he got the quantities right on the elegiac couplets. Also very Mark is his acknowledgement, in his first book, to Professor Southwood, who stood in as thesis supervisor 'when Richard Dawkins was away in the Plantations for two sabbatick terms'. 'The Plantations' meant Florida.

Mark is not to be confused with his contemporary at Oxford, Matt Ridley (no known relationship, although Matt investigated and established that they are in the same Y-chromosome tribe). I count both as dear friends; both are first-class biologists and successful writers. On one occasion, a journal editor succeeded in getting them to review each other's books in the same issue without telling them. Each was complimentary about the other, and Mark wrote that Matt's book would be 'an excellent addition to our joint CV'.

In 1984 Mark and I accepted the invitation of Oxford University Press to be the founding editors of a new annual journal, *Oxford Surveys in Evolutionary Biology*. We only lasted three years in the job, before handing our baby over to Paul Harvey and Linda Partridge. But we were pretty pleased with the distinguished authors we managed to capture during those three years (papers were commissioned rather than submitted), together with a star-studded editorial board to grace the title-page.

At the end of my undergraduates' Oxford careers, I took very seriously my role in coaching them for their final examinations. American students typically sit an exam at the end of every term in every course they take (and often a mid-term exam as well). Oxford is very different. Apart from informal tests called 'Collections', which are administered separately by the various colleges to monitor progress and have no official standing, most Oxford students have no proper exams at all between the end of their first year and the end of their third. Everything is packed into the terrible ordeal of 'finals'–an ordeal exacerbated by the requirement that they dress up in 'sub fusc' for the occasion.

In my time this dress code was dark suit and white bow tie for men, dark skirt, white shirt and black tie for women, black academic gown, black academic cap or mortar board. Since 2012 the authorities have found a neat form of words to proclaim a politically acceptable sex-blindness: 'students identifying as either sex can wear historically male or female clothes'.

Adding to the intimidating atmosphere of sub fusc is the strict invigilation. Officially, students wishing to visit the men's or women's room have to be escorted by a minder of the same sex, to make sure they don't cheat by looking things up while there, but by the time I came to be an invigilator we usually didn't bother with that rule. At least in my time there was no need to frisk examinees for internet-capable mobile phones, as there presumably must be today.

The whole performance is calculated to terrify, and nervous breakdowns are not uncommon around the time of finals. My colleague David McFarland, tutor in psychology at Balliol College, was once telephoned by the invigilator in the examination room: 'We're getting concerned about your pupil Mr—. Since the exam began, his handwriting has been growing steadily larger, and each letter is now three inches across.'

I saw it as the duty of a college tutor to see students at frequent intervals during their last term to hold their

hands through the ordeal of finals and the weeks of revision leading up to it. I would take the whole cohort
together in my room for regular coaching in exam technique, and I made them do lots of mock exam questions, each in exactly one hour. The self-imposed time
limit was important. In each of their written papers,
they would have three hours to write three essays,
chosen from about twelve on offer. In our joint counselling sessions I urged them not to deviate too far from
allocating equal time—one hour—to each of the three
essays. I was exaggerating a little, in the interests of
warning them how many students under pressure fall
into the trap of getting carried away with a favourite
topic, devoting most of their time to it and leaving no
time to answer less preferred questions.

'Suppose you were the world authority on the topic
of your favourite question,' I suggested. 'You couldn't
write down more than a small fraction of what you
knew.' With a nod to Ernest Hemingway I advocated
'iceberging'. Nine-tenths of an iceberg is submerged.
If you are the world authority on a topic, you could
write about it till kingdom come. But you've only got
one hour, like everybody else. So, cunningly exhibit
the tip of your iceberg and let the examiner infer the
huge quantity that lies beneath the surface. Say, 'Notwithstanding Brown and McAlister's objection . . .' and

convey to the examiner that you could, if you had time, go to town on Brown and McAlister. Do not actually do so because that would take too long and leave you no time for all the other iceberg summits that you need to skip on to. Just the name-drop does the business: the examiner will fill in the rest.

It's important to add that iceberging works only because we assume that the examiner knows a lot. Iceberging is a terrible technique in, say, instruction manuals where the author knows what is being imparted but the reader does not. Steven Pinker makes the point forcibly as the 'curse of knowledge' in his splendid book *The Sense of Style.* When you are actually trying to explain something to somebody who knows less than you do, iceberging is the exact opposite of what you should be doing. It works in exams solely because it is fair to presume that your reader is an examiner who knows a lot.

When I was an undergraduate myself, the wise and learned Harold Pusey took me and a group of contemporaries for similar coaching, and iceberging was one of the tips he gave us. I think his metaphor was not actually icebergs but shop windows, and it works just as well. An impressive shop window is sparsely furnished. A few tasteful items, elegantly presented, evoke the riches stashed away inside the store. A good window-dresser doesn't clutter the window with everything in the shop.

Another of Mr Pusey's tips (yes, Mr: he was a don of the old school who never bothered with a doctorate), which I passed on verbatim to my own students, was this. When you've read the exam paper and spotted a favourite topic, don't immediately start writing that essay. First, decide which three of the twelve questions you are going to tackle; then make a written plan for all three essays, each on a separate sheet, before you start writing any one of them. While you are writing your first essay, you will find that ideas for the other two continually occur to your primed mind. When that happens, jot down a note on the appropriate one of your three sheets of paper. Then, when you come to answer your second and third questions, you'll find that much of the thinking work has already been done, at almost no cost in time. I'm told that this tip is also pertinent to American students taking the Advance Placement exam.

I didn't have the courage to pass on another of Harold's pieces of advice: stop revising a whole week before your final exams start; spend the last week punting on the river and let it all sink in. I did, though, tell them yet another piece of his wisdom: during your weeks of Oxford finals, you will probably have more concentrated knowledge in your head than at any other time in your life. Your task while revising is to systematize it while it simmers: seek connections and relationships between different parts of your knowledge base.

During my years in the Zoology Department, I also took my turn as examiner from time to time, and a very heavy responsibility it is. Quite apart from the hard work involved, you cannot escape the grave knowledge that your decisions are going to affect the whole future lives of promising and eager young people. Some unfairnesses are built into the system. The students end up classified into one of three discrete classes, although everybody knows that the bottom of one class is much closer to the top of the class below than it is to the top of its own class. I wrote about this in 'The tyranny of the discontinuous mind' (in *New Statesman* when I was guest editor; see e-appendix) and I won't spell the argument out again here.

But there are other unfairnesses that an examiner can, and should, do something about. How sure can you be that the *order* in which you read the scripts is unimportant? Are you getting fatigued as you read paper after paper? Is your standard of judgement drifting– either upwards or downwards–as a result? Or, if not physically fatigued, are you getting progressively bored by the inevitable familiarity as answer after answer to the same popular question thuds into your consciousness? Does this give an unfair advantage to candidates who choose unpopular questions? *Is* that advantage unfair? Does the 'boredom or fatigue effect' give an unfair advantage to those essays that you read early? Or to those that you read late? I tried to guard against 'order effects' by using some of the elementary principles that all biologists learn in designing experiments. Don't read all three of the first candidate's essays, then all three of the second, and so on. Instead, read everybody's first essay, then everybody's second essay, then everybody's third essay. And on each of your three runs through the scripts, it might be no bad thing to read them in random order, not the same order every time.

Then again, are you seduced by the elegant handwriting of this candidate, and prejudiced against the untidy scrawl of that one: a virtue or fault that has no bearing on the quality of scholarship—or does it? My first wife, Marian, and I were both examiners at various times during our careers in the Oxford Zoology Department, and we tried the experiment of reading the essays aloud to each other. That should have gone some way towards eliminating the 'handwriting effect', and it had additional advantages. At the end of a reading, without comment to each other, we would *simultaneously* on a count of three (so as not to influence each other) sing out the mark that each of us thought the essay deserved. We were both reassured by the high concordance between our separate marks. All Oxford papers are, in any case, double-marked–marked by two separate readers without collusion–which is a good attempt at safeguarding against certain types of unfairness. Also, nowadays (this didn't happen when I was an examiner) the candidates' names are disguised and they are known only by randomized numbers. This guards against personal biases either for or against, and it matters in a small department like Zoology, where most of the students are known personally to the examiners.

I had occasion to worry about order effects in other decision-making tasks, for example when I was on committees to choose new lecturers or fellows, or in judging prizes or awards. The Royal Society awards an annual Michael Faraday Prize for success in promoting science to the general public. I won it in 1990, and was later put on the committee to choose the prizewinner. The committee has a rolling membership, and I served as chairman for the last three of my five years. In the first two years, with my predecessor as chairman, I was worried about order effects. Each candidate had a dossier consisting of *curriculum vitae* and letters of support. We had all read the dossiers conscientiously before the meeting. So far so good. But then in committee we discussed all of them in order–probably alphabetical order, which makes it worse, but that's not the point I am making here. Whatever ordering principle you choose, order effects are inevitable. It was distinctly noticeable that the first few dossiers were given lengthy discussion, and the length of discussion tailed off as the afternoon wore on. This was especially unfortunate when we would spend a lot of time near the beginning discussing the minute details of a candidate who turned out to be a no-hoper, with no support from any member of the committee.

When I became chairman, I changed the system in a way that I recommend for all such committees, and therefore think is worth explaining fully here. Before the committee began any discussion at all, each one of us, having read the dossiers before we arrived, wrote secretly on a piece of paper the names of the three candidates whom we would wish to discuss first, along with a score: three points for our top candidate, two marks for the next, one mark for the third. I then gathered all the slips of paper, added up the points, and announced the rank order. I had clearly explained to the committee that this was *not* a ballot to determine the prizewinner; it was a ballot solely to determine the *order* in which we would *discuss* the candidates. We then discussed the dossiers properly and in detail, but the order in which we did so was not alphabetical, not reverse alphabetical (which is sometimes used in a vain but futile attempt to counter the known advantage the As and Cs have over the Ts and Ws), not arbitrary at all, but determined by our preliminary secret ballot. After our full discussion, we then proceeded to our final secret ballot rounds to choose the winner. The eventual winner might turn out to be the same candidate who had topped the preliminary 'order' ballot, but it might not: the exhaustive discussion during the afternoon could well change people's minds. Under the old system, the lion's share of the discussion time was squandered on candidates who never had any hope anyway. The new system meant that we had time to discuss more thoroughly those who did have at least some support, and to discuss them in a just order.

Sub-Warden

To serve on the committees that chose new lecturers and fellows was one of the serious responsibilities of my life at Oxford. There were others–financial, pastoral, custodial. A fellowship at a typical Oxford or Cambridge college brings with it trusteeship of a large charitable institution, making investments and disbursements that, in the case of a relatively wealthy foundation like New College, could be quite substantial. In addition, the fellows were as a group responsible for student welfare and discipline, for the upkeep of the chapel and other precious medieval buildings, and for much else besides. We elected officers from among our number to oversee each of the main functions. I was happily and rightly never elected to any office (I would have been hopelessly bad). However, there is one office no fellow of New College can escape: Sub-Warden. Other Oxbridge colleges may elect a Vice-Warden (Vice-Master, Vice-Principal, Vice-Provost etc., depending on the bewilderingly variable names that Oxford colleges choose for their heads), a colleague whom all respect to deputize for the head of the college. That isn't New College's way. We don't elect our Sub-Warden: it is a one-year burden which descends inexorably down the list of fellows, and respect doesn't come into it. Every year until the black spot hit me in 1989, I was able to count off the years of its approach. That count was always a maximum figure: a year was ominously knocked off it every time a colleague ahead of me in the list died or–which was pretty common–left to take a professorship elsewhere. 'Ominously', because I dreaded it.

The onerous nature of the Sub-Warden's duties is justified by their brevity: only one year out of your life. As Sub-Warden I had to attend all committee meetings, and that meant all sub-committees and all appointing

and electing committees as well as the full college meetings where I had to write the minutes. As it turned out,
I quite enjoyed my minutes, using them as a vehicle
to try to amuse my colleagues–those who read them,
which was by no means all, as we sometimes discovered during the subsequent meeting. The Sub-Warden
has to take the chair when the Warden is unavoidably
absent from college meetings, or when the Warden has
to recuse himself from a discussion that concerns him
personally. The role is a particularly weighty responsibility when the college is electing a new Warden, for the
Sub-Warden of the day has to preside over the whole
voting procedure. Thank goodness this didn't happen
on my watch. In all the four wardenship elections in
which I participated, the Sub-Warden of the day either
happened to be well qualified or rose to the occasion.
In one case some smart footwork ensured that the year
of office of a notoriously unstable, not to say downright
misanthropic fellow was somehow postponed in favour
of a respected 'safe pair of hands'. Incidentally, my lack
of skill as a political operator is indicated by the fact
that in three of those four elections I nominated the
runner-up.

As Sub-Warden I had to preside over dinner in Hall,
and say grace before ('Benedictus benedicat') and after
('Benedicto benedicatur'). I was one of the majority

who pronounced this last word 'benedicahta'; some of the older, classically trained fellows pronounced it 'bene-dye-cay-tour', which fascinated me although I never dared follow them. I doubt that they really thought that's how the Romans pronounced it, but their justification was surely thought-out and deliberate, probably buried in some erstwhile dispute among the dominies. One of my predecessors as Sub-Warden, the ancient historian Geoffrey de Ste Croix, used to refuse to say grace on conscientious grounds (he described himself as 'an atheist, politely militant'). Equally conscientiously, however, he went out of his way to line somebody up to say it for him. Once when I was a dinner guest at King's, our sister college at Cambridge (whose chapel, incidentally, is one of the most beautiful buildings in England), the senior fellow presiding was the incomparable Sydney Brenner, one of the founding fathers of molecular genetics and winner of a well-deserved (they aren't always) Nobel Prize. Sydney gavelled everybody to stand, then solemnly intoned to his neighbour: 'Doctor—, will you please say grace?' I, however, was of the school of thought of the great philosopher Sir Alfred Ayer who, when Sub-Warden of New College, cheerfully said grace on the grounds that 'I will not utter falsehoods but I have no objection to making meaningless statements.'

I was once fiercely attacked for adopting the same stance by Rabbi Julia Neuberger, one of Britain's best-known rabbis and a well-established member of the 'great and good', both a dame and a member of the House of Lords. She sat next to me at a rather formal luncheon and furiously accused me of hypocrisy when she dragged out of me that I was prepared to say grace when presiding in New College Hall. I countered that although it meant a lot to her it meant nothing to me, so why should I object? It seemed to me a matter of simple courtesy, like removing your shoes when entering a Hindu or a Buddhist temple. Wasn't I right simply to honour an ancient tradition? (Although actually I'm not sure that 'Benedictus benedicat' is all that ancient; does it, perhaps, like so many 'ancient' traditions, go back no further than the nineteenth century?) Once, at the beginning of a dinner in Wellington College after a debate with, among others, the Bishop of Oxford, the philosopher A. C. Grayling and the journalist Charles Moore (who for some reason brought a brace of shotguns to the occasion), the Master, the justly celebrated Anthony Seldon, jovially invited me to say a secular grace. Put on the spot, I couldn't think fast enough to do any better than, 'For what we are about to receive, thank the cook.'

Most daunting of the Sub-Warden's duties was making speeches, usually for the purpose of welcoming new fellows or saying farewell to departing ones. It was the speeches that I especially dreaded as my year of tribulation approached, having heard some pretty poor Sub-Warden turns along with some good ones. As it turned out I could do it, although not off the cuff: I had to spend lots of time preparing my speeches in advance, and in this I was greatly helped by the more spontaneously witty Helena Cronin, philosopher and historian of science at the London School of Economics, who was my close collaborator at the time—we were helping each other to write our respective books, as I shall explain later.

Making speeches about new fellows is difficult because it's in the nature of their newness that one doesn't know them and has to rely for material on their written *curricula vitae*. The CV of a new law fellow, Suzanne Gibson, for example, announced a professional interest in 'the body' as a 'visual and narrative structure'. I had a bit of fun with that, hamming up the role of a hypothetical future barrister trained in law at New College:

M'lud, m'learned friend has presented evidence that my client was seen burying a body at dead of night. But, ladies and gentlemen of the jury, I put it to you that a body is a visual and narrative structure. You cannot convict a man on evidence of burying nothing more than a visual and narrative structure.

Suzy was a good sport who took it well, and we later became close friends. Another new fellow whom I had to introduce that evening was Wes Williams, a scholar of French who became a valued colleague. We already had two other Williamses, so I was able to make something of this:

> For years we had only a single Williams on the fellowship. We soldiered on but it didn't look good, and I'm afraid it was a long time before we managed to get a second Williams. I am therefore delighted this evening to welcome our third Williams, and I can officially announce that all future election committees will include at least one token Williams to see fair play.

These welcoming speeches always take place at 'dessert'. The quaint ceremony of formal dessert, a version of which is observed in most Oxbridge colleges and which I never liked, takes place in a separate room after dinner, where port and claret, sauternes and hock, have to pass clockwise around the circle, and nuts, fruit and chocolates are handed round by the most junior fellows. New College has a curious contraption called the port railway, dating, as you might expect, from the nineteenth century, which is supposed to (and occasionally does) convey bottles and decanters by a pulley system across the gap in the circle which is the fireplace. Snuff also is traditionally handed round, but seldom actually taken (at least since the days of one venerable and long-retired fellow whose consequent prodigious sneezes would reverberate companionably around the oak panelling for the rest of the evening).

Although the Sub-Warden doesn't have to seat fellows and their guests (as the presiding fellow does in some other colleges), he is expected to beam the role of genial host at dessert. I did my best, but there was one awkward evening. As I was helping people to find their seats I became aware, from a sort of ominous rumbling, that all was not well. Sir Michael Dummett, immensely distinguished philosopher, Wykeham Professor of Logic in succession to Freddie Ayer, stickler for grammar, conscientious and passionate campaigner against racism, world authority on card games and voting theory, was also famously choleric. When angered he would go even more than usually white, which somehow seemed–though this may be my fevered imagination–to make his eyes glow a menacing red. Pretty terrifying . . . and it was my duty as Sub-Warden to try to sort out whatever this problem was.

The rumble grew to a roar. 'I have never been so insulted in my life. You have the most atrocious manners.

You obviously must be an *Etonian*.' The target of this damning sally was not me, thank goodness, but our quirkily brilliant classical historian, Robin Lane Fox. Robin was hopping with anxiety and bewildered apology: 'But what have I done? What have I *done*?' I didn't immediately succeed in discovering what the problem was, but in my hostly role I saw to it that the two of them were seated as far from each other as possible. I later learned the full story. It had begun at lunchtime that day. Lunch is an informal, self-service meal and fellows sit where they like, although it is conventional to fill up the tables in order. Robin noticed that a new fellow was hesitantly looking for a place. He courteously motioned her to sit, but unfortunately the chair he indicated was the very chair for which Sir Michael was heading himself. The perceived slight rankled, simmered up through the afternoon and finally boiled over after dinner at dessert. The story had a happier ending, as Robin told me when I asked him recently. A couple of days after that distressing incident, he was approached by Professor Dummett who offered the most gracious apology, saying that there was nobody in the college whom he would less wish to insult than Robin. Thank goodness I was never the target of his ire, although I might have been vulnerable as he was a devout Roman Catholic with the zeal of the convert.

Not that it is at all relevant, Robin Lane Fox does happen to be an Etonian. You may know him as the gardening correspondent of the *Financial Times* and author of *Better Gardening*, whose chapter on 'Better Shrubs' follows the one on 'Better Trees' with this deliciously anachronistic, and utterly characteristic, opening salvo:

> Swinging down from the branches to the level of better shrubs, I will not leave the days when the world was young and Dawn Redwoods dawned among the dinosaurs. Among mastodons and dimetrodons, what could be more natural than my own dying species, the Oxford ancient-history-don? Though long declared moribund, we are far from extinct.

World authority on Alexander the Great and enthusiastic horseman, he agreed to advise Oliver Stone on filming the epic *Alexander*, on condition that he was allowed to appear as an extra, leading the cavalry charge. And he did. It has been a privilege of my career to have been surrounded by such idiosyncratically unpredictable colleagues, who would make even committee meetings entertaining. I could tell parallel stories of many such colleagues and friends, but will desist. One will have to stand for all–although I suppose that belies the very meaning of idiosyncrasy.

I have great affection for New College and for many friends made there over the years. I feel pretty sure I'd say the same if the roll of the dice had placed me in another college–or indeed a Cambridge college–for these very similar institutions are wonderful places, mixing scholars in different subjects but sharing the same academic and educational values–values from which, I like to think, the students benefit. Quirky individuality, nevertheless, abounds, and Oxbridge colleges are famously hard to govern, as many an incoming head from the big world outside has discovered. Yes, we have our share of scholarly prima donnas, clever but not necessarily as clever as their vanity persuades them. And we have the reverse: a scholar so lacking in vanity as laughingly to tell, at lunch, a story like this against himself:

I was telephoned by the student newspaper today: 'Dr ––––, have you any comment on the fact that, in your lecture this morning, one of the students yawned so vigorously that he dislocated his jaw?'

As it happens, the same student newspaper, *Cherwell* (pronounced 'Charwell', like the Oxford river after which it is named) once telephoned me when it was doing a survey of dons to see how cool we were. The student reporter put to me a list of questions to assess my street cred, such as: 'What is the price of a packet of Durex?' And then: 'What is the price of a Big Mac?' To which, in my naivety, I replied: 'Oh, about £2,000 with a colour screen.' He was laughing too much to continue the interview, and put the phone down.

In one of my speeches as Sub-Warden of New College, I had to bid farewell to the chaplain, Jeremy Sheehy, who (as was the custom in those days) was moving on to a Church of England living. We had often voted together on the liberal side of contentious issues and I mentioned in my speech a political affinity with him that I sensed at college meetings, 'catching his eye in concord, across the abyss of our differences'. At the time, the New College kitchen was in the habit of serving a rather delicious pudding, a sort of moist, black, cakey sponge topped by a creamy white sauce, but it always appeared on the menu under an unfortunate name: *Nègre en Chemise*. The Reverend Jeremy was repeatedly and rightly upset by this, and I wanted, as my parting gift to him, to get the name changed. I went to the chef (one of the few powers possessed by the Sub-Warden) and asked him to serve the dish up at dinner, but under a new name. In my speech at dessert that night, I told the story and explained that I had chosen the new name in the chaplain's honour: *Prêtre en Surplice*. Alas, after he left, it wasn't long before the delicacy started to reappear under the original name, *Nègre en Chemise*, and by then I no longer had Sub-Warden's powers to do anything about it.

Incidentally, I heard of a related problem at a care home for old people in England. One day, the menu included a traditional English pudding, the long, raisin-infested, custard-bespattered suet roll called Spotted Dick. The local government inspector demanded that it be banned from the bill of fare as its name was 'sexist'.

The arduous climax to the Sub-Warden's oratorical career at New College is the speech at the gaudy, the annual reunion dinner to which a different cohort of old members is invited each year. The choice of age group marches backwards a few years at a time with, in deference to the grim reaper, lengthening stride as vintage gives way to veteran and eventually we hit the 'old gaudy', which embraces everybody who came up to New College before some early cut-off date. Then the cycle begins again with the 'young gaudy'–those who left the college only a decade or so ago. As it happened, in my year as Sub-Warden the cycle had reached the old gaudy but their depleting numbers, alas, couldn't fill the tables, so they were afforced by new blood from the young gaudy, callow youths in their thirties. And so I faced the difficult task of appealing to guests one half of whom were separated from the other half by a world war, a great depression and about fifty years. Not an easy speech to compose. I tried to play on the contrast between the roaring twenties, when the old contingent had been undergraduates, and the 1970s which, at least by contrast with my own era, the 1960s, could, at a pardonable stretch, be regarded as somewhat staid. Describing myself as having reached 'the lunchtime of life', I cobbled together something about 'gilded age meeting crabbed youth', which I think pleased the old while not seriously annoying the young who perhaps didn't believe it anyway.

I tried to arouse nostalgia in the oldies, coupled with incredulous amusement in their young successors, by reading from the Junior Common Room (JCR) Suggestions Book of the 1920s, which the college archivist kindly lent me. Incredulity, for example, on discovering that apparently many of the baths in the 1920s were in one great hall with cubicles, as testified by the several letters that said things like: 'Would the gentleman who was attempting unsuccessfully to sing, in the fifth bath on the left this morning, please refrain in future?' Incredulity too, perhaps, at the presumptuous treatment of college servants, tell-tale badge of the haughty 'Brideshead generation' that emphatically no longer typifies the Oxford colleges (with the possible exception of the much despised 'Bullingdon Set'):

> If one wants a plate of cucumber sandwiches sent up to ones rooms for tea, I understand the kitchen requires notice before 11 am. This is most inconvenient.
>
> Would it be possible for either the boot cleaner or the bath attendant to scrape the mud off the football boots (and if necessary grease them) in the bathroom?

There were many complaints about the door of the JCR squeaking. I'd like to think that the 1970s cohort might quietly have applied a drop of oil to the hinge rather than braying for someone else to do it for them.

But mostly the appeal of my quotations was gentle nostalgia for a bygone age:

> Would it be possible for a new pair of hairbrushes (really hard) and a new comb to be provided in the old bathroom?

May I suggest that pipe cleaners be provided in JCR? These articles strike me as being more useful than tooth picks.

Wishing to telephone this morning, I was surprised to find the Telephone Box missing. What can have happened to it? May I add, as a suggestion to be conveyed to the right quarter, that there seems to be no particular reason to replace it?

I think my speech went down quite well. One of the old brigade wrote a thank-you letter to the Warden, which said it reminded him of his old tutor, Lord David Cecil. It seemed to be meant as a compliment, although Kingsley Amis's recollections, in his autobiography, of that aristocratic savant give me pause.

Lore of the jungle

Among the 115 mammal species on Barro Colorado Island in the Panama Canal there is an irregularly shifting population of *Homo scientificus*, and this includes short-term migrant visitors, invited for a month or so to interact with (and, it is hoped, refresh and invigorate) the resident population of biologists. In 1980 I was privileged to be invited as one of two birds of passage, the other being, I was delighted to hear, the great John Maynard Smith.

The densely forested Barro Colorado Island sits in the middle of Lake Gatún, which forms a substantial chunk of the Panama Canal, and is home to a world-renowned tropical research centre run by the Smithsonian Tropical Research Institute (STRI). Why such forests are so rich in species is one of the perennial questions in ecology. Their biodiversity outstrips all other major ecosystems, and the six square miles of Barro Colorado are surely (with the possible exception of Wytham Wood near Oxford) the most intensively studied, most picked-over, most analysed, most binocular-scanned, most mapped square miles of forest in the world. What a privilege, to be invited there for a month.

At the time of my visit, Ira Rubinoff, the Director of STRI in Panama who originally invited me, was going on sabbatical leave, leaving the institute in the capable and genial hands of his deputy, my old friend Michael Robinson. Mike and I had been fellow graduate students of Niko Tinbergen at Oxford in the 1960s. He was a little older than the rest of us, having gone back to university as a mature student to indulge his passion for entomology, after what some (not I) might consider a mis-spent youth as a left-wing agitator. At that stage of his life, British troops were fighting insurgents in Malaya, and Mike once spent a whole night rushing around the streets of Manchester painting slogans on wall after wall after wall: 'Hands off Malaya.' 'Hands off Malaya.' 'Hands off Malaya.' Dawn came and he prepared to crawl off to bed, unarrested and with a good feeling of a night well spent and Manchester well and truly taught a lesson. He looked up at his final daub with a sigh of satisfaction and noticed to his horror that it said: 'Hands of Malaya.' He didn't need to go back to check his earlier handiwork. A sinking hindsight told him that all his slogans throughout the night contained the same slip, mechanically repeated from the first one.

When, after completing an excellent doctoral thesis on stick insects at Oxford, Mike was offered a job at STRI, his officially organized itinerary to Panama took him via Miami. Because he had once been a member of the Communist Party, the US authorities refused to grant him a visa to touch down in Miami, even though he would never leave the airport's secure area, and in spite of the fact that his fully agreed salary in Panama was to be paid by the US government. Stalemate! I forget how it was resolved, but he did eventually make it to Panama. All must later have been comprehensively forgiven (or at least officially forgotten), because he eventually rose to become Director of the National Zoo in Washington DC, one of the most famous zoos in the world. At the time of my visit to Panama he was accepted enough to be Acting Director of STRI, and was still exactly as I remembered him: that rosily beaming face with its little red goatee and matching apical tuft on the top of the head (a young woman in Oxford, seeking to establish which one he was in a group, had once whispered to me: 'Is he the one with the little beard?' while her accompanying hand gesture cheekily indicated the top of the head).

My guide when I arrived in Panama was another old friend from Oxford days—Fritz Vollrath, the spider man. If Mike Robinson was cheerful, Fritz is world-class cheerful, but with none of the negative connotations of the 'life and soul of the party': more the life and soul of everyday life itself. I first encountered his quizzically laughing eyes when he arrived in Oxford from Germany to work as a teenage 'slave' in the Tinbergen group. He was introduced by the wildly brilliant Juan Delius, his cousin, who was at that time a leading member of the group. Fritz immediately fitted in, laughing at his own broken English even more than the rest of us did. When I met him again in Panama years later he had scarcely changed. His English was much better and his Spanish didn't seem at all bad either. We drove around the environs of Panama City, pausing to watch a sloth as it slowly descended from a tree for its weekly defecation. We climbed a peak in Darien—alas, not the very one where (as I murmured to myself) stout Cortez with his eagle eyes silently star'd at the Pacific and all his men look'd at each other with a wild surmise. Fritz was based in Panama City, while I was bound for Barro Colorado in the interior of the country, but it was a delight to see him for just that one day. He is now back in Oxford, close friend and distinguished authority on spiders, their behaviour and the unparalleled properties of their silk.

The journey from Panama City to Barro Colorado was (is?) by a small, rattling train with unpadded wooden seats. It stops by Lake Gatún, halfway across the peninsula, at a tiny halt (too small and deserted to be called a station). Hard by is a landing stage, and every train is met by a boat from the island. Or, at least, is supposed to be met. On one occasion during my month, John and Sheila Maynard Smith had gone on a day trip to Panama City. They returned late, on the last train, and were pleased to see the boat chugging towards the dock. Then, to their dismay, it suddenly turned around and headed back to the island. Apparently the boatman had decided it was so unlikely anyone would be on the last train that he might as well not bother even to look. The Maynard Smiths shouted and yelled but the wretched man didn't hear them above the engine noise. There was no telephone, and so the elderly couple were forced to spend the night at the halt, with little shelter and nothing but wooden boards to sleep on. They were surprisingly nice about it the next morning. I never discovered whether the boatman was sacked, nor what mental aberration caused him to turn the boat around without troubling to check if anyone was waiting for him; nor why, indeed, if he had no intention of going to the station dock, he had bothered embarking in the first place.

At the time of my own first arrival, all went according to plan and the boat did its duty. From the island's small dock, you climb steep steps to the main compound of the Research Institute: a purpose-built cluster of red-roofed houses and laboratories. My bedroom proved to be bare but functional, and I didn't mind the companionable large cockroaches. Two cooks provided hot meals at set times in the communal dining room, where the researchers gathered to eat and talk. There were probably about a dozen of them when I was there, mostly graduate students and post-docs (a post-doctoral fellowship is the normal next step of a bright young academic scientist after completing a PhD), working on a range of subjects from ants to palm trees. Most were from North America, one from India–and the Indian biologist Ragavendra Gadagkar was of particular interest to me as he was working on wasps, the primitively social *Ropalidia*, which plausibly might represent an intermediate on the diagram Jane Brockmann and I had constructed for our paper, published in the journal *Behaviour* the previous year, on potential evolutionary origins of insect sociality. (More of this in the next chapter.)

I don't think I imagined this: the social atmosphere in the dining room and around the compound seemed a tiny bit colder than I had become accustomed to among groups of working scientists. It thawed markedly later during my month there, and I eventually felt accepted enough to remark on it–at which point I was told that this was a well-acknowledged characteristic of the place, which the residents attributed to the fact that they were on an island. I wasn't sure how to tie this psychological insight in with my knowledge of the theory of island biogeography (the title of a famous book by two old Barro Colorado hands, Robert MacArthur, who died tragically young, and Edward O. Wilson). But after I had been on the island a month, I too found myself feeling ever so slightly territorial when newcomers arrived. I therefore made a conscious effort to counteract this by going out of my way to befriend the latest arrival before I was to leave, Nancy Garwood, at the New Year's Eve party. As it turned out she had been there before so didn't need me to go out of my way, but I'm glad I did, and I hope she was too.

This party was also memorable because of the firework display on a huge ship passing through the canal just beyond the trees. Actually falsely memorable, because for years I have been utterly convinced that we saw in not just a new year but a new decade:

1 January 1980. So detailed and full were my recollections of 'seeing the new decade in', it took multiple documentary evidence, kindly sent me by Ira Rubinoff, Ragavendra Gadagkar and Nancy Garwood, to finally convince me that what I had thought to be a crystal clear memory was faulty. It was actually 1 January 1981, not 1980. I was quite shaken to discover this, because it made me worry how many other clear memories actually never happened (and the reader of my memoir is, I suppose, duly warned).

The dreamlike presence of large tankers deep in the jungle was one of the most vivid memories I took away from the place. On several afternoons I had joined the resident scientists swimming off a raft, and it was a surreal experience to see those gigantic vessels gliding calmly, and surprisingly quietly, through the still, clear water, just a few yards away behind the trees. Some of the women scientists liked to sunbathe, and I couldn't help wondering what the tanker crews thought about the undraped feminine pulchritude diving off the raft deep in the jungle. If those mariners were Greek, did they think Sirens; or if German, Lorelei? Or—peering through the lush tropical vegetation—did they see a vision of Eve's innocence before the Fall? They had no way of knowing that these tropic nymphs had PhDs in science from some of the top universities in America.

I've mentioned the apparent territoriality of those busy and dedicated scientists whose island fastness I was briefly allowed to invade, but I mustn't exaggerate it. I learned from friendly experts most days, either in the field or in the dining room. Elizabeth Royte independently noted the same slight initial *froideur* in her book about her own visit to Barro Colorado, *The Tapir's Morning Bath*. For her, as for me, it thawed later, as she gradually became accepted as an in-group islander and was allowed to help with the research. The first to befriend her was the senior scientist on the island, the delightfully eccentric Egbert Leigh, and he was hospitable to me too. I already knew his name as the author of a thought-provoking paper on 'The parliament of genes' and was quite surprised to find this deep-thinking theorist deep in the Central American forest. But there he was, with his family, in the only permanent residence on the island, known as Toad Hall. 'Toadish', I later learned, was an epithet of high praise in Dr Leigh's vocabulary. I never quite discovered what it meant to him: I suspect something multifaceted and subtle, like 'spin' in the private vocabulary of the English mathematician G. H. Hardy (a term of approbation, derived in his case from cricket, whose exact meaning C. P. Snow struggled to elucidate in his affectionate memoir of Hardy). Egbert Leigh and

I found common ground in our admiration for R. A. Fisher, and he barked out his appreciation in accents best summed up in the phrase (I can't trace the origin of the witticism) 'irritable vowel syndrome'.

If the island housed theoretical firepower in the form of Egbert Leigh, the intellectual armoury was massively afforced again with the arrival of John Maynard Smith, the first half of whose month as visiting consultant overlapped with the second half of mine. John always was a man eager to learn as well as teach, and it was wonderful to walk the jungle trails in his company and learn biology from him—as well as learn from him *how* to learn from the local experts who conducted us. I treasure an aside from him about one young man who was walking us through his research area: 'What a treat it is to listen to a man who really loves his animals.' The 'animals' in this case were palm trees—but that was John all over, and one of the reasons I loved him. And miss him.

Among the actual animals, as opposed to photosynthetic honorary animals, were the well-named spider monkeys, with their splendid fifth limb in the form of the prehensile tail. And the howler monkeys, with their bone-amplified voice boxes, whose travelling waves of crescendos and decrescendos could easily be mistaken for a squadron of jet fighters roaring through the canopy.

On one occasion I encountered a full-grown tapir, close enough to see the ticks on its neck, engorged with its blood. It was hard to walk a day in the jungle without picking up one's own cargo of ticks. But they were always small when they first hitched a ride, and everyone carried a roll of sticky tape to peel them off. Tapirs, by the way, seem never to have occurred in Africa, so it was a bit of a solecism for Stanley Kubrick to portray them as hunted by our hominin ancestors at the beginning of that magnificent film *2001: A Space Odyssey*.

Insofar as I got any constructive work done during my time in Panama, it took the form of writing chapters of *The Extended Phenotype*, and in this I was helped by discussions with some of the scientists on the island. From the dates, I know that I spent Christmas 1980 on the island but I remember nothing about it, which suggests that not a lot of fuss was made of the occasion. I do recall some kind of a party with a cabaret, which may have been associated with Christmas; Ragavendra Gadagkar was co-opted to act as Master of Ceremonies, somewhat to his bemusement I think, as he had only just arrived.

I discovered a special affinity for the leaf-cutter ants, introduced to me by Allen Herre along with the more sinister army ants, who one night invaded a bathroom and linked limbs to hang in there, festooned like noisome brown-black curtains. Allen was not the only regular who earnestly warned me about the giant ponerine ants, *Paraponera*, whose formidable sting placed them among the most talked-about denizens of the jungle. My fear-primed eyes saw them often, and I kept my distance with the utmost respect.

I found the leaf-cutter ants more appealing and would stand for what seemed like hours watching the flowing green torrents of walking leaves: tens of thousands of workers, each with its verdant parasol held aloft en route for the dark, subterranean fungus gardens. I was filled with naive fascination that they were cutting leaves not to slake their own taste for greenery, either now or later, but to make compost on which to grow fungi which would eventually be eaten by others in their teeming colony after they themselves had died. Were they motivated by some myrmecine equivalent of an 'appetite', which was satisfied not by a full stomach but by, say, the feel of a leaf in the jaws, or something even more indirect? I didn't need John Maynard Smith to remind me that natural selection favours 'strategies' which are not assumed to be understood by the animals that act them out. It is not for us to say whether ants feel any conscious appetites or desires, wants or hungers. I felt a glow of understanding, which was repeated in an encounter with the army ants, later described in my third book, *The Blind Watchmaker*. I explained that, as a child in Africa, I had been more afraid of driver ants than of lions or crocodiles; but I quoted E. O. Wilson to the effect that a driver ant colony is 'an object less of menace than of strangeness and wonder, the culmination of an evolutionary story as different from that of mammals as it is possible to conceive in this world'. And I continued:

As an adult in Panama I have stepped aside and contemplated the New World equivalent of the driver ants that I had feared as a child in Africa, flowing by me like a crackling river, and I can testify to the strangeness and wonder. Hour after hour the legions marched past, walking as much over each others' bodies as over the ground, while I waited for the queen. Finally she came, and hers was an awesome[1] presence. It was impossible to see her body. She appeared only as a moving wave of worker frenzy, a boiling peristaltic ball of ants with linked arms. She was somewhere in the middle of the seething ball of workers, while all around it the

---

[1] Today I could not use 'awesome': the word has become so debased, it is now no more than a routine term of mild approval. Yes, yes, I know, language evolves: 'Chill out, Professor.' I still lament the loss of a valuable word: 'awe-inspiring' would spoil the rhythm of the sentence.

massed ranks of soldiers faced threateningly out-wards with jaws agape, every one prepared to kill and to die in defence of the queen. Forgive my curiosity to see her: I prodded the ball of workers with a long stick, in a vain attempt to flush out the queen. Instantly 20 soldiers buried their massively muscled pincers in my stick, possibly never to let go, while dozens more swarmed up the stick causing me to let go with alacrity.

I never did glimpse the queen, but somewhere inside that boiling ball she was, the central data bank, the repository of the master DNA of the whole colony. Those gaping soldiers were prepared to die for the queen, not because they loved their mother, not because they had been drilled in the ideals of patriotism, but simply because their brains and their jaws were built by genes stamped from the master die carried in the queen herself. They behaved like brave soldiers because they had inherited the genes of a long line of ancestral queens whose lives, and whose genes, had been saved by soldiers as brave as themselves. My soldiers had inherited the same genes from the present queen as those old soldiers had inherited from the ancestral queens. My soldiers were guarding the master copies of the very instructions that made them do the guarding. They were guarding the wisdom of their ancestors, the Ark of the Covenant . . .

I felt the strangeness then, and the wonder, not unmixed with revivals of half-forgotten fears, but transfigured and enhanced by a mature understanding, which I had lacked as a child in Africa, of what the whole performance was for. Enhanced, too, by the knowledge that this story of the legions had reached the same evolutionary culmination not once but twice. These were not the driver ants of my childhood nightmares, however similar they might be, but remote, New World cousins. They were doing the same thing as the driver ants, and for the same reasons. It was night now and I turned for home, an awestruck child again, but joyful in the new world of understanding that had supplanted the dark, African fears.

I made a half-hearted attempt at some quantitative observations on the leaf-cutter ants, but I never really got anywhere; there wasn't enough time. And I fear I am not very good at getting down to properly planned research with a specified end. I can do 'pilot experiments', flitting like a butterfly as the interest takes me, but to do real research you have to write down the time course of the project in advance and adhere to it rigorously. Otherwise it's too easy to stop when you have the result you want–and that, if not quite deliberate cheating, has been a serious source of error in the history of science.

I spent much of a day watching, in horrified fascination, a clash between two rival colonies of leaf-cutters, which made me think of the First World War. The large battlefield became littered with limbs, heads and abdomens. I hoped, and half believed, that the ants didn't feel pain or fear. They were acting out genetically programmed automatisms, wound up like clockwork in their tiny brains–Maynard Smithian 'strategies'–but that doesn't in itself mean they didn't feel pain. I'd be pretty surprised if they did, but I can imagine no way of deciding the question.

The mind of an academic needs refreshing by such an interlude as mine in Panama with dear JMS, and when I returned to everyday life in Oxford it seemed just that little bit less everyday.

Go to the wasp, thou sluggard

Evolutionary economics

Natural selection is a miserly economist, invisibly counting the pennies, the nuances of cost and benefit too subtle for us, the observing scientists, to notice. Human economists weigh up rival 'utility functions', alternative quantities that an agent, such as a person or a firm or a government, might choose to maximize: gross national product; personal income; personal wealth; company profits; the sum of human happiness. None of these utility functions is 'correct' to the exclusion of others. Nor is there any one correct *agent*. You can choose any utility function you like and attribute it to any agent you like, and you'll get the appropriate, but different, result.

Natural selection isn't like that. Natural selection maximizes only one 'utility': gene survival. If you personify a gene as a metaphorical 'agent' doing the maximizing you'll get the right answer. But in practice genes don't directly behave as agents, so we shift our gaze to the level at which decisions are actually taken: usually the individual organism which, unlike a gene, has sense organs to apprehend the world, memories to store past events, computational apparatus in a brain to take decisions from moment to moment, and muscles to execute them.

Why, by the way, do biologists find it helpful to personify at all, whether with genes or individuals, to view them as 'agents'? I suspect it's because we are an intensely social species, social fish swimming through a sea of people. So much of what happens in our environment is caused by the deliberate actions of persons: how natural, then, to generalize to inanimate 'agents'. One way this inclination manifests itself is as superstition—fear of poltergeists or spirits—and that's the downside. But the upside is that scientists, so long as they know what they are doing, can use a legitimate personification as a handy and congenial short cut to getting their sums right. I once heard the Nobel Prize-winning biologist Jacques Monod make a remark that has stayed with me for its imaginative colour: 'When faced with a chemical problem of this kind, I ask myself, what would I do in this situation if I were an electron?' Physicists can explain refraction by personifying photons, adjusting their angle to minimize their travel time through media that slow them down to different extents. A photon is like a lifesaver on a beach, optimizing his trajectory towards a drowning swimmer, off to one side of the beach. He runs (fast) most of the way along the beach, then bends his angle to swim (inevitably more slowly), choosing his two angles to minimize his total travel time. When photons move from air (fast travel) to glass (slower), you'll correctly calculate the angle of refraction if you assume that they behave like agents, although not consciously calculating like the lifeguard. A stone tossed through the air follows a trajectory as if it's 'trying' to minimize a mathematical quantity that physicists can calculate. In a chemical reaction you get the right answer if you assume that the reactants are 'trying' to maximize another mathematical quantity called 'entropy'. Of course, nobody thinks these inanimate entities are really *trying* to do anything. It's just that you get your sums right if you fancy that they are, and the human mind is geared to think in terms of purposeful agents.

So, biologists shift the focus of our legitimate personification from gene to individual organism. We leave open the question of whether the organism is a conscious agent. We know the gene isn't. The organism takes decisions calculated (unconsciously is all we need to assume) to maximize the long-term survival of the genes that ride inside it–the genes that programmed, via embryonic development, the nervous system that makes the decisions. The decisions give every appearance of being those of a shrewd economist, acting as if deploying (distributing, eking out) limited resources in the service of passing genes on to future generations. The limited resources of a potato plant flow in from the sun, the air and the soil. The shrewd economist which is the plant has to 'decide' how to apportion those resources between tubers (storage for the future), leaves (solar panels to gather more sunlight to turn into chemical energy), roots (to suck up water and minerals), flowers (to attract pollinating insects with costly nectar to pay them), stems (to hold the leaves aloft to the sun) and so on. Allocating too generously to one sector of the economy (say, the roots) and too meanly to another (say, the leaves or flowers) will result in a less successful plant than a perfectly balanced distribution to all the departments of the plant's economy.

Every decision that an animal takes, whether behavioural (when to tug on which muscle) or developmental (which bits of the body to grow bigger than others), is an economic decision, a choice about the allocation of limited resources among competing demands. So are decisions on how much of the time budget to allocate to feeding, how much to subduing rivals, how much to courting a mate and so on. So are decisions on parenting (how much of the limited budget of food, time and risk to spend on the present child and how much to hold back for future children). So are decisions on life history (how much of life should be spent as a caterpillar, growing by feeding on plants, and how much as a butterfly, sipping aviation fuel from the nectaries of flowers while pursuing a mate). It's economics everywhere you look: unconscious calculations, 'as if' deliberately weighing up the costs and benefits.

That's all theory and a bit hand-wavery. Can we go out and record the moment-to-moment behaviour of animals in the wild and calculate their time budgets as examples of their economic decisions? Yes, we can; but it requires more or less continuous observation of individually marked animals, in their natural environment. And it can only be done by a skilled and meticulous observer, with huge reserves of patience, persistence, intelligence and dedication. Allow me to introduce Dr Jane Brockmann.

I met Jane when she bounced cheerily into my office in Oxford in the summer of 1977. She had been accepted by my colleague and boss, Niko Tinbergen's successor, the idiosyncratically brilliant David McFarland, to do a post-doc. As things turned out, her arrival was delayed a year, by which time David had gone on sabbatical leave, so Jane, by default, worked with me, his deputy. I came to think it a most fortunate happenstance for me, and I like to think that Jane, too, did not regret it.

Jane's PhD at the University of Wisconsin was on the great golden digger wasp, *Sphex ichneumoneus*. She came to Oxford bearing large quantities of meticulously systematic observations of the behaviour of individually marked female wasps in two different field sites, in New Hampshire and Michigan, and it was on these measurements—originally made for a quite different purpose, more related to David McFarland's field than to mine—that we ended up working together.

Not all wasps are social, like the familiar striped vespids known to Americans as yellowjackets, and to all of us as jam-loving spoilers of teatime in the garden. Many wasp species are solitary, and *Sphex* is among them. Female digger wasps, once mated, do all the work on their own, with no workers to help. Males disappear after mating, leaving the females holding the babies. Well, not literally holding. The typical cycle is as follows. The female digs a burrow about six inches deep, slightly angled from the vertical and terminating in a short side tunnel leading to a widened chamber. She then sallies forth in search of prey, which for this species of digger wasp consists of katydids (elegant, usually green, long-horned grasshoppers). She catches a katydid, stings it to paralyse but not kill it, flies home with it, then drags it down the burrow into the chamber. She repeats this several times until she's amassed a tidy pile of up to half a dozen katydids in the burrow, then lays an egg on top of the pile. She sometimes then excavates another side chamber elsewhere in the same burrow and repeats the process with fresh katydids. Finally she seals up the burrow, and proceeds to do the same with a new burrow. Some digger wasp species pick up a small stone in their jaws and use it as a hammer to tamp down the soil–a feat which has been dramatically hailed as tool use, once thought a human monopoly. When an egg hatches inside its secure, dark chamber, the larva feeds on the paralysed katydids, fattens on their nourishment, and eventually pupates and emerges as an adult wasp of the next generation, male or female.

Although they are called solitary because they don't live in huge colonies with armies of sterile workers, these wasps are in another sense not solitary. They dig their burrows close to where they themselves hatched, so 'traditional' nesting areas naturally grow up. This generates a kind of village atmosphere in a particular patch of ground, with dozens of female wasps going about their separate business, largely oblivious of each other but occasionally clashing. This closeness enabled Jane to sit in one place with her notebook and watch all the wasps in the area, each of whom she had marked using a system of colour-coded paint spots. She knew every wasp by its code name (Red Red Yellow, Blue Green Red etc.) and she mapped where each wasp's burrow was, then her next burrow and her next and so on. Among much else, Jane had observed that if a female comes across a burrow that another wasp has dug, she may save herself the trouble of digging her own and use the already existing burrow. And thereby hangs the tale we went on to tell.

Others, by the way, have told different tales. Charles Darwin was appalled by the cruelty of stinging a prey to paralyse it, keeping the meat fresh for larval consumption, rather than killing it outright. If the prey were killed it would decay and wouldn't be so good for the larva to eat. We have no way of knowing whether the prey feels pain as its tissues are slowly devoured from within while it is paralysed and unable to move a muscle to prevent it. I devoutly hope not, but the possibility gave Darwin the horrors. According to the great

French naturalist, Darwin's contemporary Jean-Henri Fabre, there's something clinically ruthless about digger wasps' precise manner of stinging. Fabre said that they carefully aim their sting, targeting one by one the nerve ganglia strung out along the ventral side of the prey—presumably achieving paralysis with an economic minimum of venom.

Philosophers, too, have used *Sphex* to construct a narrative of their own, stimulated by some classic experiments, again begun by Fabre and repeated by others since. When a hunting wasp returns to her burrow with prey she doesn't immediately drag it underground. Instead, she parks it near the entrance, then goes into the burrow empty-handed, re-emerges and only then drags the prey down. This has been described as a burrow 'inspection', the idea presumably being to check that there are no obstructions in the hole before she pulls the prey in. It is a well-replicated finding that, if the experimenter moves the prey a few inches while the wasp is down doing her 'inspection', when the wasp re-emerges she searches for the prey. Having found it, instead of dragging it straight down the hole she does another 'inspection'. Experi-menters have repeated this tease several dozen of times in succession. Every time, the 'stupid' wasp fails to 'remember' that she has only just 'inspected' the burrow and therefore there is no need to do it again. It appears to be a kind of robotic behaviour, akin to setting a washing machine back to an earlier part of its cycle, say back to 'wash' when it is about to start 'final rinse'. No matter how many times you do it, the silly machine doesn't 'remember' that it has already washed the clothes! *Sphex* has even conferred its name, in philosophical jargon, on this kind of mindless automatism–'sphexish behaviour' or 'sphexishness'. Jane is one of those wasp-watchers who is sceptical of this interpretation. She suspects that the wasp isn't being 'sphexish' at all. The misunderstanding arises from the human assumption that it is 'inspecting' the burrow. Jane and others believe it needs to approach the prey coming out of the burrow, so that when it reverses back into the burrow dragging the prey, its abdomen is aimed in the right direction. So it goes down head first, turns round inside, and emerges head first so its abdomen is pointing down the burrow when it seizes the prey to drag it down. It's just a way of taking aim, not an 'inspection' at all.

Exploring evolutionarily stable strategies

At the time of Jane's arrival in Oxford, *The Selfish Gene* had only just been published, and my mind was dominated by one of its central ideas, the evolutionary game-theoretic ideas of John Maynard Smith: the

'ESS' or evolutionarily stable strategy. I was working on a lecture, 'Good strategy or evolutionarily stable strategy', to be delivered at a conference on sociobiology in Washington the following year (see page 131). Whenever I heard a story about animal behaviour–such as the stories Jane told me about her wasps–my mind at that time immediately leapt, with almost unseemly zeal, to translate it into ESS terms.

We need ESS theory whenever it happens that the best strategy for an animal depends on which strategy most other animals in the population have adopted. 'Strategy' doesn't imply conscious deliberation; it is simply a rule for action, like a computer applet or a clockwork mechanism. It might be something like: 'Attack first. If opponent retaliates, run away, otherwise continue the attack.' Or: 'Begin with a peaceful gesture. If opponent attacks, retaliate, otherwise continue peaceful.' Sometimes there is simply a *best* strategy, in an absolute sense, regardless of what other strategies prevail in the population, and then natural selection will simply favour it. But often there is no single best strategy: it depends upon what other strategies dominate the population. A strategy is said to be evolutionarily stable if it is the best thing to do given that everybody is doing it. Why should it matter what 'everybody is doing'? Because if something else was better than what 'everybody is doing' natural selection would favour that something. So, after a few generations of natural selection, it would no longer be true that 'everybody is doing it'–that is, the original behaviour. It would not be *evolutionarily stable*. It would be evolutionarily *un*stable, in the sense that the population would be *invaded*, evolutionarily speaking, by an alternative strategy, the 'something else' that I mentioned.

Some birds have a habit (Jane Brockmann herself later, with a colleague, reviewed the scientific literature on it) called 'kleptoparasitism': piratical stealing of food from other birds. Frigatebirds make a living by pirating fish from other species (as I later witnessed in Galápagos, and with Jane herself in Florida), but kleptoparasitism also happens within species, for example some gulls. Is piracy an evolutionarily stable strategy? We answer that by imagining a hypothetical population of gulls in which almost everybody is a pirate and hardly anybody is doing any fishing. Would it be stable? No. Pirates would starve because there'd be no fish to steal. Imagine you're the sole honest fisher in a population of pirates. Even though you'd stand to lose a fair number of fish to the ubiquitous pirates, you'd still eat better than any one pirate. So a population of 100 per cent pirates would be 'invaded', over evolutionary time, by the honest fisher strategy. Natural selection would favour honest fishing and the frequency of honest fishers would increase. But it would increase only up to the point where piracy just starts to pay better.

So, piracy is not an evolutionarily stable strategy. Is honest fishing an ESS? Now we postulate a population consisting entirely of honest fishers. Would it be invaded, evolutionarily speaking, by pirates? Yes, it might well. If you were the only pirate in a population of honest fishers, there'd be rich pickings for you. So natural selection would favour piracy, and the frequency of pirates would increase.

But again, it would increase only to the point where it no longer pays compared to the alternative. So we end up with a balance between pirates and honest fishers, balanced at some critical frequency such as 10 per cent pirates, 90 per cent fishers. At the balance point, the benefits of piracy and honesty are exactly equal. If the ratio in the population should chance to swing away from the balance point, natural selection restores it by favouring whichever 'strategy' is temporarily at an advantage because it is rarer than the critical frequency.

It's an important part of the theory that the frequencies we are talking about are frequencies of *strategies*. This doesn't have to coincide with frequencies of individual strateg*ists*, although that is the way I chose, for simplicity, to express it. 'Ten per cent pirates' could mean that every individual gull spends a random 10 per cent of its time pirating and 90 per cent fishing. Or it could mean that 10 per cent of individuals spend all their time pirating. It could apply to any combination that achieves the 10 per cent frequency of the pirating *strategy* in the population. The mathematics works out the same, regardless of how the ratio is achieved. By the way, there's nothing magical about '10 per cent'. I chose that number just for a simple example. The actual critical percentage would depend on economic factors that would be hard to measure–indeed, would take a gull-loving equivalent of Jane Brockmann to measure.

Such matters, to be discussed in my Washington conference lecture, were buzzing around in my prepared mind when Jane Brockmann breezed into my Oxford room and we sat down and talked of her wasps. Sometimes they dig, sometimes they exploit the digging efforts of others, and perhaps also the others' haul of katydids. You can just imagine the excitement in my ESS-primed brain when I heard this. Pirates and Honest Diggers! Is 'digger' an ESS? If the majority of the population dig, would the digging strategy be invaded by a rival strategy called 'parasitize the digging efforts of others'? And is 'parasitize' an ESS? Probably not, because if nobody is digging there'll be no burrows to hijack. Could there be a critical ratio, at which diggers and pirates are equally successful? What excited me is that Jane evidently had mountains of hard, quantitative data. Maybe we could use her data to actually *measure* the economic benefits and costs of the two strategies. Unlike the bird pirates and fishers of my draft Washington talk, for whom nobody had any real data, Jane Brockmann's voluminous recordings of the timed behaviour of individually marked wasps had the tantalizing potential to turn into the first real field trial of mixed ESS theory.

Jane and I decided to work together on the project, but we needed more theoretical expertise, more mathematical wizardry than either of us could muster. It was time to call in the big guns, and the biggest gun in my world was my student Alan Grafen. It might seem strange to say that my guru and mentor was my own student, but it is true. He was that sort of student. He shared my enthusiasm for ESS theory, and helped me to understand its finer points and many other aspects of evolutionary biology. He taught me some of the intuitions and instincts of a mathematician, even if I couldn't follow him through the thickets of symbol manipulation. There are mathematicians and physicists who sashay into biology, thinking they can clean up its act in a week. They can't: they lack the intuitions and knowledge of a biologist. Alan is an exception. He has a rare combination of mathematical and biological intuition (shared, I believe, with his hero R. A. Fisher) which enables him to smell out the right answer to a problem almost instantaneously—and, like Fisher but unlike me, he can then go on to do the algebra if asked to do so, to prove he was right. He is now my colleague at Oxford, Professor of Theoretical Biology and a well-deserved Fellow of the Royal Society.

I first encountered Alan in 1975, when he was an undergraduate and I was tutor in Zoology at New College and in the middle of writing *The Selfish Gene*. A tutor from another college had recommended a young man from Scotland as something way out of the ordinary, and I agreed to take him on for tutorials in animal behaviour. In those days, the custom was for undergraduates to read their essays aloud in the first part of the tutorial, and we'd then discuss them. I've forgotten what Alan's first essay was about, but I vividly remember the goosepimpling awe with which I listened to it. 'Out of the ordinary' had been an understatement.

Alan's undergraduate degree was in psychology (his course had an animal behaviour option, which was why it was appropriate to send him to me for tutorials). I hoped he would go on to do a doctorate with me, but he decided to take the hard option of an Oxford master's degree in mathematical economics, supervised by Jim (later the Nobel Prize-winning Sir James) Mirrlees, a fellow Scot and one of the world's leading mathematical economists. Economics is increasingly important in evolutionary theory, so this would be a good choice for Alan whether he came back into biology or made his career as an economist. In the event, he did come back into biology and did his DPhil with me. But when Jane Brockmann came into our lives he was still studying mathematical economics, and he was to make good use of it in the wasp research which the three of us undertook together.

First things first, however. The day following Jane's arrival, as she remembers (but I had forgotten), was the occasion of the Great Annual Punt Race. Less serious than the Oxford vs Cambridge boat race and probably a lot more fun, its competing teams were those of our Animal Behaviour Research Group (ABRG) and the Edward Grey Institute of Field Ornithology (EGI). The EGI is another sub-department of Zoology, named after the former Foreign Secretary and keen ornithologist Lord Grey (who, on the eve of the First World War, had intoned the unforgettable lament: 'The lamps are going out all over Europe; we shall not see them lit again in our lifetime'). Both teams deployed an anarchically variable number of punts (flat-bottomed boats propelled by a pole thrust against–and all too often stuck in–the river bed). The name of the game was not just speed but sabotage, and Jane's abiding memory is of John Krebs (later Sir John, now Lord Krebs FRS and one of Britain's most distinguished biologists) showing up as especially ruthless on the EGI side. Did Alan, perhaps, see scope for a little ESS modelling: the Honest Punter strategy versus the Piratical Saboteur? Probably not; he has more sense and was too busy honestly poling his punt along.

Then it was down to the serious business of wasps. In her two different field sites, the main one in New Hampshire and a subsidiary one in Michigan, Jane had spent more than 1,500 hours meticulously recording the behaviour of individually colour-marked female digger wasps. She had an almost complete record of the histories of 410 burrows and of the nest-related activities that dominated 68 complete wasp lifetimes. As I said, she had originally used these records for an entirely different purpose, which she had already written up for her PhD thesis at the University of Wisconsin. Together with Alan, we now decided to use the same raw data again, to put real, measured economic values on the costs and benefits involved in ESS theory.

In my room in the Department of Zoology, with its view overlooking Matthew Arnold's dreaming spires, Jane and I worked together every day at my PDP-8 computer, typing in the numbers from her voluminous wasp records, and pushing them through numerous statistical analyses. Alan swung by every few days for a visit, casting his swift and expert eye over our statistics and patiently teaching Jane and me how to think like mathematical economists. We all three worked together on plugging his economic ideas into formal ESS models. It was a magical time, one of the most constructive periods of my working life. There was so much to learn, and I learned from both my colleagues. I like to think that I am a natural collaborator, and one of the regrets of my life is that I haven't done more of it.

The first model we tested—colourfully named Model 1—turned out to be wrong; but, in textbook philosophy-of-science fashion, its disproof gave us the clue to inventing the much more successful Model 2. When we first proposed Model 1, we regarded 'Join' as a piratical strategy, cashing in on the digging and katydid-collecting efforts of Honest Diggers. All the predictions of Model 1 turned out to be wrong, so we went back to the drawing board and came up with Model 2. Model 2 postulated two strategies called Dig and Enter. 'Dig' speaks for itself. 'Enter' means 'Enter an already dug burrow and use it exactly as though you had dug it yourself.' This is not the same as the piratical 'Join' of Model 1, for an interesting reason.

That reason stems from an additional fact about the wasps: they quite often *abandon* the burrow they are working on. Why they do this is not always clear, and indeed the reasons seem to vary. Perhaps there is a temporary problem like an invasion by ants or a centipede; or maybe a wasp died while away from her nest. What this means is that an Enterer might find a burrow untenanted and assume sole ownership. Or, if the previous owner had not abandoned it, the two would continue to work on the nest, each ignoring the other—except that if they happened to meet in the nest at the same time (which was quite rare because they spent most of their time out hunting) they would fight.

Model 2 suggests that Dig and Enter should be equally successful at a balanced frequency. When lots of digging is going on, entering becomes more successful, because there's a good supply of abandoned burrows. But if the frequency of Enter rises too high, not enough burrows are being dug, therefore there aren't enough abandoned burrows for the Enter strategy to prosper. Now, here's an interesting complication. A wasp may abandon her nest at any time, even when she has already stowed katydids in it. So an Enterer might stand to gain not only a ready-dug burrow but a ready-caught cache of katydids too. The model assumes—with justification from Jane's measurements, as she and I showed in a separate paper–that an Enterer has no way of telling whether a burrow has really been abandoned, or whether the owner is just temporarily out hunting. And we showed in yet another separate paper that each wasp behaves as if she knows how many katydids she herself has caught, but is blind to the number that another wasp might have put in the burrow.

If a wasp is the sole incumbent of a burrow–whether or not she dug it in the first place–there is a risk that she will be joined by an Enterer. And an Enterer runs the risk that the burrow she has chosen to enter is still occupied by its original owner. Both these outcomes are less favourable than being in sole possession. This is in spite of the fact that (as the discarded Model 1 emphasized) a shared nest is likely to contain more katydids (two wasps to hunt them), and there is a 'winner take all' benefit to the wasp who ends up laying an egg to take advantage of the shared cache. To use informal, personifying language: a wasp may dig a new burrow and 'hope' not to be joined by another wasp; or she may enter an existing burrow, 'hoping' it has been abandoned by its previous owner. In Model 1, Join was a strategic decision. In Model 2, both Join and Is Joined were undesirable accidents, unfortunate outcomes of a decision to Enter. Dig and Enter, by contrast, were alternative strategic decisions: at equilibrium wasps should be indifferent between the two of them. Model 2, if correct, could be summed up in a limerick:

> There's an insect called *Sphex ichneumoneus*,
> Whose encounters are seldom harmonious.
> Twixt Enter or Dig
> They don't care a fig,
> But to Join or Be Joined is erroneous.

But how could we measure such benefits in order to compare them and test Model 2? We had to think carefully about the correct way to use Jane's data to assess the benefits and costs accruing to each strategy. The evidence showed that individual wasps didn't use the same strategy every time, so there was no point in totting up the benefits and costs to individual wasps. We had to do the totting up for strategies themselves, averaged across wasps. And to do this we recognized what we called *decisions*. The whole of a wasp's adult life was a series of decisions, where each decision committed a wasp to a definite and measurable period of *time* associated with a particular burrow. Each period ended at the precise moment when the next decision was taken, initiating an association with a new burrow, whether dug or entered. Benefits and costs were credited to each decision. Then we could average the net benefit associated with decisions to dig versus decisions to enter.

A successful decision was one which resulted in an egg being laid on katydids in the burrow. If the wasp laid two eggs in different chambers in the burrow, the decision was twice as successful. But could we refine our measure of benefit by taking account of the katydids on which each egg was laid? Presumably an egg laid on a single katydid constituted less of a success than an egg laid on three katydids, because the first larva would be less well nourished. Moreover, not all katydids were the same size, and a quirk of wasp behaviour enabled Jane to measure them.

Remember my philosophical digression on 'sphexishness' and the wasp's habit of leaving prey by the burrow's entrance while she goes briefly inside and re-emerges? This gave Jane her chance. While the wasp was down in the burrow she would quickly measure the length of the katydid, taking care to replace it exactly where the wasp left it so as not to trigger any 'sphexish' repeats. Volume is a better measure of nutritional value than length, and we approximated volume as the length cubed. In the case of a shared burrow we credited the sum of the katydids to the benefit score of the wasp who ended up laying an egg on the joint cache–winner take all.

This, then, was our measure of benefit: numbers of katydids (or estimated volume of katydid flesh) on which an egg was successfully laid. How about cost? In a stroke of insight that made a big impression on both Jane and me, Alan urged that *time* was the appropriate currency of cost. Time is a precious commodity to these wasps. The summer season is brief, they don't live long; their genetic success will depend on how many times they manage to repeat the burrowing/nesting cycle before the season–and their life–ends. This, indeed, was our rationale for recognizing the concept of a 'decision': a commitment of time by a wasp to a particular burrow, for a period terminated by the next decision. Every minute of a wasp's time, then, was accounted as a *cost* on the ledger of the decision to enter upon a strategy. The net benefit of Dig was measured as an average *rate:* the sum of the benefits from all digging decisions divided by the sum of the time costs. And we calculated the equivalent net benefit rate for Enter.

Now this is where we start to 'think ESS'. According to our ESS model, we should predict that Dig and Enter will coexist at a balanced frequency where their success rates are equal. If the frequency of Enter were to drift above this equilibrium, natural selection would start to favour Dig, because too many enterers would find themselves sharing a burrow and at risk of engaging in, and perhaps losing, a costly fight. And vice versa: if Enter drifts below the equilibrium, Enter is favoured because there are rich pickings among abandoned burrows. The observed frequency of Enter in the New Hampshire population was 41 per cent, and we conjectured that this might actually be the equilibrium frequency for the New Hampshire population. In that case the measured success rates of Dig and Enter should be equal. So we looked.

The actual rates were not identical (0.96 versus 0.84 eggs per hundred hours, with a similar conclusion for the katydid volume scores) but there was no statistically significant difference between them, and they were close enough to encourage us to test the model further. Alan did some clever algebra which enabled him to deduce from the model four further predicted numbers, which we could then compare with the observed numbers. The numbers concerned were the proportions of wasps who fell into four categories, and the predictions were the proportions that *ought* to be observed if the population was at equilibrium according to our ESS model. The table below shows the results: these observed figures from New Hampshire failed to falsify the predictions of Model 2. This pleased us.

Proportion of wasps who:	Observed	Predicted
Dig then abandon	0.272	0.260
Dig and do not abandon	0.316	0.303
Enter and find themselves alone	0.243	0.260
Enter and find themselves sharing	0.169	0.176

Nevertheless, we were mindful of the principle that a model whose predictions are not falsified by the observed data is impressive only insofar as those predictions were *vulnerable* to falsification. If you use too much of the observed data in order to deduce your predictions, the predictions almost can't help being right. We demonstrated by computer simulation (plugging in random hypothetically possible data instead of Jane's real data) that this was far from the case with our Model 2 predictions. The model could very easily have been wrong, yet it wasn't in fact. It had stuck its neck out and survived. Karl Popper would have loved it.

Well, it survived in New Hampshire. As if to ram home the point that Model 2 could very easily have been wrong, it actually *was* wrong in the other population Jane studied, in Michigan. We were disappointed, but stimulated to think constructively about why this was. We came up with various suggestions, of which the most interesting was that the Michigan wasps were adapted to a different environment from the one in which Jane studied them. Perhaps the Michigan wasps were 'out of date', their genes having been adapted to some earlier set of conditions–somewhat as our human genes are adapted to a hunter-gatherer way of life in Africa, but we now find ourselves in cities, with shoes, cars, refined sugar and other food surpluses. The Michigan wasps were working in a large, raised flowerbed, which must have been pretty different from their normal environment, and indeed was different from the more natural-seeming environment of the New Hampshire wasps.

In spite of its failure in Michigan, the success of our Dig/Enter model in New Hampshire was striking, and it remains one of only a few quantitative field tests of Maynard Smith's elegant theory of the 'mixed ESS' (in this case the 'mixture' is between Dig and Enter). For me, it exemplified the joy of working in collaboration with compatible colleagues having complementary knowledge and skills.

Consider her ways and be wise
Our work on the ESS model came to an end and was sent off to the *Journal of Theoretical Biology*

for publication, becoming 'Brockmann, Grafen and Dawkins, 1979'–the citation engendering, as always, a nice sense of accomplishment. Jane and I continued to work together on a larger paper called 'Joint nesting in a digger wasp as an evolutionarily stable preadaptation to social life'. This paper presented, and substantiated statistically, many of the background facts that we had used in the ESS paper. And it had its own theoretical aim, which was to contribute to the controversial debate over how social behaviour in insects might have originated from solitary ancestors. Could the kind of uncooperative, inadvertent nest-sharing, which we had shown to be evolutionarily stable in solitary digger wasps, have been the forerunner of the massive cooperative colonies of wasps, ants and bees which are such a spectacular feature of life on Earth? The close genetic relatedness within social insect colonies is certainly an important factor, as my friend and colleague Bill Hamilton had so persuasively argued. But could there have been other pressures predisposing to social life, and could one of these other pressures have been foreshadowed by something like our ESS model in ancient wasp ancestors? Jane and I worked hard on this paper, mostly in her lodgings in Oxford (tastes and smells are notoriously evocative, and I associate those happily productive times with the taste of Cinzano and a slice of lemon amid clinking ice cubes), and eventually we published it in the journal *Behaviour*.

The organization of the paper was unusual, in a way that I am quite proud of and would like to see emulated. The standard plan for a scientific paper was, and still is, the one I had fought a losing battle against during my four years, from 1974 to 1978, as editor of the journal *Animal Behaviour* (assisted by the vivacious Jill McFarland, wife of my then boss and predecessor as editor, David McFarland): Introduction, Methods, Results, Discussion. This plan, although dull, makes sense for a certain kind of scientific study in which a single experiment is planned, executed and discussed. But what if a series of experiments is performed sequentially as each prompts the next? Pose a question, try to answer it with Experiment 1. The result of Experiment 1 raises a further question, answered by Experiment 2. Experiment 2 needs clarifying by Experiment 3; the result of Experiment 3 provokes Experiment 4. And so on. It seemed to me that the obvious plan for such a paper would be: Introduction; Question 1, Methods 1, Results 1, Discussion 1, leading to Question 2, Methods 2, Results 2, Discussion 2, leading to Question 3, Methods 3 . . . etc. But time after time, as editor, I received papers laid out like this: Introduc-tion; Methods 1, Methods 2, Methods 3, Methods 4; Results 1,

Results 2, Results 3, Results 4; Discussion. Seriously! What an utterly bonkers way to write a paper: tailor-made to destroy the narrative flow of a story, kill the interest, emaciate the relevance to the rest of the subject! I struggled as editor to persuade authors to abandon it, but old habits die hard.

When Jane and I came to write our paper, we had a narrative flow to offer, even though it took the form of a series of observational measurements rather than experiments. Our conclusions constituted a series of factual statements about the wasps, each one needing statistical justification, and each one prompting a new question which led to the next factual statement in building up an argument about the possible origins of social life in insects. So we wrote out a *summary* of our paper consisting of thirty discrete factual propositions, every one of which we had substantiated with quantitative evidence. Each of those thirty propositions then became a *heading* in the paper itself. Underneath each heading, the text, tables, diagrams, statistical analyses etc. were directed at demonstrating the truth of the heading. You could get the gist of the paper simply by reading the headings. And indeed, since the journal required a Summary at the end of each paper, we simply reprinted all the headings in sequence as a single summary narrative. The same scheme was independently adopted by Jim Watson in his excellent textbook of molecular genetics. And I was to use a version of it again much later in the last chapter of *The Greatest Show on Earth*, where I took the famous final paragraph of Darwin's *Origin of Species* and made each phrase, in sequence, a section heading of my last chapter. The body of each section was a meditation upon Darwin's phrase.

Here is the sequence of headings from Jane's and my paper, which constitute a concise summary of the facts that we demonstrated. Bear in mind, if you read them, that each one is substantiated in the paper itself by the text, numbers and analyses that follow it.

One suggested evolutionary origin of insect sociality is joint nesting by females of the same generation.

Long before selection favoured joint nesting itself, it might have favoured some other incidental pre-adaptation such as the habit of 'entering' abandoned burrows, found in the usually solitary wasp *Sphex ichneumoneus*.

We have comprehensive economic records of individually marked wasps.

There is little evidence of consistent individual variation in nesting success.

Wasps often abandon the nests they have dug, and other individuals adopt or 'enter' these empty burrows.

'Dig/Enter' is a good candidate for a mixed evolutionarily stable strategy.

Digging and entering decisions are not characteristic of particular individuals.

The probability of entering is not conditional upon whether it is early or late in the season.

There is no correlation between an individual's size and her tendency to dig or enter.

There is no correlation between an individual's egg-laying success and her tendency to dig or enter.

Individuals do not choose to dig or enter on the basis of immediate past success.

Individuals do not dig and enter in runs, nor do they alternate.

Wasps do not choose to dig or enter on the basis of how long they have been searching.

At one study site digging and entering decisions are roughly equally successful, but at another entering decisions are perhaps slightly more successful.

Entering wasps seem not to distinguish empty, abandoned burrows from burrows that are still occupied.

As a consequence of indiscriminate entering, two females sometimes co-occupy the same burrow.

Co-occupation should not be called 'communal' because the wasps usually share the same brood cell, not just the same burrow.

One might expect that wasps would gain some benefit from co-occupying, but they do not, for a number of reasons.

Only one egg is laid in a shared cell, and obviously only one of the two wasps can lay it.

Two wasps together do not fetch noticeably more food than one alone.

Two wasps together are no quicker at provisioning a cell than one wasp alone.

Wasps sometimes duplicate each others' efforts when they co-occupy a nest.

Co-occupying wasps often have costly fights.

About all that can be said for joint nesting is that it may reduce parasitism.

The risk of joint nesting is the price wasps pay for the advantages of taking over an already dug and abandoned burrow.

A mathematical model assuming 'dig/enter' as a mixed evolutionarily stable strategy has some predictive success.

If the parameters changed quantitatively, the *Sphex* model could come to predict selection in favour of joint nesting as such.

The selection pressures would have to be very strong to overcome the demonstrated disadvantages of co-occupying.

Variants of the *Sphex* model may be applicable to other species, and may help our understanding of the evolution of group living.

The theory of evolutionarily stable strategies is relevant not just to the maintenance of behaviour but to its evolutionary change.

Our conclusion was that the Dig/Enter model which fitted the New Hampshire population of *Sphex ichneumoneus* could, if named economic parameters were to change over evolutionary time, move into any of a number of 'spaces', including 'social space' (see graph opposite). We took our Model 2 and calculated what would happen if we systematically varied two of the terms in the algebra, namely B_4, the benefit of joining, and B_3, the benefit of being joined. Would the model yield a stable ESS at different values of these two benefits?

The New Hampshire population of *Sphex ichneu-*

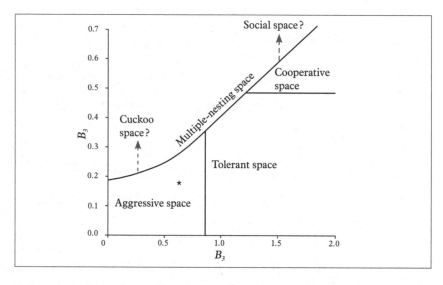

Plan view of 'economic landscape' for the evolution of social insects based on our game theory model of digger wasps' behaviour. The two economic variables, B_3 and B_4, are the benefits to a wasp of being joined, and of joining, respectively. The star represents Sphex ichneumoneus in 'aggressive space' (where a wasp is better off being solitary). The map shows that, if economic conditions change, there is a smooth trajectory from aggressive space all the way through 'tolerant space' to 'cooperative space' and 'social space'.
Source: *H. J. Brockmann, R. Dawkins and A. Grafen, 'Joint nesting in a digger wasp as an evolutionarily stable preadaptation to social life', Behaviour 71 (3), 1979, pp. 203–44.*

moneus is indicated by the star as being stable in 'aggressive space' (where the wasps are better off being alone). Our analysis showed that the model allows a smooth gradient, as the B values change (over evolutionary time), through 'tolerant space' (where wasps who are joined do better than lone wasps but joiners do worse) to 'cooperative space' (where joiners do better than lone wasps and wasps who are joined by enterers do best of all). All the way along this evolutionary gradient there are stable solutions such that both digging and entering would be favoured at the (changing) equilibrium frequency. Our analysis showed that, even without the strong kinship reasons which undoubtedly apply in many cases, social behaviour could evolve from a *Sphex ichneumoneus* type of ancestor. And of course close kinship only adds to the pressure to become social and stay social.

Interlude in Florida

In 1978 Jane's year at Oxford came to an end, and we sadly relinquished her to the University of Florida in Gainesville. But we three musketeers were to be reunited. In 1979 I took a sabbatical leave in Jane's Gainesville lab, and I arranged for Alan to join us towards the end of my time there. Jane was by then working on another species of solitary wasp, *Trypoxylon politum*.

These 'mud daubers' are related to *Sphex* and have similar habits—but, instead of digging burrows underground, they build aerial 'burrows' up on walls, under bridges, on rock faces. The aerial burrows are tubes built of mud, carried dollop by dollop from streams. The tubes often are stacked side by side, hence the name 'organ pipe mud dauber'. Having built its tube, the wasp then provisions it, like *Sphex*, except that *Trypoxylon* hunt spiders instead of katydids and they pack a succession of them into one tube, separated by mud partitions. Jane worked under a bridge on these wasps, recording the comings and goings of marked individuals just as she had with *Sphex*. Alan worked with her on the theory, and we both spent time under the bridge with Jane and some of her students, helping to observe the wasps—and dodging the cottonmouth snakes, which frightened me more than they frightened the natives.

I loved watching the building behaviour of these wasps, partly because I was writing *The Extended Phenotype* at the time, and animal artefacts played a starring role in my argument. I was especially charmed by the wasps' habit of exploiting what looked to me like the physical phenomenon known as thixotropy, as a kind of 'welding' technique. When a wasp returned to her tube with a ball of mud in her mouth, she would apply the ball to the lip of her tube. She then buzzed her wings noisily and the vibration transmitted through her jaws visibly caused the mud to 'melt', like quicksand: not just the new ball of mud but (I suspect, although I couldn't see for sure) the mud of the lip of the tube, so they melted into each other. Melting, melding, welding–it did indeed look like welding.[1] In fact, it looked to me as though the vibration was acting on the mud in the same way as the heat of a welder's acetylene flame acts on metal, temporarily liquefying the lip so that new mud bonded securely with old. Jane doesn't think anyone has published this 'thixotropy' suggestion before, so I offer it here for what it's worth.

Jane and I conducted a graduate seminar in evolution and behaviour at Gainesville, in which we were joined by two other professors. My chief memory of these weekly gatherings was the way in which they became increasingly dominated by the intellectual power of Alan Grafen. On the face of it he was just a graduate student among many (and one of the youngest), but it was remarkable how we all, students and professors alike, fell into the habit of turning to Alan to resolve our difficulties and tell us, in his sharp Scottish tones, how to think clearly about them and reach the correct conclusion.

---

[1] Just about the only skill I learned in the vaunted workshops of Oundle School, as I recounted in *An Appetite for Wonder.*

My Florida sabbatical wasn't all wasps and work. Jane, Alan and I were joined by Donna Gillis, a friend of Jane's from the Zoology Department, and the four of us set off to discover more of Florida. We went on drives to Disney World (Alan insisted on taking all the most hair-raising rides) and Sea World (Alan was the first to volunteer to be pushed into the pool by a performing seal). We went to the university's Seahorse Key marine biology research station on the Gulf of Mexico coast, where we self-catered and slept in dormitory bunks. We saw *Limulus* ('horseshoe crabs', although they aren't crabs at all but distant relatives of spiders; Jane later did research on these 'living fossils'). We saw thousands of ghost crabs (they really are crabs) scuttling into their burrows as we approached, leaving their easily traced footprints. Most memorable was the phosphorescence in the sea, caused by the disturbance of microscopic organisms in the plankton. We played 'ducks and drakes', skimming flat stones over the water and watching the glowing ripples as the stones struck the surface. Donna danced on the night beach, her toes in the wet sand inscribing patterns of glowing and then fading phosphorescent blue, charmingly singing of herself in the third person: 'She's dancing.'

On another beach, Alan and I swam naked, which alarmed Jane and Donna because–to Alan's and my surprise–apparently it was, and is, illegal, even at night. Now that I think about it, an incident many years later gives me reason to believe the illegality really is taken seriously in the United States. The anthropologist Helen Fisher and I were skinny-dipping in Lake Michigan one warm summer night after a hot day of conference speeches at Northwestern University. A police car drew up on the road a hundred yards away. It was dark, so I don't know how they had spotted us, but they trained a searchlight on us and bellowed through a bull-horn: 'You are subject to arrest. You are subject to arrest. You are subject to arrest.' Panic-stricken, without waiting to get dry, Helen and I tugged our clothes on as we ran. No such incident marred Alan's and my rather swift and brief dip in the moonlit Florida waves. I suspect, with hindsight, that we did it for reasons of bravado more than enjoyment. Jane now tells me she discourages her students from swimming off that particular beach because one often sees sharks there.

Back at Gainesville, much of my time on this sabbatical was spent writing chapters of *The Extended Phenotype*, taking advantage of the library, and consulting Alan almost daily on matters of evolutionary theory and how to think straight about it. But I also spent time collaborating with Jane (again relying heavily on advice from Alan) on a new paper called 'Do digger wasps commit the Concorde fallacy?'

The Concorde fallacy

Economists recognize the sunk cost fallacy–throwing good money after bad. Before I had heard of it, I had recognized the same error in the context of evolutionary biology, and named it the 'Concorde fallacy', first in a 1976 paper in *Nature* which I wrote jointly with an Oxford undergraduate pupil, Tamsin Carlisle, and then in *The Selfish Gene*. Here's the definition of the Concorde fallacy in the *Oxford Dictionary of Psychology*, edited by Andrew Colman:

> Continuing to invest in a project merely to justify past investment in it, rather than assessing the current rationality of investing, irrespective of what has gone before. Thus gamblers often throw good money after bad in an attempt to escape from escalating debts . . . and the length of time a female great golden digger wasp *Sphex ichneumoneus* is prepared to fight over a disputed burrow depends not on how much food there is in the burrow but on how much she herself has put there, the wasp that has carried the largest amount of prey into the burrow being generally least willing to give up fighting. The phenomenon was first identified and named in an article in the journal *Nature* in 1976 by the British ethologist Richard Dawkins (born

1941) and his undergraduate student Tamsin R. Carlisle (born 1954). Also called the *sunk cost fallacy*, especially in decision theory and economics . . . [Named after the Anglo-French supersonic airliner, the Concorde, the cost of which rose steeply during its development phase in the 1970s so that it soon became uneconomical, but which British and French governments continued to support to justify past investment.]

Another name I've used is the 'Our-boys-shall-not-have-died-in-vain fallacy'. During the mounting opposition to the Vietnam War, as I remember from my teargas-dodging days in California in the late sixties, one of the arguments against pulling out went something like this: 'Lots of Americans have already died in Vietnam. If we withdraw now, they will have died in vain. We can't let all those boys have died in vain, so we have to go on fighting' (and a whole lot more of our boys will die, but we don't mention that). Jane and I were mildly disturbed to find, on re-analysing her data, that her *Sphex* digger wasps appeared to commit their own version of the Concorde fallacy. Here's how.

It doesn't often happen that an entering wasp meets an incumbent at their shared burrow, but when they do they fight, and the loser usually flees for good, leaving

Sphex ichneumoneus *wasps fighting near the entrance to their shared burrow*
Drawing by Jane Brockmann

the winner in sole control of however many katydids the two wasps have gathered. Presumably the wasps are fighting to determine who gets the burrow, which is valuable to both of them. The value of the burrow is larger when there are lots of katydids in it, and that value should be the same, you might think, for both wasps, regardless of who caught the katydids. So, if the wasps behave like rational rather than Concordian economists, you would expect both to fight harder for a well-provisioned burrow than for a scanty one.

But that is not what happens. Instead, Concorde-fashion, each wasp fights as if the value of the nest is determined by how many katydids she herself has caught, regardless of the true future value of the burrow. This showed itself in two ways. First, there was a statistical tendency for the wasp who had contributed the most katydids to end up winning the fight. Second, the *duration* of each fight was correlated with the number of katydids that the *loser* had contributed. Here's the Concordian rationale for that result. Each fight comes to an end when one decides to flee, and this defines her as the loser. A Concordian wasp will give up early if she has contributed few katydids, late if she has contributed many. Hence the correlation between fight duration and number of katydids that the loser had caught.

Jane and I fretted, half jokingly given my role in naming it, over the fact that digger wasps appeared to commit the Concorde fallacy. Was it just that, in John Maynard Smith's jocular phrase, 'Natural selection has bungled again'? We sought advice, as ever, from Alan, who pointed out a number of things. Animal design, like design by human engineers, is not perfect in some absolute sense. Good design is always constrained. A suspension bridge is not guaranteed to withstand all conditions: the engineer designs it as cheaply as possible within a specified safety margin. What if, for some reason, the sensory and nervous apparatus that a wasp needs to count katydids in a shared burrow is costly, while the apparatus needed to measure her own effort in catching katydids is cheap? The most economical 'design' of wasp might then indeed look Concordian, especially if–as is indeed the case–sharing of burrows is not very common.

As it happens, there is some indirect evidence that the nervous/sensory apparatus a wasp might use to count prey is indeed costly to run. It comes from a related genus of digger wasp, *Ammophila*, studied in the Netherlands by Gerard Baerends, incidentally the first graduate student of my old maestro Niko Tinbergen. Unlike Jane's *Sphex ichneumoneus*, Baerends's *Ammophila campestris* is a progressive provisioner. Instead of gathering a full store of food, laying an egg on it and then sealing the burrow and leaving, *Ammophila campestris* brings food (caterpillars, not katydids in this case) daily to its growing larvae. Moreover, it has two or three burrows on the go at any one time. The ages of its larvae are staggered, and their food needs are correspondingly different. The wasp 'knows' that younger larvae need less food than older larvae, and she feeds them unequal quantities accordingly. But now, here's the surprising fact. She assesses the needs of each of her larvae only during a single early morning inspection round of all her burrows. After the inspection round is over she behaves for the rest of the day as if she is completely blind to the contents of the burrows.

Baerends showed this in a neat experiment. He

systematically switched larvae around from burrow to burrow. No matter how small the larva now in a particular burrow, the wasp would continue to feed it the large prey appropriate to the larger larva who had occupied the burrow during the morning inspection round. And vice versa. It is as though the wasp has an instrument for measuring the contents of a burrow, but this instrument is costly to run so she switches it on only once a day, during the morning inspection round; the rest of the day the apparatus is switched off to conserve running expenses. This would explain the blunders made by the wasps in Baerends's experiments, blunders compatible with the view that they were now blind to burrow contents. Of course, such 'blunders' would not normally matter because, in the absence of a Baerends, larvae don't hop from burrow to burrow.

Ammophila campestris really needs its assessment instrument because its normal habit is progressive provisioning of multiple burrows housing larvae of different ages; but even so, it severely limits the time it is switched on. *Sphex ichneumoneus*, working on only one larva at a time, and sharing burrows rather rarely, doesn't have such a strong need for this costly apparatus, so it never switches it on, or doesn't possess it at all. It therefore appears to commit the Concorde fallacy. This, at any rate, was how we rationalized our results.

And really, we shouldn't have allowed ourselves to be disappointed by the wasps' 'performance'–especially if they are as 'sphexish' as some philosophers believe. Intelligent humans in positions of power commit the Concorde fallacy, and humans make decisions a lot more silly than that when assessing risks, costs and benefits, as psychologists such as Daniel Kahneman have shown us.

The delegate's tale

D avid Lodge, in his campus novel *Small World*, compares the academic conference to a Chaucerian pilgrimage:

> The modern conference resembles the pilgrimage of medieval Christendom in that it allows the participants to indulge themselves in all the pleasures and diversions of travel while appearing to be austerely bent on self-improvement.

I warm to the analogy, having used Chaucer for a different purpose in *The Ancestor's Tale*. Here I choose six conferences out of hundreds, as representative stations along my scientific pilgrim's way.

Lodge's cynical view was not contradicted by a memorable conference during the time I was writing

The Selfish Gene, sponsored by the Boehringer pharmaceutical company in a splendidly over-the-top German castle. The subject was 'The creative process in science and medicine', and it was certainly the poshest conference I have ever attended. The main guest list comprised scientists and philosophers of immense distinction, many of them Nobel Prize-winners. Each of these illustrious figures was allowed to bring along a couple of junior colleagues–squires to their knights, as it were. My old maestro Niko Tinbergen was one of the 'knights', and he took Desmond Morris and me as his squires. Other knights (some of them knights in the literal sense) were Sir Peter Medawar (immunologist, essayist and legendary polymath), his 'philosophical guru' Sir Karl Popper, Sir Hans Krebs (the world's most famous biochemist), the great French molecular biologist Jacques Monod and several other household names in science, each with their juniors in attendance. There were only about thirty of us in all. I felt immensely fortunate to be present, and hardly dared say a word.

We sat around a large polished table (I don't think it was actually round, which is a pity for my 'knights' conceit), with our names proud in front of us (incidentally, why, at such events, does the name so often face its owner, who presumably knows who he is, rather than out into the room towards those who might make use of it?). Strewn around the table were notepads and pencils, bottles of mineral water, sweets (ugh) and cigarettes galore. These last were more than usually unfortunate because Karl Popper had a famous distaste for cigarette smoke. On one occasion at a different conference he had risen from the floor to make a special request that nobody should be allowed to smoke. Nowadays such a plea would not be necessary; it would go without saying. But those days were different, and it was a symptom of the regard in which the great philosopher was held that the chairman acceded to his request. Or almost. What he said was: 'In deference to Sir Karl and out of respect for him, please would any delegate who wishes to smoke leave the hall and smoke outside.' Sir Karl rose again: 'No, zat would not be good enough. When zey come back in, I could smell it on zeir bress.'

So you can imagine the consternation raised by the tobacco largesse scattered over the conference table in our opulent *Schloss*. Every time the hand of a smoker strayed tablewards, a flunkey would come bustling over to clutch a sleeve and whisper, 'No please, not to smoke, Sir Karl cannot stand it . . . *bitte schön.*' But, as far as I remember, the cigarettes remained on the table in full view, to tempt the unfortunate addicts for the duration.

The conference was structured loosely around a series of invited presentations, followed by questions from around the table and extended discussion. Every morning at breakfast, with German thoroughness, each of us was handed a massive pile of paper, a complete word-for-word transcript of everything that had been uttered the previous day: every last *um* and *er* and infelicitous restarting or recycling of a sentence. I pitied the red-eyed night shift of typists slaving through the small hours to produce this torrent of verbiage. But there was a problem: how to tie each pearl to its oyster–who had said what, in other words. The chairman of each session was briefed to make us preface each intervention with our name. Peter Medawar, who chaired the opening session, also asked the first question and identified himself to the tape recorder with characteristic aplomb: 'This is Medawar, shamelessly abusing the privileges of the chair.' But most people, in the cut and thrust of discussion, forgot to say their names, so an alternative solution was deemed necessary. It proved to be even more distracting than the cigarettes. On a rotating stool, perched high on top of the massive, polished table, sat a young woman in a short skirt. Every time one of the delegates started to speak she would swivel, like a gun turret on a battleship, to locate him and write down in a notebook his name and first sentence. These notes were then used by the night typists to attribute each laboriously rendered paragraph.

It was fascinating for a young scientist to eavesdrop as the giants of his profession unmasked their creative processes. Hans Krebs's recipe for how to win a Nobel Prize was too modest to be credible: it amounted to 'Go into the lab every day at 9 a.m., work all day until 5 p.m., then go home; and repeat the process for forty years.' I've already quoted Jacques Monod's engaging revelation that he was in the habit of imagining himself an electron deciding what to do next. I did something similar when I asked myself, following my scientific hero Bill Hamilton, what I would do if I were a gene, trying to get copies of myself passed on to future generations.

At the very end of the conference, one of the invited guests, a Japanese physicist who had uttered not a word throughout, timidly asked if he might finally say something. He explained that he would lose face if he went back to Japan and confessed that he hadn't spoken. It would have been technically enough if he had stopped there, but he went on to say something rather interesting. Most physicists, he pointed out, are obsessed with symmetries of various kinds. Japanese aesthetics, on the other hand, favour asymmetry, and perhaps this gives Japanese physics a different perspective. I immediately thought of Pamela Asquith, a young

Canadian anthropologist friend doing a study in what might be called meta-primatology–the comparative study of primatologists. Her thesis was that Japanese primatologists brought a different cultural perspective to their monkeys, to complement the western angle. A comparable point has been argued for female primatologists, of whom there are numbers disproportionate to other sciences.

PBM

Of all the Nobel Prize-winners I was in awe especially of Peter Medawar, who had long been a hero of mine, as much for his writing style as for his science. Badly disabled by a stroke at a disquietingly young age, he was assiduously cared for by his wife, Jean (the knot of his tie seemed softer and looser than a man would produce). The slight slur in his speech scarcely hindered his wit and erudition. Only once did I glimpse a chink in the armour of valiant bonhomie. I was hurrying along a corridor, nearly late for one of the lectures, and I passed the Medawars, also rushing as fast as Peter could, which was not very. Jean, in an urgent hiss, called me back ('Richard, Richard') and appealed for help in getting him through the door into the conference room. As I did so, I was moved by her solicitude for him and by his evident anxiety not to be late, a guard-dropping moment that belied the outward patrician nonchalance.

On another occasion he mentioned that he and my father had been exact contemporaries as schoolboy biologists at Marlborough College. 'Your father and I were united in our detestation of one A. G. Lowndes.' Lowndes had been their much loved and legendarily successful biology teacher, and I reminded Sir Peter that he had written an affectionate obituary of his erstwhile mentor. 'Oh well, I felt that when the old bugger croaked I ought to do my bit for him.'

At some point around that time I was invited by Redmond O'Hanlon, who was on the editorial staff of the *Times Literary Supplement* (*TLS*), to review one of Peter's books. I submitted a rave notice, one of the most enthusiastic book reviews I have ever written (alongside some stinkers whose style, now that I think about it, owes inspiration to Medawar himself).[1] My only slightly negative sentence was a judgement that I set down in order to repudiate it: 'Some have described Medawar as a "loose cannon on deck" but I would vigorously contest this charge . . .' I was never sent a proof to correct, and when the issue came out I was horrified to find that they had cut all my most glowing

---

[1] For a selection of these reviews, please see the e-appendix.

praise and printed the review under the headline 'Shots from a loose cannon'. I stormed round to Redmond's office above his wife Belinda's famous Annabelinda dress shop in Oxford. Surrounded by what seemed to be an overflow collection of stuffed reptiles, shrivelled monkey hands, fetish objects and other bizarre memorabilia from his travels, he listened to my prolonged tirade in silence, then left the room without a word. He returned carrying an object which he solemnly, and still without a word, presented to me. It was a double-barrelled shotgun. I shall never know whether it was loaded (Redmond's eccentric adventurism is such that it is possible) but in any case the gesture paradoxically disarmed me. I don't think Redmond was actually responsible for the mischievous sub-editing, and Peter was magnanimous when I wrote to tell him of the episode.

A decade or so later, near the end of Peter's life, Jean invited me to one of her dinner parties in their house in Hampstead, north London. Peter's physical condition had deteriorated since our meeting in Germany, but his mind was as sharp as ever, and she would invite two or three guests every week to keep him entertained. Those, like me, who scarcely knew him personally, felt especially honoured to be invited, and the evening was not to be forgotten. My fellow guest was the distinguished journalist Katharine Whitehorn, and I suspect that she entertained him a lot better than I did, awed as I was. The only concession to his malady was when he apologetically excused himself to retire early to bed: 'I'm afraid I'm a very sick man.'

I felt honoured again when, in June 2012, Charles Medawar presented me with a priceless book from his father's library: the *Festschrift* volume presented to the great Scottish naturalist D'Arcy Thompson on his retirement, edited by Peter and signed by all the authors including V. B. Wigglesworth, J. Z. Young, J. H. Woodger, E. C. R. Reeve, Julian Huxley, O. W. Richards, A. J. Kavanagh, N. J. Berrill, E. N. Willmer, J. F. Danielli, W. T. Astbury, A. J. Lotka, G. H. Bushnell, and of course the two editors, W. E. Le Gros Clark and P. B. Medawar himself. D'Arcy Thompson's own signature was there as well, glued in for good measure. Most of those authors were household names to me and my undergraduate contemporaries in zoology, and D'Arcy Thompson was a particular hero, described by Peter Medawar as

an aristocrat of learning whose intellectual endowments are not likely ever again to be combined within one man. He was a classicist of sufficient distinction to have become President of the Classical Associations of England and Wales and of Scotland; a mathematician good enough to have had an entirely mathematical paper accepted for publication by the Royal Society; and a naturalist who held important chairs for sixtyfour years . . . He was a famous conversationalist and lecturer (the two are often thought to go together, but seldom do), and the author of a work which, considered as literature, is the equal of anything of Pater's or Logan Pearsall Smith's in its complete mastery of the *bel canto* style. Add to this that he was over six feet tall, with the build and carriage of a Viking and with the pride of bearing that comes from good looks known to be possessed.[1]

If I am asked to name a single scientist whose writing style inspired me more than any other, I would say that other aristocrat of learning Peter Medawar, and maybe this short extract tells you why.

Double Dutch

In 1977 I was invited to present a lecture at the International Ethological Conference held at Bielefeld in West Germany. At that stage of my career, it was quite an honour to be invited (as opposed to volunteering) to give a lecture at this flagship conference of my then field of animal behaviour, and I took great trouble over

---

[1] P. B. Medawar, 'D'Arcy Thompson and growth and form', in *Pluto's Republic* (Oxford, Oxford University Press, 1982).

my speech, which I called 'Replicator selection and the extended phenotype'. Subsequently published in the journal *Zeitschrift für Tierpsychologie*, it was the first time I introduced the idea, and the phrase, 'extended phenotype', which was to become the title of my second book.

The International Ethological Conference is held every two years in a different country and I attended eight of them: in The Hague, Zurich, Rennes, Edinburgh, Parma, Oxford, Washington and Bielefeld. In *An Appetite for Wonder* I mentioned the 1965 conference in Zurich, where I presented my doctoral research for the first time and was rescued from a technical debacle by the Austrian ethologist Wolfgang Schleidt. The conferences were inaugurated long before my time, as rather cosily small gatherings dominated by the flamboyantly handsome Konrad Lorenz and his more quietly thoughtful and also handsome colleague Niko Tinbergen. Speeches were prolonged by the fact that these two grand old men of the subject–not really old then but already grand–took turns to translate for the benefit of the audience, both ways between German and English. By the time I started attending, the conferences had become much larger, German speeches were diminishing in frequency, and there was no longer enough time for the translations.

Language problems hadn't gone away, however. At

another one of these biennial conferences, in Rennes, an elderly delegate from the Netherlands advertised his lecture in the programme as being in German. I am sorry to say that consequently, as he rose to speak, the majority of the Anglo-American contingent in the audience shuffled shamefacedly to the exit. I remained in my seat out of embarrassed politeness. This priceless Dutchman waited at the lectern, smiling patiently, until the last ignominious monoglot had slid out. Then his smile broadened into a contented beam as he announced (the Dutch are perhaps the most linguistically gifted of Europeans) that he had changed his mind and would now deliver his lecture in English. His audience then became yet more depleted.

The leading French delegate to that conference took a straw poll, the evening before her big speech, on how many would understand her if she obeyed orders from her French masters and spoke in French. Embarrassingly few put up their hands and consequently she decided to speak in English. Her change of mind was well advertised in advance and she drew a good audience for an excellent lecture.

At that same Rennes conference, a colleague from Cambridge jabbered his lecture much too fast. At the end, a questioner stood up and furiously berated him in equally rapid-fire Dutch. Unschooled in the language as I was, I was one of many who got the point. We native

English speakers must not abuse the privilege we enjoy: through various accidents of history, our *lingua anglica* has emerged as the new lingua franca. I suspect that this clever Dutchman had actually understood my Cambridge friend perfectly well, and was complaining not on his own behalf but for the benefit of others–probably not Dutch–who would have struggled to understand high-speed Cambridge English. I have done the same kind of thing, with respect not to language but to difficult scientific matters which, I had reason to fear, might not have been understood by some students in the audience. In other words, like (I suspect) my dear mentor Mike Cullen,[1] I have sometimes pretended not to understand a point of science in order to force a speaker to be clearer. In any case, I was humbled by this Dutchman's public-spiritedness, to the extent that when I returned to Oxford I resumed my interrupted (since school) study of German, under the tutelage of the wonderful Uta Delius–only to be told, by a disgracefully insular colleague, 'Oh, you shouldn't do that. It'll only *encourage* them.' (Those colleagues and friends who can–with affection and not much difficulty I suspect–guess his identity will hear the words in his distinctive intonation.)

For my plenary talk at the Bielefeld conference I hope I spoke slowly and clearly enough to be understood by

---

[1] My eulogy at his memorial service is largely reproduced in *An Appetite for Wonder.*

all. At any rate, the only negative comment from a poly-glot Dutchman was a furious attack on the colour of my tie. Admittedly it was a lurid purple, conspicuously–and to his quivering sensibilities jarringly–out of har-mony with the rest of my attire.

I never nowadays commit such a sartorial solecism, by the way. The only ties I ever wear are hand-painted by my multitalented wife, Lalla, all to her own animal designs. The subjects include penguins, zebras, impalas, chameleons, scarlet ibis, armadillos, leaf insects, clouded leopards and . . . warthogs. This last tie, I have to admit, has been subjected to severe criticism in high places, sig-nally falling short of royal approval. I wore it on the oc-casion I was invited to one of the Queen's weekly lunches in Buckingham Palace among a bewilderingly eclectic mixture of about a dozen guests: those present included (around the table) the Director of the National Gallery, the Australian rugby captain, whose 'build and carriage' were exactly as you might imagine, a poised ballerina (ditto), Britain's most prominent Muslim,[1] and (under it) at least six corgis. Her Majesty was charm itself, but my warthog tie did not amuse. 'Why do you have such ugly animals on your tie?' Though I say it myself, my

---

[1] An affable gentleman, whom I have admired ever since learning that he offered Salman Rushdie sanctuary in his own home when hysterical mobs of his co-religionists were baying for the blood of that eminent man of letters.

reply wasn't bad for the spur of the moment: 'Ma'm,[1] if the animals are ugly, how much greater is the artistry to produce such a beautiful tie?' I actually think it is rather admirable that the Queen does not limit her conversation to meaningless *politesse*, but respects her guests enough to tell them what she really thinks. As for warthogs, my aesthetic agrees with hers: they are ugly. But there's a sprightly insouciance about the way they run with their tails pointing vertically upwards: not exactly charm, certainly not beauty, but an air of jaunty high spirits that makes me glad they are around. And it is a splendid tie. As I like to think the Queen would have thought on reflection.

Returning to earlier, purple-tie days and my Dutch critic, the idea of the extended phenotype itself escaped his anger–for which I was thankful, because he possessed a notoriously sharp intellect with a tongue to match. Though a distinguished elder of our shared field and author of an important theory of human origins, he wasn't everyone's cup of tea. One of Evelyn Waugh's minor characters, 'Uncle Peregrine', was a 'bore of international repute whose dreaded presence could empty the room in any centre of civilization'. I'm sorry to say that my neckwear critic had a similar reputation (the

---

[1] Pronounced 'Mam', by the way, as we were firmly told by the equerry who prepped us, not 'Mahm' as is commonly supposed.

mere mention of his name would clear whole corridors in the world of ethology) coupled with a finely honed persecution complex. There was a (not totally implausible) rumour that he was paid a full professor's salary by the University of Amsterdam, on the strict condition that he never set foot in Amsterdam. He came to live in Oxford.

I'm afraid he was the butt of other unkind jokes in the Netherlands. He once submitted to a Dutch journal a paper, in English, which contained a typo: 'Man is a ridicolous species.' He meant 'nidicolous', defined as a species whose young are heavily dependent on their parents (like thrush nestlings), as opposed to 'nidifugous' (like chickens or lambs, whose young leave the nest on their own sturdy legs and are much more appealing to us). The distinguished editors of the journal surely knew full well what the author meant, but they pleaded–in a later mock-apologetic *erratum*–that he had been unreachable in the African jungle and they had had to make a quick decision, trusting in the laws of probability: 'ridiculous' occurs much more frequently in English than 'nidicolous' and both involve a one-letter mutation from the misprint. So the published version reads: 'Man is a ridiculous species.' Perhaps that persecution complex was not entirely without justification. Nowadays, a spellchecking computer would have done the job for them, and almost certainly reached the same decision.

Cold water, hot blood

I next pick out a 1978 conference in Washington DC, because an incident there has become part of the folk-lore of the so-called 'sociobiology controversy' and, unlike most regalers of the tale, I was an eye-witness. The conference was convened by my old Berkeley friend the ethologist George Barlow, and the anthropologist James Silverberg, to discuss the sociobiology revolution and how to carry it forward. Edward O. Wilson, author of *Sociobiology* itself, was the star speaker, and I was also invited because *The Selfish Gene* was acquiring a following at the same time. There was much over-lap between Wilson's magisterial *opus* and my slighter volume, although neither book influenced the other. One important difference is that John Maynard Smith's powerful theory of evolutionarily stable strategies (ESS) plays a prominent role in *The Selfish Gene* but is mysteriously absent from *Sociobiology*. I regard this as the most serious shortcoming of Wilson's great book, although it was overlooked by critics at the time–and, as noted in the previous chapter, my contribution to the Washington conference was accordingly devoted to the topic. Maybe those critics were distracted by the fusillade of silly political attacks on Wilson's final chapter about humans–attacks from which *The Selfish Gene* suffered some (not very damaging) collateral damage. The whole sorry history is well and fairly treated by the sociologist Ullica Segerstråle in her *Defenders of the Truth.*

At the Washington conference I was in the audience for a panel discussion when a motley rabble of students and leftist fellow travellers rushed the platform, and one of them threw a glassful of water at Edward Wilson, who was using crutches at the time, having injured himself training for the Boston Marathon. Some journalists have described a 'pitcher' of 'iced water' being 'poured' on his head. This may have happened too, but what I saw in the confusion was water being thrown sideways from a glass in Wilson's general direction, fended off by David Barash, flaring his Bernard Shaw (or W. G. Grace) style beard towards the attacker in a classic primate agonistic display. Barash was the author of a readable student textbook of sociobiology who has grown, through his later books, into a sage and humanely prophetic voice in our field. The assailants were chanting slogans obviously inspired by the Harvard Marxist cabal led by Richard Lewontin and Stephen Gould, so it was good that Gould himself was on the platform with Wilson and Barash, and in a position to quote Lenin's condemnation of 'an infantile disorder'. In the same vein the chairman of the session, plainly upset, stood up and made an angrily passionate speech, concluding: 'I am a Marxist and I wish *personally* to apologize to Professor Wilson.' Ed Wilson himself took it with his customary good humour. I expect he knew, as we all did, that among all the hubbub he had inadvertently scored a quiet victory that day.

Northern nightingale

In 1989 Michael Ruse, founding editor of the journal *Biology and Philosophy*, convened a conference on 'The border zone between evolutionary science and philosophy'. This conference was remarkable not so much for its theme as for its location: Melbu, a village on one of the islands off the coast of northern Norway. More than the beauty of the place and the night sun, what was memorable was the–what shall I call it?–the *sociology* of the conference venue. Once a prosperous centre of the fishing industry, Melbu had fallen on hard times. In response to this change in its fortunes, a consortium of citizens led by the dentist got together to found a community centre, which would bring money into the village by building and operating a conference venue. The most unusual feature of this enterprise was that it was run entirely by volunteers, freely giving their time, money and resources out of what appeared to be pure, altruistic public-spiritedness. I may be exaggerating a little but the conversations among us delegates from abroad, at meals and on midnight walks, dwelt more on our wonderment at the idealism of the townspeople than on the official topic of the conference itself.

Two pleasant vignettes mark Melbu in memory. In a gigantic cylindrical fishmeal store–no longer needed for its original purpose because of the decline in the industry but still faintly smelling of it–we had a grand gala dinner. Come the appropriate hour, we gathered outside and stood in a long line–not just the conferees but, as it seemed, most of the inhabitants of the village who were, indeed, nearly all volunteers for the enterprise. We stood, and we stood. And we stood. Finally, a Norwegian biologist colleague left the line to investigate the delay. He returned in high humour with the perfect explanation: 'The cook is drunk!' This was so very Melbu, and so exactly like the plot of 'Gourmet Night' at Fawlty Towers, our mounting impatience dissolved in good-natured laughter. Our spirits were still high when we finally entered the giant drum to be greeted by a spectacular vision of thousands of candles all around the perimeter. The meal itself was fine.

'Still are thy pleasant voices, thy nightingales awake.' The first night of the conference, in the community centre, I was at the buffet dinner when I was suddenly struck dumb by one of the most beautiful voices I had ever heard, singing in the next room. Mesmerized, I left the dining room and gravitated towards the music as if lured by a Rhinemaiden. A lovely soprano, accompanied by a string quintet of obviously professional quality, was singing in German to a nostalgic, probably Viennese, waltz tune. I was entranced, and made enquiries. The string players were indeed professionals, who came from Germany every year to Melbu to play, pro bono, for love of the place and its idealism. The sweet soprano was not German but Norwegian: Betty Pettersen, Melbu doctor and one of the consortium who had joined forces with the dentist to found the centre. We became friends during the course of the conference, and I was sorry to lose touch as the years passed.

But there was a sequel. In September 2014 I was invited to the Blenheim Palace Literary Festival in Woodstock, near Oxford. Blenheim Palace is the magnificent Vanbrugh-designed residence of the Dukes of Marlborough (the Churchill family, and indeed Sir Winston was born there). It's a beautiful location for a literary event, and I usually go to the festival to promote each of my new books. This time, for *An Appetite for Wonder*, the format would be different: the interview was to be punctuated by music, chosen by me–as in the BBC radio show *Desert Island Discs* (on which I once appeared as castaway)–to illustrate scenes from my life. The big difference in the Blenheim Palace version was that the music would be live, played by the Orchestra of St John's under their Director John Lubbock, with a soprano, a contralto and a pianist.

Among the fifteen pieces I chose, I really wanted the haunting Viennese waltz-time song, to remind me of Betty and the Spirit of Melbu. I knew neither its name, nor that of the composer. The tune itself, however, was firmly lodged in my head, one of my morning shower repertoire. So I played it on my EWI (electronic wind instrument) into my computer microphone, and emailed the melody to half a dozen musicians in the hope that one of them would recognize it. And one–only one–did: Ann Mackay, a dear friend of Lalla's and mine who, by a lively coincidence, had been engaged as the soprano soloist for my Blenheim concert. She knew the song well, had often performed it and possessed the sheet music: *Wien, du Stadt meiner Träume* ('Vienna, city of my dreams') by Rudolf Sieczynski. It all worked out to perfection: Annie sang it beautifully in the long, sparkling Orangery of Blenheim Palace, and stirred sweet memories of my Melbu nightingale.

EWI son et psycho lumière

The EWI (pronounced 'ee-wee'), by the way, is another story. In 2013 Lalla went on the popular BBC radio show *Loose Ends* to talk about an exhibition of her art that was being mounted at the National Theatre in London. The band Brasstronaut, which provided the musical interlude on the show, included Sam Davidson, a virtuoso on the electronic wind instrument. Lalla was intrigued and got talking to him. When she reported the conversation to me, as a sometime clarinettist I was even more intrigued. I had an email conversation with Sam, and stored away in my head the idea that some day, if the opportunity arose, I would love to try my hand at the EWI.

Meanwhile I happened to be approached by the London advertising firm Saatchi & Saatchi. They had been commissioned to produce the opening tableau to get the ball rolling at the Cannes documentary film festival. They chose the theme of 'Memes' and wanted to feature me. I was to enter, stage left, and give a precisely scripted three-minute lecture on memes. I'd then walk off, giving way to a weird psychedelic film in which words and phrases from my lecture would be, as if by magic, incorporated together with spinning images of my face, loud booming music and strange light-show effects coming from all sides, my voice distorted and made vaguely musical with surreal echoes and harmonies–exploiting the whole whizz-bang repertoire of computer graphics and sound. Could this be what is known as postmodern? Who can tell?

It was all designed to seem like a conjuring trick, as if the computerized *son et lumière* had somehow picked up the words and phrases of my lecture and instantly magicked them, in fragments and distorted echoes, into the fabric of psychedelia. To the audience it would have sounded as though weird, dreamlike memories of my lecture had somehow been captured and instantly re-arranged and regurgitated. Of course, in truth the Saatchi team had recorded me giving the word-for-word identical lecture weeks earlier, in an Oxford studio, and so had plenty of time to extract snatches from the recording to make their phantasmagoric film.

Anyway, the idea was that, as the sound and light show soared to its end, I should walk back on stage, this time carrying a clarinet and playing the refrain from the music that had just stopped thundering through the wraparound loudspeakers. 'Er, it is true, isn't it, that you play the clarinet?' Well, I hadn't touched a clarinet for fifty years, no longer possessed one, wasn't at all sure my embouchure was up to it. But then Lalla reminded me of the EWI. I explained to them what it was. It was now Saatchi & Saatchi's turn to be intrigued.

Would I be prepared to learn to play the EWI in order to do the walk-on-and-play climax to their psychedelic extravagarama? How could I not? 'Just try me.' We both rose to the challenge; they bought me an EWI and I set about learning to play it.

The EWI is a long, straight thing shaped like a clarinet or oboe, with a mouthpiece at one end, a cable leading to a computer at the other, and woodwind-style keys in between. The mouthpiece contains an electronic sensor. Blow into it and a noise comes out of the computer: clarinet, violin, sousaphone, oboe, cello, saxophone, trumpet, bassoon–the mimicry of the real instrument is as good as software can make it, and that means very good indeed. If the computer is connected up to the massive speakers and boom-boxes of the theatre in Cannes, it sounds pretty impressive.

Electronic keyboard instruments purport to mimic real instruments too, but the added control you can exert when you blow into the mouthpiece of the EWI makes all the difference. You can transmit emotion in a way that you really can't with a keyboard trying to mimic orchestral instruments (you can with a piano, but that's because the keys are sensitive to how hard you hit them: hence the full name, pianoforte). The EWI is fingered pretty much like a clarinet or an oboe. This makes it much easier, startlingly easier, for a beginner to produce the sound of a cello, say, with a beautifully resonant vibrato, or a singing violin, without any of the agony we associate with those bowed instruments during the years of scraping, scratching apprenticeship. Tongue the EWI's mouthpiece hard, and the software renders it as the characteristic 'zing' of bow hitting string. Tongue the mouthpiece when the software is set to trumpet mode and you get the 'lippy' attack of that instrument; or in tuba mode and you get a satisfying oompah. Tongue it in clarinet mode and you hear exactly what you would hear from a real clarinet. In any mode, blow steadily harder and then die away, to create a soulfully burgeoning crescendo and sighing diminuendo. For the finale of the Saatchi show I played it in trumpet mode, blaring and forthright. I actually made a mistake through stage fright, but managed to recover and the Saatchi team were kind enough to congratulate me on my impromptu 'improvisation'. They tell me the YouTube video went viral.

Astronauts and telescopes

In 2011 the astronomer and musician Garik Israelian convened a most remarkable gathering on Tenerife, in the Canary Islands. This volcanic archipelago off the coast of Morocco is a major centre of astronomy because there are mountains high enough to penetrate most clouds, and important observatories take advantage of this on both Tenerife and La Palma. Garik had the inspired idea of getting scientists together with astronauts and musicians to see what they had in common, and what they could learn from one another: hence the event's name, 'Starmus'. The musicians included Brian May, former lead guitarist with Queen, a supernormally nice man; the scientists included Nobel Prize-winners such as Jack Szostak and George Smoot; and among the astronauts were Neil Armstrong, Buzz Aldrin, Bill Anders (who, though an unbeliever, had been made by the NASA PR department to read aloud from the book of Genesis), Charlie Duke (who, disconcertingly, has become a born-again Christian), Jim Lovell (captain of the nearly doomed Apollo 13), Alexei Leonov (first man to walk in space), and Claude Nicollier (the Swiss astronaut who walked in space to repair the Hubble telescope).

Halfway through the conference, several of us were flown in a small plane to the neighbouring island of La Palma, to have a panel discussion inside the housing of the largest optical telescope in the world, the Gran Telescopio Canarias, with its 409-inch mirror. Lalla and I travelled with Neil Armstrong, and it was a pleasure to see how well deserved was his reputation for modesty and quiet courtesy. This was in no way belied by his very reasonable policy of never giving autographs to random strangers–instituted (as he explained to an eager autograph hunter on the journey) when he discovered that his signature–and even his fake signature–was being sold on eBay for tens of thousands of dollars.

The La Palma giant telescope was stunning. Instruments such as this, and the similar telescopes of the Keck Observatory on the big island of Hawaii, move me deeply, I think because they represent some of the highest achievements of our human species. And, as my friend Michael Shermer has recorded, I was especially moved by the Mount Wilson 100-inch telescope in the San Gabriel Mountains near Los Angeles, once the largest telescope in the world and the one with which Edwin Hubble first decoded the expanding universe. Before Mount Wilson, the title of largest telescope was held (for the longest time) by the Earl of Rosse's 72-inch 'Leviathan of Parsonstown' at Birr Castle in Ireland, to which I have an extra emotional attachment because of its association with Lalla's family. I experienced the same swelling of the chest when I visited CERN and the Large Hadron Collider: again, the near-lachrymose pride in what humans can do when they co-operate, across nations and across language barriers.

The spirit of international cooperation was hovering over the whole Starmus conference. When Buzz Aldrin arrived late in the conference hall, Alexei Leonov was in the front row of the audience. Completely unde-terred by the fact that somebody was trying to give a lecture, this jovial Khrushchev-lookalike stood up and bellowed at the top of his voice: 'Buzz Aldrrreeen'. Arms outstretched, he strode towards the incoming Aldrin and enveloped him in a full Russian bear-hug. At dinner, Leonov showed himself to have artistic as well as astronautical talent. Lalla and I were entranced to see him dash off on the back of the menu a rapid self-portrait (see the picture section) for Garik Israelian's little boy Arthur. The tie, included at Arthur's request to match the one Leonov was wearing at dinner, adds a quirky charm to the picture—as if added charm were needed, given the beaming abundance of the commod-ity on display in another bear-hug for Jim Lovell, hero of the triskaidekaphobogenic[1] Apollo 13.

Neil Armstrong sat with Lalla on the return flight from La Palma to Tenerife. They talked of many things, including the remarkable fact—vivid demonstra-tion of Moore's Law—that the total computer memory on board Apollo 11 (32 kilobytes) was a small fraction

---

[1] Google it.

of the capacity of a Gameboy that Armstrong pointed out in the possession of a child in a neighbouring seat. Alas, that gracious and courageous gentleman was no longer present when Garik reprised the Starmus conference three years later. It was again a great experience, this time with a much larger audience and with Stephen Hawking as special guest.

Flashing back again to the 1970s and the gatherings of my earlier career such as the Washington conference on sociobiology, an element of nostalgia creeps in. In those days I was able to be simply a delegate, listening to talks with interest, approaching speakers afterwards to follow up points of interest, perhaps having dinner with them. Recent conferences, especially since publication of *The God Delusion*, have become a very different kind of experience. Although I am not a celebrity with lots of people recognizing me in the street (thank goodness), I seem to have become a minor celebrity in the secularist, sceptical, non-believing circles that convene the sorts of conferences to which I am now invited. The other major change is the arrival of the selfie. I don't think I need to elaborate, except to say that the invention of the cellphone camera is a mixed blessing. And you can take that as polite British understatement.

Christmas lectures

In the spring of 1991, the telephone rang and a pleasant voice with a gentle Welsh lilt announced: 'This is John Thomas.' Sir John Meurig Thomas FRS, a scientist of distinction and the Director of the Royal Institution (RI) in London, was ringing to invite me to give the Royal Institution Christmas Lectures for Children, and I went hot and cold as he did so. The warm flush of pleasure at the honour was swiftly followed by a cold wave of trepidation. I immediately knew I would not be able to refuse the commission, and yet I lacked confidence that I could do it justice. I was aware that this renowned series of lectures had been founded by Michael Faraday, who gave them himself on nineteen occasions, culminating in his famous exposition of 'The Chemical History of a Candle'. I knew that in recent years the BBC had been televising the series, and the lecturers had included such scientific heroes as Richard Gregory, David Attenborough and Carl Sagan. If I'd lived in London as a child, I would probably have been in the audience.

Sir John understood my fears (he'd given the Christmas Lectures himself) and kindly didn't press me for a decision there and then, but invited me to visit the RI to discuss the possibility. I went up to London and he was as quietly kind as his telephone voice promised, paying special attention, as he showed me around, to the many legacies and traditions of his personal hero, Michael Faraday. With one of these traditions I was already only too familiar. A year or so earlier I had been invited to give a 'Friday evening discourse', another regular custom of the RI dating back to the 1820s. This particular tradition bristles with intimidating formality. Both the lecturer and the audience are expected to wear evening dress. The lecturer has to stand outside the lecture theatre while the clock strikes the hour. On the last stroke, an official flings open the double doors, the lecturer strides purposefully in and immediately has to start talking science in the very first sentence, with absolutely no introduction or preamble of the 'Great pleasure to be here' kind. That's an admirable tradition. More difficult, the concluding sentence of the lecture must be uttered, with decisive finality, at the exact moment when the clock begins to strike the following hour. As if that were not alarming enough, the lecturer is literally locked in 'Faraday's Room' for the twenty minutes preceding the lecture, having been handed Faraday's own short book on how *not* to give a lecture–a bit late for that, you might think. I learned that this locking-in tradition began some time in the nineteenth century when a lecturer found the formality too much to bear and did a runner at the last minute. Sir John wasn't too sure, but he suspected it was Wheatstone (he of the eponymous Bridge). I did read Faraday's notes during my twenty minutes of incarceration, and amazingly I did manage to round off my lecture exactly as the clock was striking, despite being thrown off my stride by an illusion, only gradually dispelled in the course of repeated surreptitious glances in the dim light, that a particular dinner-jacketed gentleman in the audience was Prince Philip.

I took a deep breath and accepted Sir John's invitation to give the five Christmas Lectures, 'for a juvenile auditory' to quote Faraday's original rubric. The tradition is to make minimal use of slides (magic lantern, I suppose it would have been called in the early days, PowerPoint or Keynote now). Instead, there is a heavy emphasis on live demonstrations. If you want to talk about a boa constrictor, don't show a picture of a boa constrictor, borrow one from the zoo. If a child can be called out of the audience to wear the boa around her neck, so much the better. Such demonstrations require a lot of preparation in advance, and I soon found I had underestimated the time I'd have to commit. The rest of that year leading up to the Christmas climax was marked by frequent trips to London for planning sessions with Bryson Gore, the RI's chief technical officer, and with Richard Melman and William Woollard of the independent television company Inca, subcontracted by the BBC.

Bryson was (no doubt still is, although he has moved on from the RI) a stalwart of technical ingenuity and improvisation. His fiefdom was a large workshop chaotically littered with useful rubbish, including props from earlier lecturing seasons (which, you never know, might one day come in handy). It was his job to make, or supervise the making of, apparatus and other necessaries for the lectures–not just the Christmas Lectures but the Friday evening discourses and many more. It was a trifle unfortunate that his given name sounds like a surname, so the audiences might have thought I was being old-fashioned when I addressed him as 'Bryson' during the lectures. In earlier times, indeed, lecturers had referred to his predecessor as 'Coates'. Bryson's services, and those of his staff (a young man called

Bipin), were placed at my disposal, and I had to think hard, and discuss with Bryson and William and Richard, how to use them.

One agreeable and unanticipated feature of the Christmas Lectures was that the very name was a golden key to unlock goodwill whichever way I turned. 'You want to borrow an eagle? Well, that's difficult, I honestly don't see how we can realistically, I mean do you seriously expect . . . Oh, you're giving the Royal Institution Christmas Lectures? Why didn't you say so before? Of course. How many eagles do you need?'

'You want an MRI scan of your brain? Well, who is your doctor, have you been referred to the MRI Department on the National Health Service? Or are you going privately? Do you have health insurance? Have you any idea how expensive MRI scans are, and how long the waiting list? . . . Oh, you're doing the Christmas Lectures? Well, of course, that's different. I'm sure I can slip you into a research run, no questions asked. Can you come to the Radiography Department on Tuesday during the lunch hour?'

By just dropping the name of the Christmas Lectures, I managed to borrow an electron microscope (big, heavy, and transported at the lender's expense), a complete virtual reality system (whose owners went to the enormous labour of programming a simulation of the RI lecture theatre), an owl, an eagle, a hugely magnified circuit diagram of a computer chip, a baby, and a jactitating Japanese robot capable of climbing walls like a much enlarged, ponderously hissing gecko.

I chose, as the overall title of my series of five lectures, *Growing Up in the Universe*. I meant 'growing up' in three senses: first, the evolutionary sense of life's growing up on our planet; second, the historical sense of humanity's growing out of superstition and towards a naturalistic, scientific apprehension of reality; and third, the growing up of each individual's understanding, from childhood to adulthood. The three themes ran through all five one-hour lectures, which were entitled:

'Waking up in the universe'

'Designed and designoid objects'

'Climbing Mount Improbable'

'The ultraviolet garden'

'The genesis of purpose'

The first lecture was typical of RI Christmas Lectures in the number and variety of demonstrations. To illustrate the power of exponential growth of a population under hypothetical conditions of unlimited food and no constraints, I used the example of folding paper. Every time you fold paper you double its thickness. If you fold it a second time, your paper is four times as thick as the original. Successive foldings increase the thickness until you reach the sixth fold, sixty-four pages thick. Regardless of the size of the paper you start with, six foldings is as far as you can normally go: the wad is too thick to fold again, and by now very small in area. But if you were somehow able to continue folding, even as few as fifty times, the thickness of your paper would reach out to the orbit of Mars. This being the Christmas Lectures, it was not enough to state the calculation. It was necessary to roll out a huge piece of paper and call up a pair of children to help fold it—only as far as 64-fold thickness, after which they struggled amid laughter. I suppose it's a good way to drive home the power of exponential growth, but throughout my Christmas Lectures I occasionally worried that a simile can obscure, rather than illuminate, the simuland (look it up if you must, and as I confess I did, but you'll find that you already know the meaning, imbibed in the way we learned our words as children).

The first lecture also demonstrated what could be called faith in the scientific method. Bryson hung a cannonball on a wire from the high ceiling of the steeply raked RI lecture theatre. I stood to attention against the wall, held the cannonball to my nose, then let go. You have to be careful not to give it a shove, but if you just let it go by gravity, the laws of physics assure you that when it swings back it will stop just a tiny bit short of breaking your nose. It requires a modicum of willpower not to flinch as the black iron ball looms up towards you.

I'm told (by no less an authority than a former President of the Royal Society who happens to be Australian) that, when this demonstration is performed by Australian scientists, only wusses hold the cannonball to their face. Top blokes hold it to their lolly bags (budgie smugglers or jocks). And I heard of a Canadian physicist who was so exhilarated by the audience's burst of premature applause as the cannonball hurtled towards him that he started forwards to acknowledge it . . .

I also, in Lecture 1, borrowed a baby (Richard Melman's niece) in order to hold her in my arms while I told of Michael Faraday's famous retort to the question 'What is the use of electricity?' He said (although the story is attributed to others as well): 'Of what use is a new-born baby?' I found myself sentimentally affected to be holding the beautiful little Hannah while I spoke–with hushed voice in order not to frighten her–

of the preciousness of life, the life that stretched before her. I was delighted when, some twenty years later, Hannah announced herself on a correspondence forum on my website, RichardDawkins.net.

Another sentimental memory that I especially treasure is from Lecture 5. I was talking about a revealing difference between two ways in which the retinal image can move. If you shut one eye and gently poke the other eyeball (through the lid) with your finger–and I asked the children to do this to themselves–the whole scene appears to move, as if in an earthquake. But if you move your eyeballs using the dedicated muscles attached to them for that purpose, you don't see an 'earthquake', even though the retinal image has moved just as if you had poked your eyeball. The world seems to stay rock steady, you just look at a different part of it. One explanation, offered by German scientists, is that when the brain gives the order to swivel the eyeball in its socket, it sends a 'copy' of the order to the part of the brain that perceives the image. This copy primes the brain to 'expect' the image to move through the precise amount ordered. So the perceived world looks steady because there is no discrepancy between observed and expected. When you poke your eyeball, no copy is sent, so the world appears to move as if it really *has* moved, as in an earthquake, because now there is a discrepancy between observed and expected.

I pretended I was going to demonstrate the effect with a key experiment. I would paralyse the eyeball-moving muscles with an injection. Then, when the brain sent an order to move the eyeball, the eyeball would stay still but the *copy* of the instruction would still be issued. So the person would see an apparent earthquake even though the eyeball hadn't moved–the apparent movement being the discrepancy between expected and actual (zero) eye movement.

This being the Christmas Lectures, the next thing to do was to call for a volunteer . . . I produced a huge veterinary hypodermic syringe, fit to sedate a rhinoceros, and asked who would like to take part in the experiment. Normally, the children at the Royal Institution Lectures fall over themselves in their eagerness to assist in demonstrations. Surely nobody would volunteer in this case, and I was about to reassure everyone that it was only a joke when one little girl of seven, probably the youngest in the audience, hesitantly raised her hand. It was my darling daughter Juliet, sitting shyly by her mother. I still choke up a little at the memory of her uncomprehending loyalty and courage in the face of the monstrous hypodermic that I was brandishing. Is it irrelevant that she is now a promising young doctor?

Moving from my smallest volunteer to my largest, in Lecture 4 I was talking about our moral attitudes to animals and the history of our exploitation of them. I quoted the Oxford historian Keith Thomas on medieval beliefs that animals existed purely for our benefit. Lobsters were furnished with claws so that we could benefit from the improving exercise of cracking them. Weeds grew because it was good discipline for us to have to work hard pulling them up. Horseflies were created so 'that men should exercise their wits and industry to guard themselves against them'.

> The willing ox of himself came
> Home to the slaughter, with the lamb;
> And every beast did thither bring
> Himself to be an offering.

Douglas Adams extended this conceit to a surreal conclusion in *The Restaurant at the End of the Universe*, where 'a large meaty quadruped of the bovine type' approached the table, announced itself as the dish of the day and encouraged diners to try 'Something off the shoulder, perhaps, braised in a white wine sauce?' 'Or a casserole of me, perhaps?' It goes on to explain that people had become so worried about the morality of eating animals that it was 'eventually decided to cut through the whole tangled problem and breed an animal that actually wanted to be eaten and was capable of saying so clearly and distinctly. And here I am.' The majority of the restaurant party ordered rare steaks all round, and the animal happily trotted off to the kitchen to shoot itself–'humanely'.

I needed somebody to read this darkly humorous and philosophically profound piece, and here, yet again, was my cue to call for a volunteer from the 'juvenile auditory'. Dozens of eager hands shot up as usual, and I pointed to one. An enormous man uncoiled his near-seven-foot height, and I beckoned him to the front.

'What's your name?'

'Er, Douglas.'

'Douglas what?'

'Er, Adams.'

'Douglas Adams! what an amazing coincidence.'

The older children, at least, realized it was a plant, but it didn't matter. Douglas gave a wonderful performance as the Dish of the Day, complete with mimed gestures when he reached, 'Or the rump is very good. I've been exercising it and eating plenty of grain, so there's a lot of good meat there.'

Although most of the props for my lectures were conjured up by Bryson and his staff, I also pressed my artistic mother into service. In Lecture 1 I was trying to convey an intuitive idea of the immensity of geological time. Many analogies have been proposed, and I've used several myself on different occasions. As others have before, here I chose to represent time by distance, one pace per thousand years. My first few steps across the stage took us back through William the Conqueror, Jesus, King David and various pharaohs, but by the time we had got back to the creatures we now find as fossils the theatre was too small and so I translated the paces into miles, making the numbers vivid by naming towns that were the appropriate distance away: Manchester . . . Carlisle . . . Glasgow . . . Moscow. For each fossil I named, my mother had painted a reconstruction of it on a large sheet of cardboard. Bryson had planted these portraits on particular children at strategic places in the auditorium and they stood up as I called on them. My parents also made a lovely model of Mount Improbable, the eponymous mountain of Lecture 3 (and my later book of that name). One side of the mountain is a sheer precipice. The impossible feat of leaping from the bottom to the top is equivalent to evolving a complex organ like an eye in one fell swoop. But round the back of the mountain is a gradual slope from bottom to top. Climbing, step by step, up the slope is how evolution works: through cumulative selection.

The lecture ended with a classic RI demonstration–for which I donned a Second World War helmet: the great Bombardier Beetle Damp Squib display. The bombardier beetle is the creationist's favourite insect. It defends itself against predators by squirting hot vapour produced by a chemical reaction. No wonder the reagents are kept in separate glands and not exposed to each other until they are squirted out of the beetle's rear end. Creationists love it because they think the intermediate, ancestral stages would all explode, rendering the evolution impossible. My demonstration, carefully prepared by Bryson, showed that actually there is a gentle slope up this particular peak of Mount Improbable.

The reaction, which depends upon the reactivity of hydrogen peroxide, requires a catalyst, and there is a smooth dose–response curve. Without the catalyst, there is no perceptible reaction at all, and I made much of the anti-climax to mock the creationist alarmism. Then I had a series of beakers of ingredients on the bench, and I added increasing doses of catalyst to each. With a small amount of catalyst the peroxide gets gently warm. Increasing doses of catalyst smoothly increase the strength of reaction until, with a large dose, the audience was able to applaud a satisfying whoosh of steam towards the ceiling and the effect would surely alarm and probably burn any predator brave enough to attack a bombardier beetle. Of course, these being the Christmas Lectures, I hammed it up, putting on the safety helmet and inviting nervous members of the audience to leave the room (none did).

In all my years as a university lecturer, I never came close to matching the rehearsing and drilling–almost choreography–that the Christmas Lectures put me through. William and Richard seemed to mastermind my every move. As the months of preparation drew towards their December climax and enormous BBC outside-broadcast trucks drew up outside the RI in Albemarle Street, William and Richard were joined by Stuart McDonald, the BBC's own stage manager, whose job was to direct the actual televising, deployment of cameras etc. Stuart, William and Richard pulled my puppet strings–and Bryson's too, for he was up and down throughout the lectures bringing on and taking off props ranging from fossils to totem poles to giant model eyes, and in many cases helping me to handle them. The choreography inevitably broke down when we had live animals, and there was a moment of comedy when Bryson and I were trying to catch stick insects walking all over my ridiculously flowery shirt. Things got out of hand again when, to illustrate the power of artificial selection, we had representatives of contrasting breeds of dogs brought on by their rather forthright owner (who corrected me with justified brusqueness when I referred to her prized German Shepherd as an 'Alsatian').

The five lectures were spaced at intervals of about two days, and each one was rehearsed, in full, three times before being finally delivered: two rehearsals the day before, then full dress rehearsal the morning before the evening performance. I suppose actors get used to this, but I'm surprised I didn't get bored by the repetition. It meant that I gave each of the five lectures four times in quick succession, making twenty hours of lecturing in all. I confess that I was getting a bit weary by the end of the third rehearsal in each case, but the sight of a live audience—what Lalla tells me is called 'Doctor Theatre'—soon dispelled that.

I spent so long in the Royal Institution during 'my' year that I still have a sense of cosy familiarity, almost like coming home, whenever I visit the place. I suspect other Christmas Lecturers feel the same. I was told—and this again must be true of all RI Christmas Lecturers—that during 'my' week my face had more hours of exposure on British television than any other. Those hours were well away from peak time, however, so didn't result, I'm glad to say, in my being recognized in the street.

Islands of the blest

Japan

A tradition has grown up of exporting the RI Christmas Lectures to Japan the following summer, and I gladly went along with the custom. They were still called the Christmas Lectures, even in June, and the series was abridged from five to three. Each of my three was given twice, however, once in Tokyo and once in Sendai–a large provincial capital city two and a half hours north of Tokyo by bullet train. It was agreed that Lalla could come with me, and she helped me work on abridging the five lectures. Bryson flew on ahead of us, with a large crate of props from the London performances. In Tokyo he met up with his opposite number there, employed by the British Council as our local scientific fixer, and they set to work sourcing materials and animals for the demonstrations.

The Japan lectures were not filmed–at least not for broadcasting–so the choreography didn't have to be so perfect (and there was no William Woollard or Richard Melman to do it anyway). This was perhaps just as well, as we did not in all cases have the same props and walk-on parts–or, in one case, slither-on, because we hired a python from an animal supply house. This raised unexpected difficulties. In the first place, the snake arrived in a box labelled, in Japanese, 'Live Turtles'. This was because the supply house feared that the parcel delivery workers would refuse to handle a box if it was labelled 'Live Snake'. We were warned that it was highly unlikely any Japanese child would volunteer to handle it, so Lalla was drafted in to make a rather spectacular entrance wearing the snake coiled menacingly around her body. The python had arrived packed in frozen Brussels sprouts to keep it inactive. By the time of the lecture, and no doubt benefiting from Lalla's body heat, it had become quite frisky; it escaped and started sliding swiftly about, with Lalla, Bryson and me in energetic pursuit and the children startled into either nervous silence or cries of alarm.

Undaunted, we stuck with the RI tradition of few slides and lots of demonstrations. There was a tank full of live praying mantises, projected by video camera up on a huge screen above my head. Once I had finished talking about them I moved on to something else, having forgotten that they were still up on the screen. Somewhat later, I became uneasily conscious that I was losing my audience. Even allowing for the fact that they were listening to a simultaneous translation with a delay, they didn't seem to be responding to my words as I might have hoped. Then I noticed that their eyes were bulging in fascination towards something above my head. I looked up at the screen to see a giant female mantis happily munching (it really is a suitable word for those splendid jaws) the severed head of her sexual partner. What was left of him was still gamely copulating with her, possibly the more gamely for having lost his head. (There is some evidence that sexual behaviour in male insects is inhibited by nerve impulses from the brain. My friend and flatmate Michael Hansell was once giving a talk about his caddis fly larvae and expressed regret that he couldn't persuade the adults to breed in captivity. At this the Professor of Entomology, the endearingly cantankerous George Varley, growled, almost contemptuously, from the front row: 'Haven't you tried cutting their heads off?') The video feed from the mantis tank was too much of a distraction. Killjoy that I was, I asked the technicians to switch it off.

Compared with London children, who were falling over themselves with eagerness to volunteer for the demonstrations, the Japanese children were far more shy. Perhaps, too, they were intimidated by the vast size of the auditoriums, which in both Tokyo and Sendai were frighteningly much bigger than the RI theatre in London. And I suppose it would have been difficult to handle the language problem. Anyway, for whatever reason, hardly any Japanese children volunteered, in either Tokyo or Sendai. I don't remember how we managed this in Sendai, but in Tokyo the volunteers on almost every occasion were the same: the three delightful daughters of the British Ambassador, Sir John Boyd.

Sir John and Lady Boyd invited Lalla and me and Bryson to dinner in the Residence. Afterwards, Julia Boyd and the three girls took us for a night-time swim in the Embassy pool. Sir John was visibly uncomfortable at this because it was against the rules and, as a newly appointed Ambassador, he feared he was not setting a good example to his staff by letting his family break them. On the other hand, his guests were obviously having a nice time and he is a wonderfully generous host.

This was the beginning of a lovely friendship with the Boyd family, which has continued to this day. Two years after the Japanese Christmas Lectures I was awarded the valuable Nakayama Prize for Human Sciences and

Lalla and I went back to Tokyo for the presentation. The Boyds invited us to stay at the Residence and we were delighted to accept, although the hotel would no doubt have been very luxurious too. There happened to be an earthquake while we were there. Lalla and I were in our bedroom and were somewhat alarmed to see the walls shake and the chandeliers swing. We were reassured when His Excellency himself came through the door with the broad grin of a man who has seen it all before, waving a pair of safety helmets for us. At breakfast next morning, a visiting British Member of Parliament who was also staying at the Embassy indulged the obvious wisecrack when Lalla and I walked in: 'And did the earth move for you last night?'

The Boyds graciously came to the Nakayama prize-giving ceremony. I don't remember much about it, except for the group photograph which was taken afterwards. The photographer had an assistant, impeccably neat and bustling smart in a little black suit. It was this petite young woman's job to line us all up for the photograph, and she took it very seriously. Those of us sitting in the front row had to have our hands precisely folded in our laps, all with the same hand on top. Our knees had to be held tight together and our shoes precisely aligned. John Boyd and I, having had our limbs adjusted in the middle, became aware of suppressed giggles and snorts of laughter to our right. We dared a quick look away from our regimented eyes-front position and were rewarded with a sight to remember. Our wives, sitting together, were being sorted out by the photographer's assistant. But whereas we males only had to have our shoes and knees lined up, the ladies had to have their tights straightened too. And to do this the photographer's assistant was reaching up inside their skirts. Hence the poorly suppressed giggles.

By the time of Lalla's and my next trip to Japan in 1997, John Boyd's ambassadorship had come to an end,[1] so we were denied the pleasure of another stay in the Embassy compound, although the new Ambassador did kindly host a reception for us. I was back in the country to collect another award, the even more lucrative International Cosmos Prize, and it was an enormous honour, the ceremony to take place in Osaka in the presence of the Crown Prince and Princess. I was asked to choose a piece of music which the Court Orchestra would play in my honour. There was a strict time limit for the music, which restricted my choice. I sought advice from Lalla's old friend Michael Birkett, and after a lot of thought he suggested a suite of exactly

---

[1] He went on to become a very successful Master of Churchill College, Cambridge, and I'd love to take some credit for calling the college's attention to his excellent qualities.

the right length by Schubert who, by happy coincidence, happens to be my favourite composer. It had
the advantage of a provocative change of mood halfway
through, and the orchestra performed it beautifully,
contributing to making the whole occasion, including
a private tea with the Crown Prince and Princess, and
the prizegiving ceremony itself, extremely gracious.

Here are the opening paragraphs of my formal acceptance speech. You may guess from the wording how
much help I must have received from the professional
diplomats at the British Embassy:

Your Imperial Highnesses, ladies and gentlemen. It
is a great pleasure to be here and I'd like to start by
expressing my sincere gratitude to their Imperial
Highnesses, the Crown Prince and Princess for attending today's ceremony. I am especially grateful
to the Crown Prince for his gracious and very
thoughtful words [his speech had recalled his two
years at the University of Oxford]. I'd also like to
express my appreciation to the Prime Minister for
sending his message of congratulations today.
[Three paragraphs of diplomatic thanks cut here.]

Anyone with a passing interest in Japanese
history and culture is aware of the importance that
the Japanese place on harmony with nature. The

traditional Japanese arts, whether it be archery, calligraphy or tea-making, all have at their core the endeavour by the individual to achieve harmony with the world. The four seasons are each celebrated in their own way, and provide much of the inspiration for Japanese art and design. I myself feel positively Japanese when I think of the delight that you take in the pleasures of viewing cherry blossom in the spring, or gazing at the autumn moon.

On the other hand, the world's view of Japan in recent decades is of a country driven by technology and wealth-creation. We have looked on in admiration, and some envy, as a seemingly endless stream of impressive new products has flowed from Japan's factories. In the process you have built up the second largest economy in the world. But I know that the Japanese government is also moving actively to promote basic curiosity-driven science. I confidently expect that the next century will see a great flowering of basic scientific research in Japanese universities and institutes, including–in accordance with the aims of the Foundation–research on the environment and its problems. Impressive as Japanese achievements have been to date, I have a feeling that–to borrow an English colloquialism–'We ain't seen nothin' yet!'

For the public lecture that I later had to give on a scientific subject I chose as my topic 'The selfish cooperator'—and later expanded the speech to become the chapter of that name in *Unweaving the Rainbow*.

I love going to Japan, although I confess to being squeamish about some of the raw food—such as the raw holothurian guts to which I was introduced on my very first visit to the country in 1986. I was there as one of half a dozen scientists invited to give supporting talks at a conference in honour of Peter Raven, distinguished botanist and very nice man whom I hadn't met before, and who was being awarded the International Prize. On this occasion I was also introduced to karaoke (which did not impress me any more than the raw fish) and to the contemplative peace of Kyoto's temples (which did).

To my shame, I have never achieved any dexterity with chopsticks. How does even an expert tackle a dish consisting of nothing but a whole large turnip sitting solitary and proud in a bath of water? I was completely baffled by this very conundrum at one formal dinner where I was the guest of honour under the eyes of some twenty other guests at long, low tables arranged in a hollow square around two chalk-white geishas performing the tea ceremony. I'm afraid I simply gave up. But as far as I could see, none of the other diners made any significant inroads into their turnips either.

My most recent visit to Japan was in quest of a prize of a different kind: the giant squid. I had become friendly with Ray Dalio, brilliant financier and enthusiast for science. In pursuit of his passion for marine biology, he had bought a beautiful research vessel, the *Alucia*, and had now teamed up with two television companies, one in Japan and one in America, in a search for the giant squid, fabled sea monster of yore, in the deep seas off Japan. Dead or nearly dead specimens, or bits of them, had been brought up in trawls. But Ray was inspired by the small band of dedicated biologists, from Japan, New Zealand, America and elsewhere, who for decades had been trying to find a giant squid, alive and swimming in its natural habitat, the deep ocean. The *Alucia* was prepared for action, expert biologists from all over the world were mustered, and, to my great joy, Ray invited me along for the ride. The expedition was highly confidential and I was sworn to secrecy, because the two television companies, in the event that they succeeded in filming a live giant squid, wanted to save the news up for maximum impact.

Unfortunately, that trip was postponed; I forgot about it and went about my normal business. Then, some months later, in the summer of 2012, I received a telephone call out of the blue. It was Ray. Characteristically, he didn't beat about the bush.

Ray: 'Can you get on a plane to Japan tomorrow?'
Me: 'Why, have you found the giant squid?'
Ray: 'I am not at liberty to say.'
Me: 'Right. I'll be there.'

And I was, although it wasn't literally the next day, more like a week (fact struggles to keep up with fiction). Ray explained that I would have to take a 28-hour ferry ride from Tokyo to the Ogasawara archipelago, where the *Alucia* was anchored. These volcanic islands are sometimes known as the 'Galápagos of the Orient'. Like Galápagos, they've never been part of a continent, and they've evolved their own unique flora and fauna. But they are much older than Galápagos, and the plate tectonic forces that created them placed them near the Mariana Trench, where the sea floor is further below the surface than anywhere else on the planet.

I still didn't officially know that they had already found a giant squid, and I scrupulously kept silent about the suspected reason for my precipitate summons. Only Lalla knew why I had flown so suddenly to Japan, and she too maintained strict secrecy. To no avail, at least on one occasion. She met David Attenborough at a social event, and he asked after me. Lalla replied that I was on a ship in Japanese waters. 'Oh,' said Sir David without hesitation, 'he's obviously after the giant squid.' So much for our careful reticence.

After the long flight I spent a night in a Tokyo hotel before boarding the ferry together with Colin Bell, an Australian friend of Ray's who was also bound for the *Alucia*. We shared a cabin. Most of the very numerous passengers slept on futons on the floor in large dormitories. I can't remember how we passed the time; reading, I suppose. When we docked, we were met by members of the Discovery Channel team from the *Alucia*, and a little later we were in a small boat speeding out to where the ship was riding at anchor. The *Alucia* has a large, wet loading bay at the back, on which are perched its two submersibles, a Triton and a Deep Rover, and there was a large party of rather wet people standing there as we arrived. These included Ray, who greeted us warmly. We still didn't officially know that they had found a giant squid, but Ray tipped us the wink when we arrived, and said that they had organized a seminar on board, that very evening, to talk about the momentous discovery and how it was made. Meanwhile, would we like to go to the bottom of the sea? Of course we would. Right then, be ready in ten minutes.

I was to go down in the Triton three-person submersible, Colin in the two-person Deep Rover. The highly skilled pilot in the Triton was Mark Taylor, an

Englishman, and my fellow passenger was Dr Tsunemi Kubodera of Tokyo's National Museum of Nature and Science. He had been the scientist mainly involved in the live sighting of a giant squid and I think this wasn't the only reason Mark treated 'Dr Ku' with enormous respect under water, as everyone did on the surface.

The three of us climbed through the top hatch of the Triton while it was still on board the *Alucia* and took our seats in the spherical, transparent bubble, Mark on a raised seat, behind Dr Ku on the left and me on the right. The hatch was securely battened down, and then the Triton was lifted on a hoist and deposited into the sea, where we bobbed around waiting for the similar launch of the Deep Rover. I was entranced by the sight of the blue water the other side of the bubble as we danced around in the waves. Mark gave us a routine safety briefing and explanation of how our life-preserving hydrostat worked, including an interesting technical difference between our own craft and Deep Rover. He explained that we would be at normal atmospheric pressure throughout, despite the megapascals that would soon assail the outside of our bubble. We would therefore need no special precautions against the bends when resurfacing, even though we were going down 700 metres.

It would have been too much to hope that Dr Ku would see a giant squid *again* on the dive that I did with him, but we did see some ordinary squid, and lots of fish including sharks, jellyfish, rainbow-shimmering ctenophores and much else that might populate a zoologist's dreams. That evening, in the saloon of the ship, the zoologists of the expedition gave us the promised seminar on the science behind the successful filming and sighting of the giant squid. There were two illustrated talks. The first was by Dr Edith Widder, a marine biologist good enough to have won a MacArthur 'genius award'. She's an expert on bioluminescence and she knew that at the sort of depths giant squids favour the only light is generated by living creatures, often actually by bacteria that they carefully cultivate in luminescent organs for the purpose. Unlike the whales with which they share the deeps, giant squid have giant eyes, so they probably hunt, at least partly, by sight. Such considerations led Edie to invent the Electronic Jellyfish, a luminous lure designed to appeal to giant squid. It succeeded brilliantly. Lowered together with an automatic camera and towed behind the ship on a 700-metre cable, it bided its time—with ultimate, climactic success. The spectral, almost nightmarish shape of the giant squid pouncing on the luminous bait is a sight I cannot forget.

Also unforgettable is the film of Edie's face as she later scanned her way through the huge computer files of empty images and suddenly saw the legendary sea monster powering in from the side of the frame. She and her colleagues were filmed by the television crew as they stared at the computer screen and their facial expressions and exultant cries made me tremble with the joy of vicarious discovery (even if, as killjoys can be relied upon to suspect, the scene was re-enacted later).

The second talk at that extraordinary seminar in the *Alucia*'s saloon was by Steve O'Shea, a New Zealand marine biologist who, like Tsunemi Kubodera, had devoted much of his life to the quest for the giant squid. His ingenious idea for a bait focused on a different sense from Edith Widder's electronic jellyfish: smell. He made a purée of ground-up squid, in the hope that the smell, especially sex pheromones, would lure the giant through the darkness. It wafted in an alluring cloud from a pipe attached to the submersible, and did indeed prove to be an effective squid magnet–but unfortunately for ordinary, smaller squids only. No giant squid joined them. The final success in actually seeing a live giant squid fell to Kubodera, as O'Shea went on to describe (Dr Ku himself, although he speaks English, didn't feel confident enough in the language to present the talk). Kubodera's lure was more like a traditional angler's bait. A diamond-back squid, pretty large in its own right though not in the same class as a giant squid, baited the line tethered to the submersible. And, *mirabile dictu*, it worked. Dr Ku himself was in the submersible, the very Triton that I was to share with him a few days later, and saw the Kraken seize the bait and hold on to it for long enough to provide splendid footage for the cameras. His return to the surface was an emotional moment, as the television broadcast later showed. It seemed that the whole ship's crew turned out to pipe Dr Ku aboard and cheer this climax to a lifelong quest, generously congratulated by Edie and Steve. And, damn, I missed it by a mere couple of days.

A minor misfortune supervened. My expected week on the *Alucia*, with further dives promised, was cut short by the news that a dangerous typhoon was revving itself up in the vicinity and on its menacing way towards us. I was present when the captain advised Ray Dalio that we had no choice but to hightail it to the shelter of Yokohama harbour, two days' sailing away. That was a big disappointment for Colin and me, who had only just arrived. Nevertheless, those two days fleeing the typhoon were great fun. I gave a seminar on evolution in the saloon one evening, and Ray himself treated us to an informal breakfast tutorial on the truth behind the financial crisis, which I found riveting–as I always do when I listen to somebody who really knows his own subject and can talk about it from first principles.

Arthur C. Clarke, best known for his mind-stretching stories about outer space, has suggested that the deep sea is almost as mysterious although it's on our very doorstep. My brief forays into that alien world, both then and in a later trip in 2014 to Raja Ampat, off New Guinea, again as Ray Dalio's guest on the *Alucia*, have been among the great privileges of my life. That second trip was not aimed at any particular biological discovery like that of the giant squid, but Raja Ampat is one of the great unspoiled marine areas, breathtakingly beautiful and with a marine fauna as rich as anywhere in the world. This time I enjoyed numerous dives in the Triton, sometimes with Mark Taylor as pilot, sometimes with one or other of his two colleagues. I was delighted that my fellow guests on this second trip included Larry Summers, the immensely distinguished economist and former President of Harvard, with his literary wife Lisa New. Mealtime conversations were an intellectual feast, with Larry the academic economist and Ray the pre-eminent practitioner of the markets playing off each other.

Not that such subjects dominated: there were world-class experts on conservation aboard, and theirs was a topic that preoccupied us all. One of these experts was Peter Seligmann, chairman of Conservation International (and my cabin-mate on this trip); another was the

American biologist Mark Erdman. Mark knew these islands like the palm of his hand, and he was invaluable as
our Indonesian interpreter. He was on a quest for a particular rainbow fish in a river deep in the forest of West
Papua, which he suspected was a hitherto undescribed
species. He also suspected that it would turn out to be
unrelated to other fish in the same region but closely
related to fish from the other side of the great island
of New Guinea. If true, this would have great zoogeo-
graphic significance, as it would tell us something about
shifting tectonic plates carrying these freshwater fish
with them. The *Alucia* anchored offshore, and the ship's
on-board helicopter ferried relays of us inland and up-
river to take turns in assisting Mark in searching for his
rainbow fish. The repeated drill was as follows. Mark
would wade out into the fast stream holding one end of
the net. One of us (Ray, or I or whoever was on that
shift) would also wade out, a bit further downstream,
holding the other end of the net. We crouched down
in the (pleasantly cool) water and then, on a word of
command from Mark, we stood up and pulled the net
rapidly towards the bank, trapping whatever fish might
have swum into it. We'd then lay the net out on the shore
and Mark would inspect the contents for his rainbow.

When my turn came around on the second shift,
Mark was in place with his net when we landed on

a sandbank in the river. We set about the search and–success! The day's catch amounted to some fifteen of the little fish, and Mark's expert eye confirmed that they were, indeed, of the hitherto unnamed species that he had suspected. They were carefully kept alive in a tank pending his formal description of their details, including DNA analysis–and, of course, the all-important conferring of a scientific name on the species.

As it happened, I had a personal interest in the naming of fish: I had felt highly honoured when, in 2012, a team of Sri Lankan ichthyologists conferred the scientific name *Dawkinsia* on another genus of freshwater fish, this one from Sri Lanka and South India. There are now nine recognized species in the genus. The beautiful one shown in the picture section here (it too might seem to deserve the name 'rainbow fish') is *Dawkinsia rohani*.[1]

Galápagos

If Ogasawara is the 'Galápagos of the Orient', for me part of its attraction lay in my love affair with the Galápagos archipelago itself. It's a place of pilgrimage for Darwinians such as me, so it was perhaps not surprising

---

[1] There's a nice movie of *Dawkinsia filamentosa* sparring in a tank here: https://www.youtube.com/watch?v=FnWprpFYJhQ.

that Victoria Getty, when she happened to meet Lalla at a grand dinner in Windsor Castle, was shocked to learn that I had never been there–so shocked, in fact, that she immediately promised to put this right by arranging a trip to the islands and inviting us along as her guests.

This serendipitous conversation had come about at a gala occasion, hosted by Prince Michael of Kent, which featured the performance by a visiting Russian orchestra of a symphonic work composed by Gordon Getty, younger brother of Victoria's late husband Sir Paul Getty. Prince Michael is a prominent Russophile–I was impressed by his speech of welcome to the orchestra in Russian. He and Princess Michael are friends of Charles Simonyi, my benefactor at Oxford, through whom we had been invited. The dinner stands out in my memory not only because of Lalla's conversation with Victoria Getty but because I sat next to Susan Hutchison, former television news anchor in Seattle and executive director of Charles's charitable foundation. I found her charming, and delightful company until I discovered that she was a shamelessly enthusiastic supporter of George W. Bush, whereupon my gentlemanly manners were sorely tested. We didn't quite come to blows, and by the end of the meal we had kissed and made up. Meanwhile, Victoria had asked Lalla what Galápagos was like, and

Lalla said she hadn't been there, and I hadn't either. There and then, Victoria promised Lalla she would arrange a trip and invite us along as her guests. The very next day she telephoned Lalla to say she was looking at chartering a boat called the *Beagle* (a facultative sailing boat but otherwise not like the original one) and to fix the date of the voyage. We were overjoyed.

Then something embarrassing happened: well, an *embarras de richesses* you'd have to call it. Completely independently I was approached by an American shipping magnate, Richard Fane. One of his ships, the *Celebrity Xpedition*, plied the Galápagos Islands, and he had chartered it to celebrate an anniversary with his wife Colette in the company of ninety friends and relations. Would I like to come on board as guest lecturer to regale his guests with talk of evolution in the very place where Darwin had his first stirrings of inspiration, the place about which Darwin wrote those haunting words: 'One might almost fancy that from an original paucity of birds in this archipelago, one species had been taken and modified for different ends.' Lalla was invited too; and so also, when I told Mr Fane that I couldn't go because I'd miss Juliet's birthday, was Juliet. It was too tempting and generous an offer to pass up.

But what to say to Victoria? She had arranged the trip on the *Beagle* as a consequence of discovering that

I had never been to Galápagos. But now, if we accepted Richard Fane's invitation our participation in the *Beagle* voyage would be founded on false premises: it would be my second visit, not my first. We decided we had to come clean. Lalla telephoned Victoria to confess all. She responded extremely generously, simply saying: 'So much the better, you'll be able to explain things to us.' Her generosity continued the next time we met, which was at one of the cricket matches that she organized to honour the tradition of her late husband. That Anglophile American–he became, indeed, a naturalized British citizen–was so passionate about cricket that he carved out from a hillside in his Buckinghamshire estate a first-class pitch, where county teams would come to play against, among other teams, the Getty Eleven. After his death in 2003 Victoria continued the custom, and we were invited each summer to one of the Getty matches, in glorious sunshine with Getty Red Kites circling overhead. I call them that because Paul Getty was largely responsible for reintroducing these magnificent birds to our part of England, they having been driven extinct over most of the British Isles by gamekeepers. The cricket matches always included a sumptuous luncheon in a marquee for the guests, and we were honoured to be placed at Victoria's table where she introduced us to Rupert and Candida

Lycett-Green, who were to be our fellow passengers on the *Beagle*. I found an immediate bond with Candida as a lifelong admirer–that's putting it mildly–of her father, the quintessentially English poet John Betjeman.

The two Galápagos voyages were both marvellous but rather different. The *Celebrity Xpedition* holds ninety passengers and we got the full experience of the luxury cruise liner, but without the gruesome casinos and 'entertainments' that draw passengers' attention inwards to their floating hotel rather than out to port or starboard. The *Beagle* had just nine passengers, all guests of Victoria, and we ate round one big table, together with Valentina, our cheery and knowledgeable Ecuadorian guide.

Both boats followed the standard pattern of visits to Galápagos, anchoring off one island after another and ferrying passengers ashore in inflated Zodiacs, strong mariners handing us on and off the little craft using the 'Galápagos Grip' . The *Celebrity Xpedition* had about a dozen Zodiacs, each bearing one of the excellently knowledgeable Ecuadorian naturalists who then supervised our walk over the island, never straying far from the beaten track. Their English, though fluent, was usually strongly accented, with one notable exception: a rakishly bearded Che Guevara lookalike who dumbfounded us with his decorous and perfectly modulated

Oxford hightable English. He had apparently been educated by missionaries.[1]

The overwhelming impression I took away from Galápagos was of the tameness of the animals and the almost 'Martian' weirdness of the vegetation. There are regions of the world where, in the case of most members of the fauna, you feel privileged to catch a fleeting glimpse of one in the distance. In Galápagos the tourists have to be told that you are not allowed to touch the animals. It would be absurdly easy to do so. You have to take care not to tread on the sunbathing marine iguanas and the nesting boobies and albatrosses.

The *Beagle*, being a much smaller boat, was able to anchor off smaller islands, for example the uninhabited Daphne Major, site of Peter and Rosemary Grant's epic, long-term study of evolution in the medium ground finch. Our landing on Daphne Major was a little perilous, and I wondered how the Grants and their colleagues and students managed to offload their supplies, for everything has to be taken with them on to this deserted islet, even water. The *Beagle*'s single Zodiac was always

---

[1] A whimsically parallel case to an experience of my friends Stephen and Alison Cobb. While in western Uganda, pursuing Steve's vocation in wildlife conservation, they stopped their Land Rover in a small village whereupon, as is the way in Africa, it was instantly surrounded by a circle of smiling children. 'How are you?', the Cobbs politely asked. 'Moostn't groomble' came the chorused reply, presumably learned from a Yorkshire missionary.

supervised by Valentina, a member of the Cruz family, which seemed almost to have populated the archipelago all on its own. At well-nigh every island, or so we remarked jokingly to one another, we were greeted by a different one of her brothers. Another of her brothers was the captain of the *Beagle*. His English, although not as good as Valentina's, was probably better than he pretended. On a particularly exciting occasion I got his meaning from the Latin: '*Mola mola!*', he exulted from the helm, '*Mola mola!*' One of the most extraordinary fish in the sea, the sunfish, *Mola mola*, hung floating at the surface like a huge, vertically suspended disc, easily visible from the deck of the boat. Captain Cruz halted the *Beagle*, and in a frenzy Valentina and the rest of us seized our masks, snorkels and flippers and plunged into the sea. The sunfish didn't stick around long, but it was wondrous to see it at close quarters before it vanished into its mysterious world, which was not ours.

There were many lovely people on the *Celebrity Xpedition*, including the Fanes themselves and their extended family of many talents, but because there were so many we didn't get to know any of them really well. The *Beagle* trip with Victoria and her friends felt more intimate. Candida was unusual in that, where others would carry a camera, she carried a notebook and sat on a rock amid the scuttling Sally Lightfoot crabs recording her thoughts, observations and impressions. I was charmed by this habit and regret that I didn't follow it myself.

There is a particular poignancy in these recollections, for as I write this Candida has just died, of cancer. Each summer, she and Rupert have run the ironically named 'Great International Croquet Match' in their beautiful and wistfully English garden, hard by Uffington's thirteenth-century church with its hexagonal tower, overlooked by the White Horse prancing over the chalk downs straight out of the Bronze Age. At the 2014 tournament, just a few weeks ago, Candida, knowing it would be her last, was a model of cheerful good-hostess bravery. Rest in peace, quizzical celebrant of England, an England that Charles Darwin might still recognize thanks in part to your father. Rest in peace, enigmatically sweet shipmate and fellow explorer of the blessed islands of Darwin's youth.

Whoso findeth a publisher findeth a good thing

I have been well served by my publishers–through nearly forty years, not one of my twelve books has ever been allowed to go out of print in English–and it is therefore a bit surprising to realize that I seem to have had so many of them: Oxford University Press, W. H. Freeman, Longman, Penguin, Weidenfeld, and Random House in Britain, and an equally long list of different publishers in America. There's no single reason for this promiscuous infidelity. It began because of the opposite, loyalty: loyalty to a particular editor, Michael Rodgers, who changed employers–as is rather common in the publishing world–with disconcerting frequency.

Early books

In *An Appetite for Wonder* I told the story of my first meeting with Michael and his cautiously understated eagerness to publish *The Selfish Gene*: 'I MUST HAVE THAT BOOK!' he bellowed at me down the telephone after reading an early draft. He has now given his own account of the same episode in his memoir of a publishing career, *Publishing and the Advancement of Science: From Selfish Genes to Galileo's Finger.* Michael's book also quotes a speech I made at a 2006 dinner in London organized by Helena Cronin in collaboration with Oxford University Press to commemorate the thirtieth anniversary of *The Selfish Gene* (see page 252). I'll quote it in full because it helps to explain why my loyalty was to him more than to OUP:

> Soon after *The Selfish Gene* came out, I gave a Plenary Lecture at a big international conference in Germany. The conference bookshop had ordered some copies of *The Selfish Gene*, but they sold out within minutes of my lecture. The bookshop manageress swiftly telephoned OUP in Oxford, to beg them to rush an additional order by airfreight to Germany. In those days OUP was a very different organization, and I am sorry to say this bookseller was given a polite but cold brush-off: she must send in a proper order in writing, and, depending on supplies in the warehouse, the books might be shipped some weeks later. In desperation, the

bookseller approached me at the conference, and asked if I knew anyone at OUP who was more dynamic and less stuffy . . . I telephoned Michael in Oxford, and told him the whole story. I can still hear the thump with which Michael's fist hit the desk, and I remember his exact words. 'You've come to the right man! Leave this to me!' Sure enough, well before the end of the conference, a large box of books arrived from Oxford.

That was, of course, the English edition of *The Selfish Gene*. *Das egoistische Gen* appeared a little later. I soon received a letter from a reader in Germany who said that the translation was so good, it was as though author and translator were 'twin souls'. I of course looked up the name of the translator–Karin de Sousa Ferreira–and its surprisingly un-Germanic sound made it easy to remember. A little later, at his home university of Zurich, I met the distinguished primatologist Hans Kummer. At dinner I started to tell him the anecdote about my German translator. I had got no further than 'twin souls', without ever mentioning the name of the translator, when he spontaneously interrupted me, pointing his finger at me in a pistol-shooting gesture as he asked the question: 'Karin de Sousa Ferreira?' After two such splendidly independent testimonials, when the time came for *The Blind Watchmaker* to be published in German, I put in a strong request for the same translator, and I am delighted that my German twin soul with the Portuguese name graciously came out of retirement to render the book into *Der blinde Uhrmacher.*

I haven't always been so lucky with my translations. One Spanish edition (I won't say of which book) was so bad that three separate Spanish-speakers approached me to say it should be withdrawn. English idioms were translated word for word, in the same way as 'He gave her a ring [telephone call]', in an English novel, is said to have been translated into the Danish equivalent of 'He gave her a ring [for her finger]'. That Danish story may be an urban legend, but it is true that, in my Spanish case, the phrase 'with a vengeance' (meaning 'in a big way') was translated as 'con una venganza' which, I am assured, means what it literally says and makes no sense of the idiom. That's just one example among many. It's one reason (again among many) why computer translation is so difficult. The translator needs not merely a lexicon of words but a look-up table of idiomatic phrases like 'with a vengeance', even a table of clichés like 'at the end of the day' (meaning 'when all's said and done', yet another such cliché). Isn't language fascinating? I'm glad to say the Spanish publishers took full responsibility and commissioned a whole new translation, which is now published.

The danger of relying on computers to perform human functions reminds me of a lovely story told me by my friend Felicity Bryan, reputed to be Oxford's only literary agent. One of her clients wrote a novel whose hero was called David. At the last moment, when the book was finally edited and ready to print, the author had second thoughts about her lead character. He was more of a Kevin, she decided, than a David. So she set her computer to do a global search, replacing 'David' with 'Kevin' throughout. This worked well, until the action of the novel moved to a certain art gallery in Florence . . .

One more brief translation story. I was at a conference on evolution in Japan, listening on headphones to simultaneous translation. The lecturer was talking about early hominin evolution, *Australopithecus*, *Homo erectus*, archaic *Homo sapiens*, all that kind of thing. But what was this coming through the headphones? 'Early evolution of Japanese.' 'Fossil history of Japanese.' 'Evolutionary history of Japa . . . HUMANS.'

Michael Rodgers moved to W. H. Freeman in 1979, and a couple of years later, when my second book, *The Extended Phenotype*, was ready for publication, I took it to him there. As I have already remarked, the publishing world is a fluid one, and when Michael moved yet again, this time to Longman, again I followed him with *The Blind Watchmaker* in 1986. A couple of stories about *The Blind Watchmaker*. Near the beginning of the book I recounted a dinner-table conversation with 'a distinguished modern philosopher, a well-known atheist'. I said that I couldn't imagine being an atheist at any time before 1859 and the publication of *Origin of Species*. The philosopher demurred. Citing Hume, he couldn't see why living complexity needed any special explanation. I was dumbfounded and devoted substantial portions of the book to rebutting him, though never mentioning him by name. I'm not sure why I chose not to reveal his identity. It was, in fact, Sir Alfred 'Freddie' Ayer, Wykeham Professor of Logic and Fellow of New College, a formidably clever man of whom I was in awe. Many years after *The Blind Watchmaker* was published, he approached me to say he had just read it. He apologized (totally unnecessarily) for not having read it before, and said he was delighted to have inspired it–so he at least had identified himself. I asked him whether I had rendered our conversation correctly and he said, 'Perfectly correctly.'

My second story about *The Blind Watchmaker* I tell for no better reason than that it's quite funny. First, a little background. Many evolution-sceptics have been puzzled by an aspect of the perfection of animal camouflage. They reluctantly accept that bird eyes are sharp enough to put the finishing touches to a resemblance of intricate perfection, such as that of a stick insect to a stick, complete with buds and leaf scars. Or, another example, there are caterpillars that resemble a bird-dropping. But then, the sceptic says, how can you accept selection for the final detailed perfection of mimicry of a stick or a bird-dropping on the one hand, while also believing that the same kind of selection shaped the ancestors of those insects on their first tentative fumbling steps towards a resemblance? I quoted Stephen Jay Gould on the bird-dropping mimic: 'Can there be any edge in looking 5 per cent like a turd?' I answered the question in a way slightly different from Gould's. The very same eyes are presented with prey under a great variety of seeing conditions: dim versus bright light, corner of the eye versus full frontal, far away versus close up. A minuscule resemblance to a bird-dropping might be enough to save a caterpillar's life when seen from a distance or in twilight. But a strong resemblance would be necessary to save its life when seen close up and in strong daylight. And there is a continuous gradient from poor seeing conditions to good, which provides a smooth selection pressure for every degree of improvement in mimicry from crude to perfect. The same 'gradient' argument works for all complex adaptations–eyes, wings, all the chestnuts of creationist literature, and is of immense importance for the whole theory of evolution.

That's the background to the story. Now, Stephen Gould's name appears several times in *The Blind Watchmaker* and therefore also in the index. The index to a book, with its military-style backwards formatting, is a fine place to conceal a joke. It will not be noticed by many, but those who see it will share with the compiler a smile of secret complicity. The official history of *New College, Oxford, 1379–1979*, edited by my late colleagues John Buxton and Penry Williams, has an index compiled by a third colleague, the medieval historian Eric Christiansen (whose own memoirs of New College life will not be, and by all accounts should not be, published until after he and all his victims have died). Eric smuggled into the index of the college history some delightfully characteristic little jokes. Under 'Fellows', for example, we find 'comforts of', 'drunkenness of', 'execution of', 'expulsions of', 'factions among', 'obscurity of', 'provenance of' and, my favourite, 'philistinism of'. Turn to the pages indicated under 'philistinism of' and you find no mention of the word itself, just accounts of three building projects which obviously offended Eric's taste: two in the nineteenth century and an especially egregious one in the twentieth.

Anyway, as I said, I put a little joke, for Steve Gould's benefit, in the index of *The Blind Watchmaker*. It duly appeared in the original British edition. However, when the American publishers saw it they were aghast. They thought it in terrible taste, and they may just possibly (although I was too discreet to ask) have been conscious that Stephen Gould was one of their most bankable authors. Publication of the US edition was held up while the joke was expunged. Then, unintentionally,

Gould, S. J.,
   five percent eye, 81,
      (quoted in 41)
   five percent resemblance to
      turd, 82, (quoted in 41)
   mentioned, 275, 291
   punctuated equilibrium,
      229–52, (36)
   revealing *faux pas*, 244,
      (36)
   revealing flaws, 91, (34)
   writes off synthetic theory,
      251, (35)

Gould, S. J.,
   five percent eye, 81,
      (quoted in 41)
   on dung-mimicking in-
      sects, 82, (quoted in 41)
   mentioned, 275, 291
   punctuated equilibrium, 2
      29–52, (36)
   on Darwin's gardualism,
      244, (36)
   The Panda's Thumb, 91,
      (34)
   writes off synthetic theory,
      251, (35)

as a result of a simple oversight, the microfilm version containing the censored index was the one used in subsequent printings of Longman's British edition and Penguin's paperback. Michael Rodgers had intended that the joke should remain in the British edition. The fact that it didn't may, for all I know, confer on the first British printing a collector's value, along the lines of the 'unperforated' stamps prized by some philatelists. Here are the two versions of the index entry under dispute. Spot the difference (differences, rather, for there were some smaller jokes that gave supplementary offence to the American publisher).

Meanwhile, Oxford University Press had bought from W. H. Freeman the paperback rights to *The Extended Phenotype*, and they have remained its publishers ever since. So, although I had moved to other publishers, I still retained good relations with OUP. When, in 1989, they approached me for a new edition of *The Selfish Gene*, it seemed natural that it should include, as a new chapter, a summary of the thesis of *The Extended Phenotype*.

The editor OUP put in charge of the new *Selfish Gene* was Hilary McGlynn. I enjoyed working with her, but my most decisive influence while planning and working on this project was my friend Helena Cronin. She helped me with it while I helped her with her own beautiful book, *The Ant and the Peacock*. Everyone concerned agreed from the start that the original text of *The Selfish Gene* should remain unchanged, warts and all. The publishers felt that the original edition had acquired a kind of iconic status which should be preserved. Arthur Cain, quoting a critic of A. J. Ayer's *Language, Truth and Logic*, had called *The Selfish Gene* 'a young man's book' and the publishers wanted to keep that feeling. Amendments, second thoughts and embellishments would be confined to a large section of endnotes. And I proposed two new chapters: one called 'Nice guys finish first', on the theme of my BBC *Horizon* documentary of that name (see page 256); and one called 'The long reach of the gene', which would be the condensed version of *The Extended Phenotype*. These additions combined to make the 1989 edition of *The Selfish Gene* about half as large again, compared to the original 1976 edition.

Literary agents

I have said that I followed Michael Rodgers to Longman for *The Blind Watchmaker*. By then I had acquired a literary agent, Caroline Dawnay, at Peters Fraser & Dunlop in London, who struck a hard bargain with my new publisher (as rather dramatically described by Michael in his memoir). Caroline had contacted me after

The Selfish Gene and persuaded me, over lunch in the Randolph Hotel in Oxford, that an agent was a good thing to have, and that she was a good representative of the genus. And so she turned out to be. However, after *The Blind Watchmaker* I received increasingly pressing overtures from the New York literary agent John Brockman.

John was, and is, legendary in the publishing world as a ruthlessly tough bargainer–albeit an honest one who never pretended to be anything else (a journalist once said you could see Brockman's fin circling from a long way off). But what attracted me to him was his single-minded devotion to science and its place in our intellectual culture. This mission has grown, until now all his clients are scientists (or philosophers and scholars writing on scientific subjects), members of the fraternity that he himself has labelled 'the third culture', consciously going beyond C. P. Snow. It has reached the point where rather few authors in this category are *not* clients of Brockman Inc. His 'Edge' website has been rightly described as an 'on-line salon' for scientists and associated intellectuals. Like some blogs, it has many authors. The important difference is that Brockman's contributors are there by invitation only and they are a carefully creamed-off élite. I have written that he has the finest address book in America, and he uses it relentlessly to push science and reason, for example in his annual 'Edge Question'.

Around Christmas time each year, John ransacks his address book and cajoles its contents (both those who are and those who are not his clients) to give personal answers to the year's question. For example, a typical question was: 'What was the most important invention of the past two thousand years?' I especially remember my friend Nicholas Humphrey's answer: spectacles, for without them anyone beyond middle age would be unable to read and thus, in our verbal culture, disempowered. My own answer was the spectroscope: not that I thought it really was the most important, but I was rather late sending in my entry and by the time I got around to it all the more obvious inventions had been snapped up. Nevertheless, the spectroscope turns out to be a pretty good candidate. It goes far beyond what Newton could imagine, being the instrument by which we know the chemical nature of the stars, and by which we know–via the red shift of light from receding galaxies–that the universe is expanding, that it began in a big bang, and when.

Over the years, the annual Brockman questions have included: 'What is your dangerous idea?', 'What have you changed your mind about, and why?', 'What questions have disappeared, and why?', 'How is the internet changing the way you think?', 'What is your favorite deep, elegant or beautiful explanation?', 'What should we be worried about?' and 'What do you believe is true even though you cannot prove it?' (My answer to that last one was my belief that life, wherever it may be found anywhere in the universe, will turn out to be Darwinian life; see page 572.) Every year, John edits the answers into a book–superficially not all that different from many another annual anthology until you look at the cast list, and count the Nobel Prize-winners, Fellows of the National Academy of Sciences or the Royal Society, and general household names (at least in households where books abound and intellectuals foregather).

Much of that still lay in the future when John first approached me, but he was already well embarked on his zealous crusade for science and I was impressed. Though reluctant to leave my happy association with Caroline (and naively unaware that when authors part from agents the trauma can feel like a divorce), I agreed to meet John to hear his pitch. I was already planning a lecturing tour of the United States, so I added to the itinerary a visit to the farm in Connecticut where the Brockmans spend their weekends away from New York. But as it happened, the 'I' turned into a 'we'. Here's how.

This was 1992, when Douglas Adams reached his fortieth birthday, and his party was memorable for a particular reason. It was there that he introduced me to the actress Lalla Ward, whom he had known from the days when *Doctor Who* was at its wittiest because he was the script editor and she and Tom Baker gave added value to the wit by their inventively ironic playing of the two leading roles. At the birthday party, Lalla was talking to Stephen Fry when Douglas led me over and introduced us. Both Douglas and Stephen being absurdly taller than Lalla and me, it was natural that she and I should find ourselves facing each other under a Gothic arch formed by Douglas and Stephen as they exchanged lofty witticisms high above us. Through the archway I shyly offered to refill Lalla's glass, and when I returned we rapidly reached agreement that the party was too noisy for conversation. 'I suppose, by any faint chance, it wouldn't just possibly be a good idea to go out for a quick meal and–of *course*–return later?' We discreetly slipped away and found an Afghan restaurant off the Marylebone Road.

That Lalla had read *The Selfish Gene* and watched my Christmas Lectures was gratifying. That she had read *The Extended Phenotype* (and Darwin) as well was too good to be borne. I subsequently discovered that, in addition to Doctor Who's companion, she had played a beautiful Ophelia to Derek Jacobi's Hamlet in the BBC TV production, and was also a talented and versatile artist, published author and book illustrator. As I said, too good to be borne. We didn't return to the party.

I mentioned to Lalla that I was about to embark on my American journey, having added to the itinerary a visit to John Brockman. She said she was about to set off for a holiday in Barbados, with a girlfriend from the theatrical world. Impulsively she asked if I would take her to America with me, although it would mean letting down her friend in Barbados. Equally impulsively I agreed.

Slight embarrassments then opened up. I was due to stay with Dan and Susan Dennett on first arriving in Boston, and later with the Brockmans in Connecticut. In both cases one house guest was expected, not two. How could I broach the subject? Lalla and I fretted that our hosts would ask–it is, after all, a perfectly normal question to ask of a couple–'How long have you known each other?' and we would have had to answer, 'A week.' As it turned out, they didn't ask, and it was only years later that Lalla confessed to Dan the truth. 'Really?' said Dan, with possibly mock innocence. 'I thought you'd known each other for years.'

After leaving the Dennetts we flew to South Carolina, where Duke University boasts the largest population of lemurs outside Madagascar. Lalla (having done careful drawings of most of the lemur species long ago) already knew their Latin names, which greatly impressed not only me but also the lemur experts who showed us round (and I rather think I caught a couple of lemurs giving each other knowing winks as they witnessed me clocking yet more hidden depths). Highlight of our visit was the Aye-Aye (*Daubentonia*), an anomalous and literally extraordinary lemur with an extravagantly elongated, bony middle finger adapted to probing for insect prey. First there was just a cardboard box, betraying no contents. Then a single, long twig-like finger protruded. It was followed by a satanically comic face, peeping over the rim. Then, with splendid deliberation, the finger to end all fingers was deployed, not to winkle out an insect from a hole in a tree but straight up the nostril. Like most graduates of Oxford or any other university, I have forgotten most of what I learned in lectures. But Harold Pusey on lemurs has stuck in my memory solely because of one multiply repeated phrase. After every generalization about lemurs, out would come the inevitable refrain in Harold's deep voice: 'Except *Daubentonia*.' That's why I just said *literally* extraordinary.

From South Carolina we flew to La Guardia, where John Brockman had 'sent a car' to greet us. We saw an enormous stretch limo. 'That'll be for us,' Lalla joked to me. But no joke, it was. It was so big, the poor driver couldn't get out of the car park without a lot of back-and-forthing, and he actually bumped into a pillar on one of these manoeuvres. That was my first experience of an American-style stretch limo, and the drive through the darkness to Connecticut was a surreal experience of double-bed-sized leather seat, polished wood cocktail cabinet and crystal decanters, all bathed in glowing blue interior lights.

Claire Bloom lived in Connecticut not far from the Brockmans, and Lalla, who had played Ophelia to her Gertrude, was keen to see her again. I had never met her, nor had the Brockmans, and they invited her to lunch. She drove over and proved to be as charming off screen as on. After lunch she and Lalla worked on me to accept John's hard sell, and I eventually agreed to sign with Brockman Inc. as my new literary agency.

River, mountain, rainbow: a digressive tour

At this point, as I have recounted in an earlier chapter, I had just given the Royal Institution Christmas Lectures, and my first book signed up by John had the same working title, *Growing Up in the Universe*. The publishers were to be Penguin in Britain and Norton in America. The title was later narrowed to *Climbing Mount Improbable*, which had been the title of the third lecture in the series of five, while the content was broadened to include lots of things that did not feature in the lectures at all, and other parts of the lectures spilled over into a second book, *Unweaving the Rainbow.*

I had made a start on *Climbing Mount Improbable* when John contacted me with a new idea, a major diversion. He and his friend the distinguished British publisher Anthony Cheetham (who had been my contemporary at Balliol College, although we didn't know each other) had come up with a scheme–I suppose you might say 'business model'–to produce a series of twelve short books to be called the Science Masters. Each slim volume was to have a different author and was to be a personal account of the author's science. What was different about the business model was that the twelve authors were to be financially united in a co-operative: a collective. That is, from a business point of view we twelve were treated as a single author–client of John Brockman–and would each take an equal share in the royalties of all twelve books. This would mean that those of us whose books sold better than average would end up subsidizing those of us whose books sold less well. I liked the idea–can't now remember exactly why, maybe it appealed to the socialist part of my brain–and I signed up to write the short book that became *River*

Out of Eden. My fellow tillers of the Collective Book Farm included Richard Leakey, Colin Blakemore, Danny Hillis, Jared Diamond, George Smoot, Dan Dennett, Marvin Minsky . . . and Stephen Jay Gould who, unfortunately for the collective, never actually produced his book.

A pleasure that resulted from association with the Science Masters was to get to know Anthony Cheetham, whose idea, with John Brockman, it jointly was. Lalla and I met Anthony at the launch party for the series at the Cheltenham Literary Festival, and we are still good friends with him and his more than delightful wife, the literary agent Georgina Capel. We've spent several weekends at their idyllic house in the Cotswolds, watching the sun set over the roses and next day admiring the forest that Anthony has planted as a token of confidence in the future. During one of those weekends of golden Jurassic stone, the outspoken Roman Catholic apologist Cristina Odone, a fellow guest, went out of her way at dinner to pick a fight with me: good-humoured on both sides but unresolved–and presumably never to be resolved except in the unlikely event that it is posthumously resolved in her favour.

As it happened, Lalla and I were staying with the Cheethams the weekend after *River Out of Eden* was published, in the summer of 1995. Anthony went, as usual, to the nearby market town to buy the Sunday newspapers before breakfast, and we opened the *Sunday Times* to discover that my book—well, *our* book, for Lalla drew the pictures and Anthony's publishing house was a thirteenth member of the cooperative—had entered the bestseller list at number one. I don't remember whether Anthony opened champagne for breakfast, but it would have been characteristic of his ebullient generosity to do so.

River Out of Eden came out soon after the death of my uncle Colyear, youngest brother of my father, whom he resembled. I dedicated the book to his memory:

To the memory of Henry Colyear Dawkins (1921–1992). Fellow of St John's College, Oxford: A master of the art of making things clear.

By universal consent he was a brilliant teacher, humorous, lucid, fluently intelligent, who managed to get the principles of statistics across to generations of grateful Oxford biologists—no mean feat. Along with most other college fellows in biology, I used to ask him to give statistics tutorials to my own New College students. On one occasion, I had gone to see him for that purpose in his office in the Forestry Department, then called the Imperial Forestry Institute, which is relevant to this story. I was describing the young man to him ('Quite clever, a bit lazy, you'll need to keep an eye on him . . .' etc.). Colyear was taking notes as I spoke, but not in English (he was a fine linguist). I said: 'Oh, that's very confidential of you, taking notes in Swahili.'

'Good God, no,' he expostulated. 'Swahili? No, no, everyone in this department speaks Swahili. This is Achole.'

Another brief anecdote to sum up his character. At Oxford railway station the car park was guarded by a mechanical arm, which rose to allow each car to leave when the driver inserted a token of payment in a slot. One night Colyear had returned to Oxford on the last train from London. Something had gone wrong with the mechanism of the arm and it was stuck in the down position. The station officials had all gone home, and the owners of the trapped cars were in despair as to how to escape the car park. Colyear, with his bike waiting, had no personal interest; nevertheless, with exemplary altruism he seized the arm, broke it, carried it up to the stationmaster's office and plonked it down outside the door with a note giving his name, address and explanation as to why he had done it. He should have been given a medal. Instead, he was prosecuted in court and fined. What a terrible incentive to public-spiritedness.

How very typical of the rule-obsessed, legalistic, mean-spirited dundridges of today's Britain.

And a little sequel to that story. Many years later, after Colyear's death, I chanced to meet the distinguished Hungarian scientist Nicolas Kurti (a physicist who incidentally happened to be a pioneer of scientific cookery, injecting meat with a hypodermic syringe, all that sort of thing). His eyes lit up when I spoke my name.

'Dawkins? Did you say Dawkins? Are you any relation of the Dawkins who broke the arm at the Oxford station car park?'

'Er, yes, I'm his nephew.'

'Come, let me shake your hand. Your uncle was a *hero.*'

If the magistrates who imposed Colyear's fine should happen to read this, I hope you feel thoroughly ashamed. You were only doing your duty and upholding the law? Yeah, right.

Climbing Mount Improbable (1996) was the book in which my colour biomorphs made their debut (see pages 554–5), and the book was also illustrated by Lalla's beautiful drawings of real animals. But her contribution didn't end there. This was the book that initiated–by accident–our by now long tradition of joint readings. We'd been promoting the book in Australia and New Zealand . . . but wait (I'll come back to the joint readings): pleasantly triggered memories are worth recalling in another digression. And even, for good measure, digression nested within digression.

> What is a Life if, full of stress,
> We have no freedom to digress?
> But if the prospect you enrages
> You'd better skip the next few pages.

Lalla and I flew via Hong Kong and Sydney to Christchurch (dear Christchurch, has your nostalgically dated Britishness survived the earthquakes?). In between my talks promoting *Climbing Mount Improbable*, we hired a car and drove over the Southern Alps, via the Franz Josef glacier to the rainforest on the western side of the South Island with its unique tree ferns. Unfortunately we didn't get down to Fiordland (of which Douglas Adams said that one's first impulse 'is simply to burst into spontaneous applause'). Crossing back to the east side through sensuously rolling 'sheep may safely graze' pastures and tall hedges, we came to Dunedin, where I gave another talk and where we were looked after by my former New College colleague Peter Skegg. A professor of law, Peter is also a published ornithologist, and he gave us an expert tour of the protected royal albatross colony on the Otago peninsula. The sight of the great birds laboriously taking off along their runways like Boeings at an airport was familiar to Peter but new to Lalla and me, and we were entranced.

After more lectures in Wellington (where we had dinner with the philosopher Kim Sterelny) and Auckland, we flew back to Australia. In Melbourne we were met by Roland Seidel of the Australian Skeptics, wearing the differently coloured socks which, along with his pink suit, were his trademark fashion statement–not to be confused with the Stephen Potter 'woomanship' ploy in which odd socks are worn to arouse maternal instincts ('Buy our patent Oddsox brand'). Roland took us to his house in the eucalyptus forests of the Dandenong Hills outside the city. From the wooden veranda Lalla was delighted to have kookaburras swoop down to feed from her hand with their insolently truculent beaks.

We spent a few days on Heron Island (see picture section) in the Great Barrier Reef, where I was taken snorkelling by the wife of the research station manager. She soothed my panic when I suddenly came face to face with a shark by saying: 'It's OK, it's quite harmless.' But she then rather spoiled it by adding: 'But I wish it would go away and be harmless somewhere else.'

In Canberra the Australian National University gave me an honorary doctorate–and even let me keep the robe. Its colour scheme is almost identical to the Oxford DPhil, which I suppose is convenient, although a bit coals to Newcastle. While on the subject of honorary degrees, I long coveted a Spanish one, because you get a wonderful hat like a tasselled lampshade. Unlike the subject of Peter Medawar's characteristic joke, I don't aspire to collect honorary degrees through the alphabet ('Yale and Zimbabwe are unaccountably dragging their feet') but I was delighted when Valencia came through, and I now wear the enviable lampshade hat annually to the Vice-Chancellor's Encaenia Garden Party in Oxford: a splendidly anachronistic lekking ceremony for colourful academics. Among other honorary doctorates, I take particular pleasure from the ones at Juliet's two alma maters, St Andrews and Sussex, the latter presented to me by Lalla's dear friend Richard Attenborough, as Chancellor (see picture section). When my friend Paula Kirby saw the photograph she said: 'Very nice, but why have you come dressed as Liquorice Allsorts?'

To return, finally, to the entry point of this multiple digression, after Australia Lalla and I flew to California to continue the promotional tour for *Climbing Mount Improbable*. But the many speaking engagements down under had conspired with the cold that typically follows a long flight to give me laryngitis, and I could scarcely speak: so Lalla stepped in and read selected passages from the book in her beautiful voice (it's not for nothing that the BBC cast her in Shakespeare). After her readings we turned the amplifier up so that I could croak out answers to a few questions from the floor. As we wended our way back east, my voice gradually returned. However, so well were Lalla's readings received that they had to continue, and thus the tradition became established—one that we continued when promoting subsequent books, the two of us reading alternate passages. We have now recorded most of my backlist of books as a double act, under the erudite guidance of Nicholas Jones of Strathmore audio publishing. It seems to work well: the change of voice every few paragraphs serves to stop the listener nodding off, and it's especially useful for distinguishing quotations from the surrounding prose, dispensing with the intrusively spoken word 'quote'.

I recorded Darwin's *Origin of Species* on my own, and also *An Appetite for Wonder*, except that Lalla read the extracts from my mother's diaries. Recording the *Origin* was a very interesting experience. I made no attempt to play the part of a Victorian *paterfamilias* but read it entirely in my own voice. My aim was to work hard at understanding every sentence so completely as to give the correct stresses to both words and syllables, and thereby aid my listeners' understanding. It was quite difficult, as Victorian sentences are often longer than modern ears are acclimatized to. I came away from the experience with an even deeper admiration for Darwin's wisdom and intellect than I had had before–and that's saying a lot.

I think I have learned something of the art of reading aloud from Lalla, and in doing so may have deepened my lifelong love of poetry. It was Lalla who persuaded me that there was a book to be written on the poetry of science, and that I should write it. *Unweaving the Rainbow*, a reply to Keats's Romantic hostility to Newtonian science, came out in 1998, two years after *Climbing Mount Improbable*, and is dedicated to her. *Climbing Mount Improbable* was dedicated to Robert Winston, who had so generously helped Lalla and me in our four attempts–alas, unsuccessful–to have a child by IVF. Before publication, it was a pleasure to announce the dedication ('A good doctor and a good man') at a debate on religion in London organized by a rabbi, where Robert (one of the most respected members of England's Jewish community) and I were speaking on opposite sides.

I think of *Climbing Mount Improbable* as my most under-rated book, but I can't blame the publishers for failing to push it. They sent pre-publication copies out to a starry list of readers who supplied lovely warm quotes for the jacket. Perhaps the puff that pleased me the most was the one from David Attenborough, who said, among other things, that he enjoyed it so much it was all he could do to stop himself waking the perfect stranger slumbering next to him in order to read out favourite passages. This the publishers refused to print, reducing his recommendation to the single word 'Dazzling'. What were they afraid of? They had only to explain that he was reading it on a long-haul night flight.

Let me digress a little on this marvellous man. Whenever the possibility comes up of Britain having an elected rather than a hereditary head of state, the awkward question is raised: it's all very well getting rid of the Queen, but just think what we might get as an alternative: King Tony Blair? King Justin Bieber? Such dire speculations are brought to a swift end when somebody points out that there is one potential figurehead behind whom everyone could unite: King David Attenborough.

Everyone knows how charming and friendly he is. Less well known is that he is a hilarious raconteur with a gift for mimicry. He could have been an actor like his brother Richard. Put him in the company of that other priceless storyteller, his friend and fellow antiquities collector Desmond Morris, then sit back and enjoy the cabaret. Also unforgettable is David's imitation of the older fellows of the Zoological Society when Desmond's glamorous wife Ramona appeared in their clubhouse and walked across their field of view. They slowly swivelled, agog in their chairs as their eyes followed her progress. David holds an imaginary coffee cup in his hand, and as he turns in hilarious imitation of Ramona's goggle-eyed admirers, the mimed cup slowly inverts until its contents spill all over their trousers.

On one occasion the *Guardian* interviewed David and me together. I can't remember the pretext, perhaps some regular spot involving double interviews. Before the interview itself, a photographer had been commissioned to take a picture of us together. For this purpose we sat outside in David's garden and talked, while the photographer snapped away. It was a marvellous conversation. At a conservative estimate we were both roaring with laughter 95 per cent of the time, and the photographer must have taken well over a hundred photographs. So when the picture editors came to choose the one to print, what did they go for? We were shown facing each other like a couple of prize-fighters, jutting out our chins in a classic primate aggressive display, exactly as though we were about to come to blows. It must have been a really tough job to find that one grim picture among at least a hundred smiling, friendly, laughing ones. Well, that's journalism. Perhaps there was a fashion at the time for 'edgy'.

Lalla reminds me of a *Sunday Times* journalist (whom I shall not name) who came to interview me in our home. Lalla was working upstairs and she could hear almost non-stop friendly laughter coming from the room below throughout the interview. But when the interview was printed, the first sentence was: 'The trouble with Richard Dawkins is, he has no sense of humour.' He's an atheist, you see, and everyone knows they have no sense of humour. (Actually, that journalist is probably an atheist too, and so are most of his colleagues on the paper: they just don't come out and say so.) Perish the thought that the public face of atheism should be allowed to laugh and smile; no, the trademark snarl must at all times be preserved.

Nor are atheists supposed to have any poetic sensibility. And this brings me back to *Unweaving the Rainbow*, the book in which, more than any other, I attempted to extol the poetry of science. As I've already hinted, this was the book where Lalla's influence on my writing first made itself strongly felt. She urged me, as the newly appointed Simonyi Professor for the Public Understanding of Science, to reach out to poets and artists. Although it contains sentences that originated in the Christmas Lectures, the real spirit of the book germinated in my 1996 Richard Dimbleby Lecture, which opened with words suggested by Lalla, and it carried her inspiration through to the end. Indeed, the title of my Dimbleby Lecture became the subtitle of the book: *Science, Delusion and the Appetite for Wonder.*

The annual Richard Dimbleby Lecture, televised by the BBC, commemorates a great broadcaster, and luminary of that once great institution. It was an honour to be invited to be the Dimbleby Lecturer for 1996, and I accepted with my customary misgivings and trepidation. My early drafts of the lecture seemed to be going nowhere, which only served to enhance those misgivings. Lalla rescued me from my despond with an inspired opening which I adopted, word for word, and which immediately set the tone for the rest of the lecture: 'You could give Aristotle a tutorial. And you could thrill him to the core of his being.'

The British publisher of *Unweaving the Rainbow* was again Penguin. In America, John Brockman switched me to Houghton Mifflin and they sent me on a book tour to promote it, of which the high spot was an event in the Herbst Theater, San Francisco. John Cleese agreed to interview me on stage and he did it wonderfully. His copy of the book was a thicket of yellow post-it notes: he had really done his homework. You can't be as funny as he is, in the particular way of being funny that he epitomizes, unless you are very intelligent. And his intelligence shone through that evening on stage. I got the impression the audience were expecting him to be funny, to the extent that, no matter how serious his words and his intentions, they simply laughed at everything he said. Admittedly they didn't get many clues in his tone of voice, for his serious voice is indistinguishable from the voice he uses when doing serious comedy, as for instance in the Argument Clinic sketch, or the deadpan voice in which he said to Michael Palin as the aspiring silly-walk developer, 'That's it, is it? It's not *very* silly, is it?' I much enjoyed the San Francisco audience's laughter and probably joined in. But afterwards I wondered whether John was a bit frustrated that people laugh at everything he says, even when he's actually being serious.

He really does seem to be permanently funny, as Lalla and I discovered when he and his wife invited us to stay, for a holiday. Here's just one of many wonderful stories he told. He overheard a woman saying, on the top deck of a bus (and he has no idea of the context):

'I washed it for her when she was born. I washed it for her when she got married. I washed it for her for Winston Churchill's funeral. And I'm not going to wash it for her again.'

Do very funny people have more than their fair share of funny things happen to them? It's hard to see why things should work out like that, but it's a question I'm prompted to ask, not just about John Cleese but other humour magnets I have known such as Douglas Adams, Desmond Morris, David Attenborough, Terry Jones. Perhaps they just have a good ear and eye for the humorous and notice funny things more than the rest of us.

The Ancestor's Tale and A Devil's Chaplain

The next book I proposed to John Brockman was *The God Delusion* but he was not enthusiastic. You can't sell a book attacking religion in America, was his view, and in those days (the 1990s) he might have been right. George W. Bush later came along to change his mind. But meanwhile, back in 1997, at another of those idyllic Cotswold weekends, Anthony Cheetham had a proposal for me which was exciting and intimidating at the same time. A complete history of life, on a grand scale: an evolutionist's equivalent, as he put it, of Ernst Gombrich's *The Story of Art*.

I was aghast at the ambitiousness of the project. It would require a huge amount of reading, rekindling knowledge that had lain dormant since undergraduate days (and I ruefully recalled Harold Pusey's remark, quoted earlier, to the effect that at the time of Oxford finals you hold in your head more concentrated knowledge than you ever will again). Moreover, much of that undergraduate knowledge would now be out of date, superseded by the mass of new information flooding in especially from the world's molecular biology labs. Would I have the stamina to do justice to Anthony's proposal? It seemed a tall order. On the other hand, I was now two years into my professorship of Public Understanding of Science (of which more in a later chapter), which had freed me of the burden of tutorial teaching. Didn't I owe it to Charles Simonyi, my benefactor, to produce something big, something worthy of the extra time granted me each day by his generosity, an *opus* sufficiently *magnum* to give my successors something to live up to?

I dithered sleeplessly for several days and nights. On bright mornings I thought I could do it and even made some rough notes of a plan. Dark nights raised the spectre of a millstone that might weigh me down for years. Lalla was in favour of my taking the plunge. I could pace myself over several years, she pointed out, break the book down into chapters and take it one chapter at a time, that's the way to tame a task. This stiffened my resolve, and in March 1997 I signed the contract with Anthony. At the same time, John made a deal with Houghton Mifflin as the American publishers, where my editor was Eamon Dolan.

I began writing in fairly high spirits, cheerfully facing the long and winding road ahead while not underestimating its length or the burden I would have to tote along it. However, two years later the sheer magnitude of the task before me pushed me back into despair. Lalla tried to encourage me. She vacated her art room, where she was wont to produce her lovely creations, so that I could have a whole wall on which to pin a huge map of the book–a plan of the history of life. That change of scene revived my flagging spirits, but only temporarily, and the contractual deadline loomed oppressively. I fell back into a cowardly yearning to abandon the project and return the publishers' advance. I was on the point of doing so when Lalla, in what amounted to a mercy mission to rescue me from my abject mental state, dashed over alone to see Anthony in the Cotswolds, and it was as a result of this crisis meeting that he wrote to me as follows, in February 1999 (I should explain that *Ancestral Voices* was the working title of the book at that stage; we later abandoned it because Coleridge's evocative phrase had been used too often already):

Dear Richard . . .

Re: *Ancestral Voices.*

I don't want you to lose a moment's sleep or suffer a pang of regret over this project. If the deadline is troubling you we shall change it. This book is too important to me and I am sure to you as well to treat it as if it was an article for a Sunday newspaper. I suggest we have a private agreement between us that, provided it is to be your next book, the delivery timetable is up to you rather than to us or to the date shown in the contract . . .

Best wishes

Anthony

Now that's the letter of a great publisher and bookman. The other thing that pulled me up from despair was the realization that the generous advance John Brockman had negotiated for the book would enable me to pay a full time post-doctoral research assistant to work on the book. That, after all, is the kind of thing advances are designed for. And the ideal candidate was spiritraisingly obvious—and, even better, right on my doorstep. Yan Wong, one of my very best pupils since the glory days of Mark Ridley and Alan Grafen, was just finishing his doctoral thesis under Alan's supervision

(I suppose that made him my grandstudent as well as my student). Yan was keen to take the assignment, for something like the same reason I had been reluctant to start the book in the first place: it would require a huge amount of work and the reading up of lots of facts. What I had seen as a deterrent, Yan, thirty years younger, saw as a challenge.

Yan began work with me in early 1999. My professorship was nominally based in the Oxford University Museum of Natural History, and Yan was given a small office in that splendid edifice (its architecture calls to mind the gothic style of the dinosaur skeletons inside it), where he worked surrounded by bones, fossils, dust and crystal cabinets. We met frequently to discuss every detail of the book, and to plan its structure. Anthony had originally thought of the history of life as proceeding in a conventional direction, forwards in time. But he happily came round to seeing the virtue of doing this history, as Yan and I preferred, backwards. Our reasons were cogent. Too many evolutionary histories culminate in man. The first chapter of *The Ancestor's Tale* is entitled 'The conceit of hindsight', explained as follows:

What of the second temptation, the conceit of hindsight, the idea that the past works to deliver our particular present? The late Stephen Jay Gould rightly pointed out that a dominant icon of evolution in popular mythology, a caricature almost as ubiquitous as lemmings jumping over cliffs (and that myth is false too), is a shambling file of simian ancestors, rising progressively in the wake of the erect, striding, majestic figure of *Homo sapiens sapiens*: man as evolution's last word (and in this context it nearly always is man rather than woman): man as what the whole enterprise is pointing towards; man as a magnet, drawing evolution from the past towards him.

We wanted to avoid the human conceit; but at the same time we had to recognize that our readers, being only human, would surely be most interested in human evolution. How could we satisfy that pardonably anthropocentric interest while not pandering to the myth that evolution marches ever upwards towards the pinnacle of humanity? By doing our history backwards. If you start at the origin of life and work forwards, your history can end up, with equal legitimacy, at any of the millions of surviving species. *Homo sapiens* should not be privileged, nor should *Ranunculus repens*, nor should *Panthera leo*, nor should *Drosophila subobscura*. But if you do your history backwards you can respectably privilege any modern species and trace its ancestry back to the same unique origin, an origin shared by all. This frees us to choose as our recent starting point the species we are most interested in–our own.

Yan and I dramatized the backwards journey in Chaucerian terms as a pilgrimage, a human pilgrimage back to the origin of all life. We had human pilgrims being joined successively at discrete 'rendezvous points' by close cousins, then more distant cousins, then very distant cousins. This had the added benefit of emphasizing that modern species are not ancestors of other modern species but their *cousins*. Surprisingly, there turned out to be only thirty-nine of these rendezvous points. The reason this number is so small is that at many of the rendezvous, very large numbers of cousins join. At Rendezvous 26, for example, most of the invertebrates join the pilgrim throng, including the insects–and, as Robert May (distinguished physicist-turned-biologist who became the British government's chief scientific adviser and President of the Royal Society) once quipped, to a first approximation all species are insects.

We needed a word for the dead ancestors that our modern pilgrims shared at their successive rendezvous points. Dredging up my school Greek I suggested 'phylarch' but it wasn't catchy enough. Eventually Yan's wife Nicky came up with the perfect word: 'concestor', a natural contraction of 'common ancestor'. Concestor 15, for example, is the common ancestor of all modern mammals.

Our other Chaucerian touch was to let a few of the pilgrims, as they joined the human journey to the past, tell a 'tale'. These tales were frank digressions, excuses to tell interesting biological stories, relevant to the whole book and not limited to the particular creature telling the story. The Grasshopper's Tale, for instance, is about race, especially the vexed topic of human races, and the grasshopper tells it because of a particular piece of research on grasshopper races. The Velvet Worm's Tale is about the Cambrian Explosion. The Beaver's Tale is about the extended phenotype. The tales were told in my voice: it would have been too twee to have the animals talking in the first person.

It is certainly fair to say that the book could not possibly have been finished without Yan Wong. He is credited with joint authorship of several of the chapters, and I am delighted to say that I have just negotiated with the publishers, through Brockman Inc., that there shall be a new edition, brought up to date with new material by Yan, and with his name now firmly on the cover as joint author.

In 2002, during one of my crises of confidence, I sought to appease the publishers and stave off deadline pressure by offering them another book–the one that would eventually be called *A Devil's Chaplain*. Anthony was keen to publish a collection of my already published essays and journalism, and so was Eamon Dolan of the American publishers, Houghton Mifflin. I knew exactly the right person to help me edit them. Latha Menon, originally from India but a long-time resident in Oxford and an Oxford graduate, had been the resourceful and astonishingly knowledgeable editor of *Encarta*, the encyclopedia produced under the auspices of Microsoft. I was on *Encarta*'s editorial board for several years and attended the annual meetings in Somerville College, under the chairmanship of the distinguished historian Asa Briggs, with Latha leading most of the detailed discussion. I was extremely impressed by her, and, when the work on *Encarta* came to an end, I successfully recommended her for a job editing science books for Oxford University Press. Could she do some moonlighting, editing my anthology? She could. She was already familiar with almost everything I had ever written and she set to work to help me choose a suitable list of writings and arrange them into seven sections. I named the sections mostly with poetic allusions such as 'Light will be thrown' (on Darwinism),

'They told me Heraclitus' (obituaries and memorials), 'Even the ranks of Tuscany' (various papers connected with Stephen Jay Gould), and 'There is all Africa and her prodigies in us' (on African matters). The last section, 'A prayer for my daughter', had a single chapter, the open letter that I had written to my daughter Juliet when she was ten. This formed a climax to the book and, as she was now just eighteen, the book itself was dedicated to Juliet on the occasion of her coming of age.

A prayer for my daughter

It might seem odd that I should have written a long letter to my ten-year-old daughter on the subject of 'Good and bad reasons for believing'. Why not just talk to her? The reason is the sad but not uncommon one that we didn't see very much of each other. Juliet lived with her mother, my second wife Eve. Eve was attractive and amusing and very good company, but we didn't have much in common apart from our love of Juliet herself. Separation became increasingly inevitable and we parted when Juliet was four–at an age when, we hoped, it might be less upsetting for her than if we left it till later. Juliet and I then saw each other regularly but more briefly than I would have wished (such visits are settled by lawyers, with their 'our-side their-side' mentality–need I say more?), and our time together was too precious for heavy discussions about the meaning of life. In her early years my limited times with her went all too quickly, reading her beloved gorilla book, or *Mogg the Forgetful Cat*, or *Babar the Elephant*; or playing the piano with her, or walking down to the river with Pepe the dear little whippet.

But I did want to communicate something deeper, and the fact that we saw so little of each other erected barriers. I was even a little shy of her: in awe of her sweet nature and her beauty from the day she was born. I was strangely tongue-tied in her presence. Religious parents send their children to Sunday school, or talk to them about their faith. I suppose I wanted to do something vaguely equivalent. She was intelligent and did well at school, and I thought she might appreciate a long, thoughtful letter. I hasten to add that the very last thing I wanted to do was indoctrinate her with my own beliefs. The entire thrust of my letter was towards encouraging her to think for herself and come to her own conclusions.

She read the letter and said she liked it, but we didn't discuss it. As it happened, John Brockman was at the time editing a book of essays for children, which he wanted to present to his son Max as a bar mitzvah present. I was one of those whom he asked to contribute, and the obvious essay for me to submit was my letter to Juliet. So it became an open letter. The published version was well received by parents around the world, who gave it to their children or read it to them. And, as explained above, I later reprinted it as the last chapter of *A Devil's Chaplain*, and dedicated the whole book to Juliet on her eighteenth birthday.

Juliet was seven when I met Lalla and eight when I married her. From the start, they seemed to get on very well. We settled down to a regime where Juliet spent alternate weekends with Lalla and me in our house, and we had some lovely holidays in the far west of Ireland, with Juliet and her friend Alexandra, in the house which my parents had restored among the dunes looking towards the Twelve Bens of Connemara. Those were happy times, commemorated by a lovely embroidery that Lalla made and presented to my parents.

But when Juliet was twelve, Eve started to develop ominous symptoms which were diagnosed as adrenal cancer. She had a large operation, which saved her life for a while, but then metastases set in and she entered the treadmill of chemotherapy, with all its gruelling side-effects. She bore it with immense fortitude and courage, buoyed up by her own characteristic black humour, which was one of the things that had attracted me to her in the first place. For example, on one occasion Lalla was taking Pepe to the vet, my nephew Peter

Kettlewell, and Eve said: 'While you're about it, just ask Peter for something to put me down: I should think the dose for a medium-sized Alsatian would be about right.' And so she laughed courageously in the face of death.

Lalla and Eve, during this time, forged a remarkable friendship, and I think this solidified Lalla's bond with Juliet. Lalla accompanied Eve on all her visits to the oncologist, took her out to lunch in a pub every week and, I think, raised her spirits while her health was declining. Lalla and I employed professional carers, friendly and competent young women from New Zealand and Australia, to help look after Eve and Juliet. And, all of us knowing that the prognosis was dire, we sent Eve on a holiday with Juliet, a fine Mediterranean cruise, which I think she enjoyed.

I suspect that the seeds of Juliet's ambition to become a doctor were sown during those terrible two years of her mother's decline. Rightly or wrongly (actually rightly, I am sure), we decided to keep no secrets from her. She knew exactly what was going on with each hospital visit. I find myself almost in tears as I write this, remembering how that lovely little girl matured far beyond her years, caring for her mother through the ordeals of successive chemotherapy cycles, hiding her own forebodings and grief in a way that no child should be expected to do, keeping calm and sensible when the rest of us were not doing so well at that. And when the end came, in the old Radcliffe Infirmary, Juliet was–what can I say except–a fourteen-year-old hero.

For the funeral, I asked Edward Higginbottom, the distinguished organist and choirmaster of New College, to find me a singer to do Schubert's *Ave Maria*. He found a sweet soprano whose pure voice really did reduce me to tears at such an affecting moment, and Juliet turned to me and hugged me. I supported Eve's mother down the aisle at the end and we all returned to our house for the wake afterwards.

Juliet had been brave for so long, it is hardly surprising that her grief struck hard after the tragic loss of her mother. Lalla held us together through those difficult years with her matchless gift for intuitively sympathetic psychology and–well, for holding things together. But Juliet's school work had suffered and she was set back in the face of the notoriously relentless pressure of Oxford High School. We took her out and sent her to D'Overbroeck's tutorial college, which suited her better and gave her a taste, I think, of what true education can be. She temporarily lost heart in her medical ambition and went to the University of Sussex, on the south coast of England, to read human sciences.

This is a mixture of biological science and social science, with which I was familiar because I had been peripherally involved in the inauguration of a similar degree at Oxford, and had been in charge of the human scientists at New College.

Juliet loved the science at Sussex. John Maynard Smith had retired by then, but he was still around, and as her biology tutor Juliet had a marvellous young Australian woman, Lindell Bromham, who taught evolution in the spirit of JMS's still fresh legacy. On the other hand, Juliet didn't enjoy the social science and found it difficult to reconcile with her own intelligently scientific approach. For her, the last straw was when one of her lecturers said: 'The beauty of anthropology is that, when two anthropologists look at the same data, they come to opposite conclusions.' Perhaps the remark was tongue in cheek but, together with the anti-Darwinian spirit of some of the social science lecturers, it still dampened the spirits of a keen young scientist!

Her interest in medicine was rekindled and the big break in her young career came when she succeeded in transferring to St Andrews, in Scotland, after only one year at Sussex. Here at last she was able to read medicine. St Andrews is one of Britain's great universities (and its third oldest after Oxford and Cambridge), and it was marvellous for her. I fondly think she was pretty good for St Andrews too. She was popular, made lifelong friends, edited the medical students' magazine, went to balls and parties, and still ended up with a top first-class degree. St Andrews doesn't have a clinical school, so its medical students disperse after their first degree. Most go to Manchester, but Juliet set her heart on Cambridge and she qualified there as a doctor in 2010. Eve would have been deeply proud of her, as I am.

The God Delusion

In early 2005, soon after *The Ancestor's Tale* was published, John Brockman signalled that his original objections to my attempting to publish *The God Delusion* in America had evaporated. George W. Bush's lurch towards theocracy–he literally said that God had told him to invade Iraq–surely had something to do with this comprehensive shift. John asked me to write a proposal in the form of a letter to him, which he could hawk around to publishers. I reproduce the opening paragraphs of the letter here.

New College, Oxford OX1 3BN
21st March 2005
John Brockman
Brockman Inc. New York

Dear John

The God Delusion

As you know, I am about to embark on writing and presenting a major television documentary, attacking religion as 'The Root of All Evil' (the working title, which will change). It has been commissioned by the religious department (!) of Channel Four, who want something hard-hitting, going after religion with all guns blazing, rather than a balanced, moderate, gentle treatment like the history of atheism series recently presented by Jonathan Miller. In my discussions with the producer, it is I who am the voice of restraint!

Channel Four will broadcast it either as two one-hour shows, or (which I and the producer would prefer) one two-hour blockbuster. Filming will begin in May or June of 2005, and the documentary will be shown presumably late in 2005 or early 2006. No doubt Channel Four will make strenuous efforts to sell it outside Britain as well. Meanwhile, the producer is working hard setting up locations for filming in various parts of the world, including America, Europe and the Middle East.

It seems sensible to write a book on the same general theme while it is in the front of my mind, and I propose *The God Delusion*. I don't see it as a straight TV tie-in.

The chapters that I went on to list in the letter bear some slight resemblance to the ones that I eventually wrote—more so than is usual for my proposals, actually. Although I presented the book to John by way of the TV documentary, it was not a television tie-in. Far from it. The documentary and the book both stand on their own, and overlap only slightly.

In America, John sold the book to Houghton Mifflin, the publishers of *The Ancestor's Tale* and *A Devil's Chaplain*. In Britain, he broke new ground. The book was bought by Transworld, a division of Random House, where my editor was Sally Gaminara. The relationship turned out a happy one and she has published all my books since then. Sally has recently written to me of her reaction when John first sent her the above letter. 'I shared it with my colleagues who were equally enthusiastic and we entered, and won, the auction for UK publishing rights.' She went on to describe her reaction when she received the typescript of the book itself. I'm especially pleased that it made her laugh:

I also hadn't bargained for the wonderful humour in it. I expected to smile a little, but not laugh out loud again and again. It was a gloriously thrilling experience.

Her response makes a sharp contrast with a reputation the book has acquired–perhaps among those who have read only secondhand accounts–for shrill and savage stridency. I'll return to this in a later chapter. Sally's letter continued:

> . . . one never knows whether one's own taste chimes with others and so the nail biting began again in the run-up to publication (September 2006). I solicited pre-publication quotes from a wide group of writers and thinkers, many of whom came up with fabulous 'puffs', far more than usual, so I allowed my excitement to spill out again. But it wasn't until you gave your first interview on publication with Jeremy Paxman on *Newsnight*, arranged by Patsy Irwin, that we got the first signs that something 'big' was on its way.
>
> From that moment on we could barely keep the book in print as the publicity it provoked spread, more and more people started to read it and the reviews came out thick and fast, almost all hugely complimentary. I remember calling you at home and speaking to Lalla (whom I hadn't yet met)

because you were out, and jabbering to her with excitement, trying to explain that something exceptional was taking place. It wasn't just the sales that were exceptional, it was the fact that the book had struck a vital chord with the public. I think it is no exaggeration to say that it started a whole new debate, certainly for this generation, about religion and its place in society and became a game-changer.

Game-changer? Well, it is true that *The God Delusion* has sold more than three million copies so far, well over two million in English, the remainder in thirty-five other languages, including sales of a quarter of a million in German. Perhaps another litmus test is the remarkable collection of 'fleas' that the book has gathered. My website, RichardDawkins.net, started to collect books with titles like *The Dawkins Delusion, The Devil's Delusion, The God Solution, Deluded by Dawkins, The Richard Dawkins Delusion, God is no Delusion*. We called them 'fleas' after a W. B. Yeats poem that I had going round in my head at the time:

> You say, as I have often given tongue
> In praise of what another's said or sung,
> T'were politic to do the like by these.
> But was there ever dog that praised his fleas?

I've included a choice selection of eleven of these fleas in the picture section.

But never mind the sales figures and the fleas. Did the book feel like a 'game-changer' at the time? Yes and no. I don't know where the phrase 'New Atheists' originated. One suggestion is a 2006 article in *Wired* by one of its contributing editors, Gary Wolf.[1] He lists under the rubric Sam Harris, Dan Dennett and me. Presumably he would have added Christopher Hitchens if *God is Not Great* had been published by then. And probably Victor Stenger too, whose books, written from the point of view of a physicist, are a little less well known but none the less powerful. Vic coined the memorable aphorism, often wrongly attributed to me: 'Science flies you to the moon. Religion flies you into buildings.' His death was announced as I was revising this book for publication. His strong voice will be much missed.

Wherever the phrase came from, 'New Atheists' seems to have stuck, as has 'Four Horsemen', which apparently took over from the earlier 'Three Musketeers' when Christopher's book appeared. I don't object to any of these phrases. It is, however, necessary to disclaim any suggestion that 'new' atheism is philosophically different from earlier versions espoused by, say,

---

[1] http://archive.wired.com/wired/archive/14.11/atheism.html.

Bertrand Russell or Robert Ingersoll. Nevertheless, although it isn't really very new, as a journalistic coining 'New Atheism' has its place because I think it's true that something really did happen in our culture between *The End of Faith* in 2004 and *God is Not Great* in 2007. *The God Delusion* was published in 2006, as were Dan Dennett's *Breaking the Spell* and Sam Harris's powerful short book, *Letter to a Christian Nation.* Our books do seem to have hit the proverbial nerve, in a way that the many excellent books that preceded them did not, at least since Russell's searingly limpid *Why I am not a Christian* (which inspired me when I read it in the Oundle school library in the 1950s).

Was it that our books were especially outspoken and uninhibited? Maybe that had something to do with it. Was it something in the atmosphere of the first decade of this century: wings of a *Zeitgeist* hovering in the air waiting for an updraft from the next four books that came along? Possibly, and George Bush's leanings toward theocracy, in parallel with the menace of the Mosque Militant, doubtless had something to do with it.

I can certainly say that the four of us didn't plan anything together. We surely read each other's books, those that were available before we wrote our own. And inevitably we must have been influenced, at least somewhat. To mention just the earliest of these books, I'd never heard of Sam Harris until I opened *The End of Faith*. In a chillingly accomplished piece of writing on the very first page, Sam sets up the scene for a horrific suicide bombing of a bus, by a young man. You know what's coming from the start. When the dust and the nails, the ball bearings and rat poison have cleared, the young man's family, though sad to lose him, rejoice in the certain knowledge that their son is in the martyrs' heaven; rejoice, too, in the material comforts of food and money showered on them by neighbours honouring his accomplishment. The punch-line for the story is like a body blow which, if anything, paradoxically *gains* in devastating force because we can see it coming all through the build-up. What do we know about the young man? Was he rich or poor, popular or unpopular, clever or not, a promising young student? An engineer, perhaps? We know next to nothing about him. But here's the kicker.

> Why is it so easy then, so trivially easy–you-could-almost-bet-your-life-on-it easy–to guess the young man's religion?

And, sure enough, Sam doesn't bother to tell us the religion. There was, and is, no need.

For my fiftieth birthday, my mother painted a cupboard (*below right*) with scenes depicting my life . . .

. . . my room in New College, with biomorphs on the computer screen and a view of the Oxford skyline (*above left*); my African childhood (*far left*); and my daughter Juliet with her dog and two cats, building castles in the air (*above*).

Some of my heroes, presided over by the greatest of them all, Charles Darwin: on this page (*clockwise*), Peter Medawar, Niko Tinbergen, Bill Hamilton, John Maynard Smith . . .

42

... and here (*from top down*), Douglas Adams, Carl Sagan (who provided part of the title of this book) and David Attenborough.

UF's 'Wasp Lady' Is Branching Ou

By The Associated Press

Dr. Jane Brockmann's been watching wasps so long that she is known as the University of Florida's "wasp lady."

Again this summer she's heading for more of the same, this time joined by Dr. Richard Dawkins, an Oxford University professor who studies small animals and insects.

They believe their work will help explain how behavior evolves in wasps, crickets, frogs and other animals — perhaps even humans.

Ms. Brockmann, 32, worked with Dawkins in England last year while she was on a North Atlantic Treaty Organization fellowship.

"The big question is, 'Why is there such diversity of behavior?'" said Dawkins, author of "The Selfish Gene."

As an example, he cited a behavior pattern in frogs: a male frog may sit in a pond and croak beguiling love songs, awaiting females to come to his call. Meanwhile, other male frogs lurk in the dark around him in hope of intercepting the females.

Another evening, Dawkins said, the croaker may become one of the "sneakies" and another act as "caller."

"We tend to think of one strategy as successful and one as a loser's approach, but obviously they're not, or else evolution would favor one above the other," Ms. Brockmann said.

She began watching wasps in preparation for her doctoral degree and estimates she put in 3,500 hours at it from 1973-75.

While working with Dawkins in England, she said, "we analyzed my

Dr. Brockmann at One Point Spent Close to 3,500 Hours Studying Wasps

(AP Wirephoto to the Su

wasp data in a way I never thought of before.

"Variability in animals' response to new situations has long been considered the province of higher animals, yet recent studies, mine included, using individually marked animals, show that insect behavior can be surprisingly variable too."

Females among Ms. Brockmann's golden digger wasp subjects may dig new nests or move into old ones to lay eggs, and she sees no way to tell which they will do.

"You'd think thee's got to be some little cue to guide her decision, but maybe there isn't," the scientist said.

She found her wasp studies helped in teaching a basic biology course on population genetics and ecology.

"As I prepared my lectures came to realize the field is very re vant to understanding adaptive beh ior, such as that of my wasps. An had to go back to basic concepts l 'fitness,'" she said. "To teach peo about a concept like that, you have know what it really means and not j be using it as a piece of jargon."

Florida. Jane Brockmann (*top*) and the subjects of her research: a *Sphex* digger wasp at its burrow entrance (*above left*) and the 'organ pipe' nests made by *Trypoxylon* mud-daubers – an example of the 'extended phenotype'.

Panama. *Top:* arriving at Barro Colorado island: first sight of the Smithsonian Tropical Research Institute from the landing stage. *Above:* the ever-cheerful Fritz Vollrath (*centre*), and the institute's deputy director, Mike Robinson (*right, with friend*). *Left:* I was fascinated by the leaf-cutter ants carrying materials to make compost for their underground fungus gardens.

Conferences. *Top*: the splendid German castle that housed the poshest conference I've been to, and the determined non-smoker Karl Popper, one of the presiding geniuses of that meeting. *Right*: inside the Gran Telescopio Canarias at the 2011 Starmus event, where we saw the splendidly affable Alexei Leonov, first man to walk in space, give a proper Russian greeting to his fellow astronaut Jim Lovell (*below right*), and dash off a self-portrait (*above*) for the organizer's son, who demanded that as he was wearing a tie at the conference, he should also be wearing one in space.

Above: with some of the so-called 'founders of sociobiology' at Evanston, Illinois, in 1989: *left to right*, Irenaeus Eibl-Eibesfeldt, George C. Williams, E. O. Wilson, me and Bill Hamilton.

Far right: looking lost at Michael Ruse's 1989 conference on the spectacularly beautiful Melbu island in northern Norway (*below*). Perhaps I had already been entranced by the voice of the 'Northern Nightingale', Betty Pettersen (*right*).

Kindred spirits. *Top*: The 'Four Horsemen' – left to right, Christopher Hitchens, Daniel Dennett, me and Sam Harris. Tragically, a few years later, I had to say 'Farewell, Hitch' (*above left*). Mutual tutorials have proved enjoyable and illuminating in equal measure, from discussion of 'the poetry of science' with Neil deGrasse Tyson (*above right*) to travels with Lawrence Krauss – with whom I am pictured below in a hermetically sealed and stiflingly hot limo during the filming of *The Unbelievers*.

I think Sam's stylish boldness in *The End of Faith* was one of the factors that pushed me into deciding to write *The God Delusion*. That and John Brockman's changing his mind, as I have already related. I'd like to think that the Horsemen books are generally as well written as *The End of Faith*, and that this quality—lending a fair wind to the shifting *Zeitgeist*—is partly responsible for the successful impact of the 'New Atheism'.

Christopher Hitchens' *God is Not Great* was another landmark event in publishing. The subtitle of the American edition, *How Religion Poisons Everything*, is powerful, and I am at a loss to understand the decision of the British publishers to change it to *The Case against Religion*. What a pedestrian decision. In fact, it looks as though the publishers later thought better of it, because they reverted to the American subtitle for the paperback edition. To release a familiar bee from my bonnet, why do publishers mess around with the titles of books as they cross the Atlantic?

Christopher Hitchens' death from cancer in 2011 robbed the atheist movement of its most eloquent spokesman, probably the finest orator I ever heard on any subject. Good public speaking is more than just decibels—a point that's often overlooked by demagogues, evangelists and—unfortunately—gullible audiences.

Christopher had a beautiful baritone, reminiscent of Richard Burton speaking Shakespeare, and he used it to perfection. But his rhetorical effectiveness stemmed more from his intellect, wit, lightning repartee; from his formidable inventory of factual knowledge, literary allusions, and personal recollections from some of the world's most dangerous places—for he had physical courage as well as intellectual armament.

God is Not Great complements rather than competes with *The God Delusion*. Where I, as a scientist, am most concerned with religious faith as a rival to science in the role of explainer, Christopher's objections were more political and moral. He found repugnant the very idea of a celestial dictator who demands total obedience and devotion, and is prepared to punish you for ever if you default—or even if you so much as doubt his existence. As he said of the tyranny of North Korea, you can at least escape it by dying. But with the *divine* 'Dear Leader', dying is only the beginning of your torment. I'll say more about Christopher in a later chapter.

Opposition from religious apologists was predictable, and I've already mentioned the flea books. But attacks came, too, from fellow atheists, sometimes in outspokenly belligerent terms. One well-reputed reviewer went so far as to say that *The God Delusion* made him ashamed to be an atheist. His reason seemed to be that I didn't take 'serious' theologians seriously. I dealt fully with those theological arguments that purport to support the existence of a deity. But I was entirely right not to bother with those that *assume* the existence of a deity as a starting point and go on from there.

I have tried but consistently failed to find anything in theology to be serious about. I certainly take professors of theology seriously when they use their expertise to do things other than theology: jigsaw the fragments of the Dead Sea scrolls, for instance; or minutely compare Hebrew and Greek texts of scriptures, or sleuth out the lost sources of the four gospels and the other gospels that didn't make it into the canon. That's all genuine scholarship, fascinating to read and deserving of respect. It's even true that historians need to study theological logic-chopping in order to understand the disputes and wars that have stained European history, for example the English Civil Wars. But the vacuous deepities (Dan Dennett's splendid word) of 'apophatic theology' (Karen Armstrong's obscurantist smoke-screen), or the expenditure of precious time arguing with other theologians over the precise 'significance for us today' of Original Sin, Transubstantiation, the Immaculate Conception, or the 'mystery' (sorry, 'Mystery') of the Trinity, none of that is scholarship in any respectable sense of the word, and it should have no place in our universities.

Theological gymnastics over the 'significance for us today' of nonsensical ideas from the past like transubstantiation lend themselves to satire—positively beg for it. A gem that I recently met: 'Of course we don't literally believe the story of Jonah and the whale. But it is symbolic of Jesus' death and resurrection . . .' Suppose science worked like that. Suppose that (to take a most unlikely hypothetical) future scientists were to find that Watson and Crick were completely wrong, and the genetic molecule is not a double helix at all. Ah well, of course nowadays we no longer *literally* believe in the double helix. But what is the *significance* of the double helix for us today? The way the two helices twine intimately around one another, though not *literally* true in the crude, materialistic sense, nevertheless *symbolizes* mutual love, don't you feel? The precise, one-to-one pairing of purine with pyrimidine is not *literally* true, nothing so crude as that, but it *stands for* . . . When you contemplate the Watson–Crick model, don't you get an overwhelming feeling–I know I do–. . . etc. etc.

For the paperback edition I wrote a new preface, in which I identified a revealingly recurrent stereotype: 'I'm an atheist *but* . . .' As with the equally common

'I *used* to be an atheist but' (as popularized by C. S. Lewis), the speaker imagines that what comes after the 'but' somehow gains credibility from what comes before. In my preface I named and responded to seven species of 'I'm-an-atheist-buttery'. (More recently, in the context of western liberal apologists for terrorist outrages, Salman Rushdie has popularized the name 'but brigade'.) I won't repeat myself here, but will return to a couple of examples in the later chapter on 'Unweaving the threads'.

Later books

My next book after *The God Delusion* was not really my own. Oxford University Press publishes a highly regarded series under the title 'Oxford Book of . . .', usually edited by an academic in the field in question. Latha Menon, whom I have already mentioned as the editor of *A Devil's Chaplain*, invited me to edit the *Oxford Book of Modern Science Writing*, which appeared in 2007. 'Modern' was deemed to stretch back one century, and the eighty-three authors were drawn from those writing in English (with the single exception of Primo Levi). I wrote connecting paragraphs between each author and the next, saying a little bit about them and adding personal colour where I could. For example, I was able to paint an affectionate verbal miniature of Sir Alister Hardy, the great marine biologist, because he had been my professor when I was an undergraduate.

> Nobody had a better feel for the great rolling pastures, sunlit green meadows and waving prairies of *The Open Sea* than Alister Hardy, my first professor. His paintings for that book still adorn the corridors of the Oxford Zoology Department, and the images seem to dance with enthusiasm, just as the old man himself danced boyishly around the lecture hall, a strabismically beaming cross between Peter Pan and the Ancient Mariner. Yea, slimy things did crawl with legs upon the slimy sea—and across the blackboard in coloured chalk with the old man bobbing and weaving in pursuit.

Latha tried to persuade me to include in the anthology something from my own books, but I couldn't bring myself to do so.

My next book was *The Greatest Show on Earth* (2009). Although most of my books had been about evolution, all of them had assumed it tacitly; none of them had systematically laid out the evidence. The British publisher was again Sally Gaminara at Transworld. In America, John Brockman negotiated a new deal with

Free Press, an imprint of Simon & Schuster, where my editor was Hilary Redmon. The book was illustrated with both drawings and coloured photographs, the pictures being skilfully assembled and arranged by Sheila Lee at Transworld. The title comes from a famous American circus, but I first saw it on a T-shirt that an anonymous donor kindly sent me: 'EVOLUTION, The Greatest Show on Earth, the Only Game in Town'. I still have it, although the lettering has faded from much wearing and washing. I wanted the whole slogan for my title, but the publishers unanimously decreed that it was too long. I managed to smuggle 'the only game in town' into the last sentence of the book. Unknown to either of us, Jerry Coyne and I were both working on books with the same aim, and they came out around the same time. I suppose the two books must have competed for the same market, but–maybe I should say 'and'–both of us published highly favourable reviews of the other's.

I stayed with the same publishers, in both Britain and America, for *The Magic of Reality* (2011), my first and only (so far) book specifically aimed at young people. Each chapter poses a question that a child might ask, such as: 'What is an earthquake?', 'Why do we have winter and summer?', 'Who was the first person?', 'What is the sun?' Before coming on to the true, scientific answer to the question, each chapter begins (this was the inspired idea of my colleague the psychologist Robin Elisabeth Cornwell) with *mythical* answers to the same question, drawn from all around the world. I put the myths in, not only because they are colourfully entertaining in their own right, but because my young readers could observe that the particular myths of their own culture (biblical, Koranic, Hindu or whatever they might be) have no special status, no privileged position over any of the rich variety of myths from other cultures. I never said that explicitly. It was just left up to the child's own observation. In the case of the Noah's Ark myth (for the chapter on 'What is a Rainbow?'), I told the story in its original Babylonian version, with the legendary shipwright being Utnapashtim rather than Noah, and the warning to build the ship coming from one member of the polytheistic pantheon, but all the other details the same. The book was illustrated by Dave McKean, a highly original artist whose striking pictures had already won him a large fan base among readers of graphic novels. His arresting style was an ideal vehicle for the world's myths, as well as for the science.

After the book was published, Sally and her team at Transworld commissioned a software company, Somethin' Else, to make an app version for the iPad. I think they did a marvellous job. It might have been better to call it an e-book rather than an app, because every word of the book is in there, along with every one of Dave's illustrations (many of them animated). Apparently, though, there are reasons to do with the recondite mysteries of Marketing why it's better to call it an app than an e-book, even if the content is literally (and pictorially) identical. In addition to the text and illustrations, each chapter of the *Magic of Reality* app has a game. For example, the chapter on gravity and orbiting planets has a description of 'Newton's Cannon', and the app has a game in which you can fire cannonballs with varying velocities. Too slow and they splash into the sea; too fast and they hurtle into space; just right (the Goldilocks velocity) and they go into orbit.

My next book was *An Appetite for Wonder* (2013), the predecessor to this one as the first volume of my memoirs. Again I stayed with Sally at Transworld, but in America Hilary had meanwhile been lured to HarperCollins and I followed her there out of regard for her abilities, just as I had followed Michael Rodgers from one publishing house to another earlier in my career. Since the book was about my childhood and boyhood, culminating in my early career as a truth-seeking scientist, Lalla suggested for the title *Childhood, Boyhood, Truth*, a clever rhyming play on Tolstoy's *Childhood,*

Boyhood, Youth. Sally and Hilary both liked it but 'Marketing' were worried that not enough readers would take the Tolstoy allusion. So Hilary suggested *An Appetite for Wonder,* which picks up the subtitle of *Unweaving the Rainbow.*

Festschrift

In 2006, Oxford University Press celebrated the thirtieth anniversary of publication of *The Selfish Gene.* Together with Helena Cronin, they hosted a commemorative dinner in London. Helena and OUP also put on a wonderful conference at the London School of Economics, chaired by Melvyn Bragg, at which four colleagues spoke to the title '*The Selfish Gene*: thirty years on'.[1] Dan Dennett, representing philosophy, began with 'The view from Dawkins' mountain'. Two biologists then followed: John Krebs on 'From intellectual plumbing to arms races' and Matt Ridley on 'Selfish DNA and the junk in the genome'. Ian McEwan spoke as a scientifically literate novelist on 'Science writing: towards a literary tradition'. I rounded the meeting off with my own response to the day's proceedings.

OUP also published a thirtieth anniversary edition of *The Selfish Gene,* for which they reprised the

---

[1] http://edge.org/documents/archive/edge178.html.

original Robert Trivers foreword, and the original Desmond Morris cover design, both of which had been absent from most of the editions, hardback and paperback, of the intervening years. The Trivers foreword is especially important because that mercurial genius chose it as the place to launch his celebrated idea of 'self-deception', which he later (2011) expanded into a great book, *The Folly of Fools.*

Moreover–a source of special joy to me–Latha Menon commissioned and OUP published a Festschrift volume of tributes, edited by Alan Grafen and Mark Ridley, called (I feel quite bashful repeating the subtitle here) *Richard Dawkins: How a scientist changed the way we think–reflections by scientists, writers and philosophers.* The book was launched at the same London dinner, where the guests, including many of those who contributed to it, signed my presentation copy, which I treasure.

The Festschrift contains 25 chapters divided into seven sections: 'Biology', '*The Selfish Gene*', 'Logic', 'Antiphonal voices', 'Humans', 'Controversy' and 'Writing'. Reading the book again, I am struck by how well and entertainingly most of the chapters are written. I coyly confess to a warm glow of (possibly wishful) thinking that my friends and colleagues really pulled out the stops for me. The warm glow extends to the content, which is consistently interesting; in some cases critical of my work (for example the warm chapter by the then Bishop of Oxford, Richard Harries), in all cases original and thought-provoking (for example, the beautiful chapter on my literary style by Philip Pullman). I would like to write a detailed response to every one of these marvellous chapters but it would require another book to do it justice.

Television

On the *Horizon*

Apart from lots of interviews here and there, my first intensive exposure to television cameras was in 1986 when I was approached by Jeremy Taylor, one of the resident producer/directors of the BBC's 'flagship' (as it was then justly called) *Horizon* series of science documentaries. Americans at the time frequently knew *Horizon* programmes as *Nova* because WGBH in Boston put out a parallel series of similarly good documentaries, many of which were rebranded, sometimes re-presented, *Horizon*s, occasionally even with an American voiceover.

I never verified the rumour that the reason for the latter change was a fear that Americans can't understand–or at least don't enjoy listening to–British English. It seems unlikely in view of the popularity of dramas such as *Upstairs Downstairs*, or the anachronism-prone *Downton Abbey*. On the other hand, I was astounded to be told by a scandalized American friend, Todd Stiefel, that the BBC's *Life*, perhaps the most ambitious wildlife documentary series ever screened, and narrated by none other than David Attenborough himself, was doctored for an American audience, with Attenborough's narration replaced by the voice of Oprah Winfrey! I'm happy to say that the verdict of American Amazon reviewers who compared the two versions has been overwhelmingly in favour of the genuine article. I can't help wondering why Oprah Winfrey agreed to do it. Was she not afraid of the inevitable comparisons with the peerless Sir David?

I was certainly afraid when Jeremy Taylor approached me–because of the formidable reputation of *Horizon/Nova*, and because I doubted whether I would be good enough to do television. I had been approached ten years earlier by another *Horizon* producer, Peter Jones, to present a documentary on *The Selfish Gene*. I declined through sheer nerves, and recommended John Maynard Smith instead. He did an excellent job.

I should say that, although my memory is that I turned down the *Selfish Gene* documentary through nerves, Jeremy Taylor, who kindly read this chapter for me in draft, has a different recollection based on his friendship with Peter Jones.

The story I recollect is that *Horizon* (not necessarily Peter) thought you too youthful in appearance to credibly present your own ideas! Rather like the choirboy delivering the sermon! Indeed, when I raised the idea [a decade later] of approaching you to present *Nice Guys*, the then *Horizon* editor, Robin Brightwell, was dead against the idea–again citing the fact that you looked 'too young', and would viewers have confidence in you? I insisted and Brightwell told me 'Well I won't forbid it–but on your own head be it!' So, if you felt a bit jittery presenting it, imagine the feelings I was concealing (I hope) from you! Of course, *Nice Guys* [the documentary that Jeremy and I went on to make] went down a storm with *Horizon*, BBC2 and the controllers and attitudes to the idea of *The Blind Watchmaker* [Jeremy's next proposal for a documentary with me] changed dramatically!

When Jeremy approached me to do *Nice Guys Finish First* I was (and presumably looked) ten years older and a bit more confident, but I was still jittery about it. What swung me over was his enthusiasm for the topic he proposed. He had read a book called *The Evolution of Cooperation*, by the American social scientist Robert Axelrod, and thought the game-theoretic approach to cooperation would form the basis of a great *Horizon*.

I knew Axelrod's work well because, long before he published his book,

I received, out of the blue, a typescript from an American political scientist whom I didn't know, Robert Axelrod. It announced a 'computer tournament' to play the game of Iterated Prisoner's Dilemma and invited me to compete. To be more precise–and the distinction is an important one for the very reason that computer programs don't have conscious foresight–it invited me to submit a computer program that would do the competing. I'm afraid I didn't get around to sending in an entry. But I was hugely intrigued by the idea, and I did make one valuable, if rather passive, contribution to the enterprise at that stage. Axelrod was a professor of political science and, in my partisan way, I felt that he needed to collaborate with an evolutionary biologist. I wrote him an introduction to W. D. Hamilton, probably the most distinguished Darwinian of our generation. Axelrod immediately contacted Hamilton, and they collaborated.[1]

---

[1] From my foreword to the second edition of Axelrod's *The Evolution of Cooperation* (London, Penguin, 2006).

Hamilton was actually a professor in the same institution as Axelrod, the University of Michigan at Ann Arbor, but they didn't know each other until I introduced them. Their collaboration resulted in a prize-winning paper called 'The evolution of cooperation', later incorporated as a chapter of Axelrod's book of the same title. So I felt a tiny bit possessive about my back-seat part in the book's genesis. In any case I loved it. I quote again from the foreword that I wrote for its second edition:

> I read it as soon as it appeared, with mounting excitement, and took to recommending it with evangelical zeal to almost everyone I met. Every one of the undergraduates I tutored in the years following its publication was required to write an essay on Axelrod's book, and it was one of the essays they most enjoyed writing.

Understandably, then, when I received the overture from Jeremy Taylor and heard how he shared my enthusiasm for Axelrod's book, I couldn't resist.

We met and, picking up on that enthusiasm, I immediately liked him. I found him faintly reminiscent of my New College friend, the intelligible philosopher Jonathan Glover. Jeremy soothed my fears about television, saying that we would start slowly and see how it went. He preferred not to script my pieces to camera, with the reservation that, if it should prove necessary, we'd change to a more scripted format. Fortunately that didn't happen. Instead, the way it eventually worked was that he and I would rather intensively talk over each piece immediately before I delivered it. Then, after each one was safely in the can, we'd discuss the next piece until I had that one clear in my mind, then record that, and so on.

The film ended up being called *Nice Guys Finish First*, and I'll call it that here although we didn't hit on the name till it was nearly finished. It's a play on 'Nice guys finish last', an aphorism which, despite the frisson of sexual innuendo, is said to have originated in the world of baseball. The first scene we shot was on Port Meadow, the great flood meadow between Oxford and the Isis river (as this stretch of the Thames is called). Port Meadow has been unploughed common land since Domesday Book, granted as grazing to the Freemen of the City of Oxford and the Commoners of Wolvercote. The Wolvercote house I used to live in with my first wife Marian overlooked its spreading acres, and it was easy to fantasize it into a kind of wetter, English version of the Serengeti Plain with wandering herds, not of wildebeest and zebra but of cattle and horses.

The relevance of common land to *Nice Guys Finish First* was 'The tragedy of the commons', the topic—and title—of a famous paper by the American ecologist Garret Hardin. Common land is ruined by overgrazing. The system of the commons works as long as everybody exercises restraint. If an individual commoner is greedy and puts too many cattle on the land, all suffer. But the selfish individual suffers no more than anybody else, and he gains disproportionate benefits because he has more cattle. There is therefore an incentive for everybody to behave selfishly: and that is the tragedy of the commons.

A more familiar example: a group of ten people go to a restaurant and agree in advance to share the bill, each one paying a tenth. One person orders a much more expensive dish than anybody else. He knows he'll be paying only 10 per cent of the increased cost on the bill, but he'll receive 100 per cent of the benefit of the more expensive dish. So there is little incentive for each individual to exercise restraint in ordering, and the bill escalates beyond the likely total if everybody bought their own.[1]

Jeremy wanted me to do a piece to camera about the tragedy of the commons and, this being television,

---

[1] To say nothing of the bodmin (Google it, together with 'Douglas Adams').

there had to be a visual illustration in the background. Port Meadow, ancient medieval common land literally on my own doorstep, was perfect. Here opportunity for gentle humour presented itself and Jeremy, as a good television producer will, seized it. It is the responsibility of the holder of the ancient office of Sheriff of the City of Oxford to attend an annual roundup of all the animals, the exact date of the roundup being kept secret. At least, it's supposed to be kept secret; Jeremy apparently got wind of it. Or maybe he didn't, but just struck lucky and took serendipitous advantage.

Owners of animals grazed illegally on Port Meadow used to be fined, the object being to reduce the tragedy of the commons, but more recently the annual roundup has become a toothless ritual in which the animals are temporarily corralled but no attempt is made to establish ownership or responsibility. This would theoretically allow the tragedy to run its course. We filmed scenes of the roundup interposed with my pieces to camera explaining the principle of the tragedy.

As I watched Jeremy directing his camera crew, I couldn't help noticing that part of his intention seemed to be comedic: he was sending up the Sheriff's men and their pet tradition. I was a little concerned, and I asked him about it. He grinned and said they wouldn't notice, and even if they did they wouldn't mind: people love being on television for any reason. I learned a lesson about the subtle wit that characterizes the best documentary directors, a trait I was to see several more times during my years of occasional television presenting. To be truly witty it mustn't be laboured, and this was another lesson I learned from Jeremy.

Jeremy was even capable of laughing at the conventions and clichés of his own medium of television while using them himself. He had me explaining the restaurant example while driving a car and talking to a non-existent passenger by my side. Simon Raikes, who directed a later documentary that I did for Channel Four, *Break the Science Barrier* (more on that below), made explicit the joke against this particular cliché by cutting from me addressing my 'passenger' to a shot of the car from outside, which clearly showed that there was no passenger to address (not even the cameraman, of course). When I protested about this, Simon laughed and said nobody would notice: it had become part of the grammar of television, an accepted convention.

A similarly accepted convention of television documentaries is the 'walking towards the camera shot', in which the presenter is filmed talking to a non-existent person who must unrealistically be presumed to be walking backwards in retreat. The cameraman really is walking backwards (at some hazard to himself and passers-by, were it not for solicitous steering by the sound man hanging on to his shoulder). I have always drawn the line at this televisual cliché. My refusal to do it has been accepted, sometimes reluctantly, by every director I have worked with. Yet another convention of television, 'the speeded-up clouds shot', often deployed to indicate passage of time, can actually be rather beautiful, so I have no objection to it. Playing tricks with time, both speeding things up and slowing them down, is something David Attenborough's marvellous documentaries frequently exploit: to great effect, although I wish he would tell us explicitly that he is doing it, at least on those occasions when it isn't obvious. His wonderfully entertaining autobiography, *Life on Air*, has a fascinating discussion of the earliest days of television documentaries, where he and his colleagues had to invent the conventions, the 'grammar' of documentary television, from scratch: when to do a fade, when an abrupt cut; when to use a voiceover, when to show the speaker's face, and so on.

After *Nice Guys Finish First* was broadcast, I enjoyed a brief honeymoon period when my name was associated with niceness instead of selfishness—as is commonly the case because so many people read my first book by title only. Three prominent corporations approached me. The chairman of Marks & Spencer,

Lord Sieff, made contact through his daughter Daniela, who happened to be a pupil of mine at New College, inviting me to lunch in the company boardroom in London. Daniela and I were the only guests and her father explained to us, plausibly enough, that Marks & Spencer was a very nice company which treated its employees well. I saw no reason to doubt him, but I'm not sure he really got the point of the *Nice Guys Finish First* documentary. Maybe Daniela explained it to him afterwards.

Then, rather less plausibly, a young woman from the publicity department of the Mars Corporation took me to lunch in order to explain to me that her company sold chocolate bars not to make money but in order to sweeten people's lives. She herself was sweet and I enjoyed having lunch with her, but I found her corporate message as cloying as her chocolate.

Finally, a British senior executive of IBM Europe, who really did understand the message of our documentary, flew me to the company's Brussels headquarters to supervise a training game for middle-management executives. The purpose was to help them bond, and thereby improve the atmosphere of the workplace. These dynamic young suits were divided into three teams, the Reds, the Blues and the Greens, in order to play a modified version of 'iterated prisoner's dilemma'

(I won't spell out the details of this game-theory classic here: the details are all in Axelrod's book and in the second edition of *The Selfish Gene*). Each team was shut up in a separate room, and they communicated their moves by runner. Good cooperative rapport between all three teams was built up and maintained during the long afternoon, exactly as Axelrod would have predicted. But alas, the theory also predicts that if a game of iterated prisoner's dilemma is known to be going to end at a fixed time, the temptation to defect rises. This is because the final round, if it is known to be the final round, is equivalent to a one-off prisoner's dilemma—where the rational strategy is to defect. And if you know that your rational opponent is likely to defect on the last round, a pre-emptive strike in the last-but-one becomes rational: and so on back. Axelrod coined the phrase 'the shadow of the future', meaning the expected time to the end of the game. The shorter the shadow, the higher the temptation to defect.

And unfortunately, in the case of the IBM game, it was known that it was going to end at 4 p.m. We should have anticipated the resulting catastrophe and, instead of announcing the termination time in advance, blown a whistle at a random, unpredictable moment. As things were, it was not at all surprising with hindsight that, just before the witching hour of teatime, the Reds massively defected against the Blues, betraying a long-standing trust that had been painstakingly built up throughout the afternoon. Far from helping these executives to bond, our game, even though it was played for tokens rather than real money, caused so much ill-feeling between the Blues and the Reds that they had to have counselling before they could work together again on the serious business of running IBM. It seems quite funny now, but I didn't feel good about it on the journey home.

Nice Guys Finish First was followed, not long after, by another *Horizon* documentary, again directed by Jeremy Taylor. This time the title came first: *The Blind Watchmaker.* Like the book after which it was named–which had just been published–the documentary was a response to creationism, and that was a sufficient reason to do much of the filming in Texas. Jeremy and I flew to Dallas, hired a car and drove to the small and sleepy town of Glen Rose. The nearby Paluxy river flows shallowly over sensuously smooth, flat limestone in which are elegantly preserved dinosaur footprints. Well, some of them are elegantly preserved, showing the characteristic three toes of dinosaurs. Others, however, are sufficiently ill-formed to be seen, by the eye of faith–and I do mean faith–as human footprints. In the 1930s, the Paluxy became a mecca for creationists eager to believe that the world was young and humans walked with dinosaurs (the 'behemoth' mentioned in the Book of Job). A market in fake dinosaur footprints together with giant human ones made of cement was set up in Glen Rose, and the 'evidence' became part of the stock-in-trade of creationist lore and literature.

Jeremy hired a local Texan film crew and we tramped through the wilderness from Glen Rose to the Paluxy river, where we spent a lovely day wading and paddling in the warm, shallow water with its sympathetically smooth limestone bottom. We were accompanied by Ronnie Hastings, a local science teacher, and Glen Kuban, the two men who had done the most to uncover the real truth about the Paluxy river 'man tracks' (they are really dinosaur spoors, but from heels so the three toes don't show). Looking at the film again today to remind myself while writing this, I am a little embarrassed by the brevity of my shorts, and they have indeed been the subject of some ribaldry on the internet. Short shorts are not fashionable today, but I must say I still can't help finding Bermudas rather ridiculously long, even ungainly. Besides, they would have got wet as I waded in the Paluxy.

My friend Jeremy Cherfas, himself experienced in television, told me a shorts story about another documentary presenter, the distinguished South African anthropologist Glyn Isaac. He was being filmed as he crouched down to pick up a fossil, which he then turned over to show to the camera. His shorts were very short and he didn't realize that his penis was visible. The director scrupulously called 'Cut' but, in Cherfas's words, 'the cameraman, being a great cameraman, just kept on rolling'. No such embar-rassment happened to me, but I have to admit that very short shorts wouldn't be Wardrobe's natural choice (to quote Lalla) for re-citing Shakespeare as I had to do–the passage where Hamlet notes the ease with which the human eye is fooled by superficial resemblances (in his case of clouds to animals, in my case of dinosaur heelprints to human footprints).

'Methinks it is like a weasel' is one of the sentences uttered by Hamlet in his cloud comparisons, and I had used it in *The Blind Watchmaker* (the book, that is) to illustrate the difference between cumulative selection and one-off selection. An infinite number of monkeys, bashing away randomly at typewriters for an infinite amount of time, will write the complete works of Shakespeare, together with an infinite quantity of other poetry and prose in an infinite number of languages. But this is no more than an illustration of the ungraspability of the very idea of infinity. Even the short sentence 'Methinks it is like a weasel' would take a regiment of monkeys more billions of years than anyone can contemplate. If you programmed a computer to simulate a monkey, typing character strings of the right length at random, and even if it took only one second to type each string of twenty-eight characters, you would have to wait about a billion billion billion times as long as the world has so far existed to achieve any likelihood of hitting 'Methinks it is like a weasel'.

I jokingly wrote in *The Blind Watchmaker* that I didn't know any monkeys but fortunately my eleven-month-old daughter Juliet was 'an excellent randomizing device and she proved only too eager to step into the role of monkey typist'. Eager was an understatement. She would visit me in my rooftop eyrie overlooking the Oxford Canal and pound away at the keyboard with her little fists, loyally trying to help me meet the deadline for finishing my book. After listing some of the random strings she typed, I went on: 'She has other important calls on her time, so I was obliged to program the computer to simulate a randomly typing baby or monkey.'

Anybody familiar with the medium of television will not be surprised to learn that Jeremy wanted to recreate the scene. Juliet's mother, Eve, brought Juliet to my room in New College where we were filming. Perhaps it was the intimidating presence of cameras and cameramen, lights and huge silvered umbrellas, and a director shouting 'Turn over' and 'Cut', but poor Juliet, even when sitting on her mother's knee, was overcome with stage fright and refused to show her virtuosity at the keyboard. So in the end the film went straight to the computer, which simultaneously compared the simulated monkey to a 'Darwin' algorithm which used cumulative selection. Partially successful 'mutant' strings were selectively allowed to 'reproduce' through successive generations, and the whole process of 'breeding' the sentence 'Methinks it is like a weasel' took only a minute or so.

The weasel program was, of course, a simulation of Darwinian evolution only in a very limited sense. It was designed only to illustrate the power of cumulative selection, as opposed to single generation randomization and selection. Moreover, it homed in on a distant target (the preordained phrase, 'Methinks it is like a weasel'), which is very different from how evolution works in real life. In real life, that which survives survives. There is no distant target, tempting as it may be to imagine one with hindsight. This was why I went on to write the much more interesting and lifelike series of 'biomorph' programs, which I will discuss in another chapter, and which also played a prominent role in the film.

A later scene in the documentary took us to Berlin.

The purpose was to film a German engineer, Ingo Rechenberg, who was pioneering Darwinian selection as a method of perfecting the design of windmills and diesel engines, but we took the opportunity to visit the Berlin Wall and watch the East German guards waiting to shoot anybody attempting to escape the Orwellian oppression of the Stasi. At this bleak and depressing spectacle Jeremy's habitual cheerfulness deserted him, and I have never forgotten his heart-cry of despair, not aimed at anyone in particular but roared up anonymously into the rain-grey sky.

I'm glad I did those two BBC *Horizon* films but, looking at them again in order to write this chapter, I am struck–even a tiny bit embarrassed–by a nervous hesitancy in my delivery of pieces to camera. Possibly one reason was an awareness that every mistake I made was costly. In those days our filming was done on 16-mm film which was expensive and couldn't be reused. Today's digital recording media have zero cost. Mistakes cost only the extra time needed for another take. Although Jeremy was very nice about it and never mentioned the cost of film and his limited budget from the BBC, every time I made a fluff in those *Horizon* films, I felt the need to apologize.

Jeremy actually denies the halting lack of confidence to which I have just confessed, and suspects that I am being oversensitive to my own shortcomings. In any case, perhaps because of the drop in the monetary cost of mistakes due to the shift to digital media, or perhaps because I was ten years older, I don't seem to notice the same hesitancy when I look today at *Break the Science Barrier*, a documentary for Channel Four that I presented in 1996.

Break the Science Barrier

Channel Four doesn't have its own internal production staff or facilities. Instead (and the BBC increasingly follows this model) they commission work from any of a large number of independent production companies that have sprung up in London and all around the country. So the initial approach for *Break the Science Barrier* came to me not from Channel Four but from John Gau Productions Ltd. It didn't take me long to discover that John Gau was one of the most respected figures in British television, a BBC veteran who had left to found his own independent company, widely revered for his experience in the world of television and his success in winning commissions and awards. I had little hesitation in agreeing to my name going forward in his bid to Channel Four. The bid was successful and John hired a freelance director, Simon Raikes, to direct the film, which John himself would produce. I got on well with both Simon and John, and I ended up very pleased with the film—an impression confirmed when I watched it again recently.

Break the Science Barrier combined, in roughly equal measure, a paean to the scientific method and the wonders it reveals, and a lament for the neglect of science in our world. To illustrate the latter, we featured the story of Kevin Callan, a British lorry driver who was sentenced to life imprisonment for murder, and sub-sequently released for reasons that we went on to tell. The jury were convinced, by expert testimony from doctors ignorant of the science of head injury, that Kevin had shaken his four-year-old stepchild Mandy to death.

The point we were making was that ignorance of science, on the part of not just the judge and the prosecution but also Kevin's own defence team, had led to an unjust conviction. When he had ventured to ask his own lawyer what experts they were going to call in his defence, the lawyer told him to shut up—and called no witnesses. The reason was that the doctors the defence had thought to call agreed with the prosecution experts. Kevin was on his own, the sole witness in his own defence, and he was sent down for life.

He was on his own but indomitable. Prison rules allowed him to order books, and he systematically set about teaching himself the recondite subject of neuropathology. Long after his release, in his small house on the Welsh coast he showed our television camera the massive files of notes on the subject that he had accumulated while in prison. These notes seemed to me as full and as detailed as those of any first-class university student swotting for a final examination—with the difference that Kevin was faced with a rather more serious interpretation of 'final'. Can you imagine how soul-destroying it would be to contemplate a life-time in prison stretching ahead of you, knowing you were innocent?

Eventually he found a book by a New Zealand neu-ropathologist, Professor Philip Wrightson, which described symptoms identical to poor Mandy's. Kevin wrote to Wrightson and sent him all the details from the court records. Wrightson studied them exhaustively and became convinced that Mandy's injuries could not possibly have been caused by shaking. They were caused by a fall, which was what Kevin had said all along.

The case was reopened on the strength of Wrightson's new testimony, and Kevin was released without a stain on his character. But, as my spoken commentary said, 'An innocent man had spent four years in jail.' If this disturbing history had taken place in an execution-happy jurisdiction such as Texas, Kevin would presumably be dead. And even in Britain, but for his astonishing tenacity and the integrity of a good doctor from New Zealand, he would still be languishing in jail and probably hideously ill-treated by the other inmates.

Our documentary included a damning indictment, by one of Britain's most eminent barristers, Michael Mansfield QC, of the scientific ignorance of the judge and all the lawyers involved. I was moved by Kevin's story, and I came away with a fierce respect for this relatively uneducated truck driver who had, by sheer force of will and intelligence, educated himself in the relevant science, and also in the scientific method of thinking. His lawyers were far better educated than this heroic young man; but they were educated in the wrong subjects, and they let him down.

We also, in the film, deplored the widespread public descent into superstition and gullibility–a theme to which I would return in *Unweaving the Rainbow* and in a later documentary for Channel Four, *Enemies of Reason*. For *Breaking the Science Barrier* we filmed Ian Rowland, a professional conjuror who performed tricks of the kind alleged by spoon-bending mountebanks to be 'paranormal' or 'supernatural'–but who was himself at pains to emphasize that he was doing nothing but tricks: 'If somebody else does it supernaturally he's doing it the hard way.' In America the same role of honest, fraud-busting magician has long been filled by that veteran sceptic James 'the Amazing' Randi. Other spectacular illusionists who go out of their way to promote scientific reason and debunk the charlatans are Penn and Teller, and Jamy Ian Swiss, all of whom I am proud to call my friends.

I have never performed a conjuring trick in my life, but (perhaps 'and' is the better conjunction) I am fascinated by what the best stage magicians can do. It might almost be said to have philosophical implications. When I watch a world-class conjuror like Jamy Ian Swiss in America, or Derren Brown in Britain, my sense of the miraculous is so strong that it takes a strenuous effort of will to persuade myself that there really is a rational explanation. Contrary to all apparent evidence, what I have seen isn't a miracle. Turning water into wine, or walking on water, would seem child's play compared to what these remarkable performers can do. I have to keep telling myself it really is only a trick, even though all my instincts are screaming 'miracle', 'supernatural', 'paranormal'. Honest spell-breakers like James Randi, or Ian Rowland, or Jamy Ian Swiss, or Derren Brown, or Penn and Teller don't need to come clean about exactly how they do it: they cannot, it would be breaking their professional code to do so. It is enough that they reassure us that it really is a trick.

A shameful confession. I was no longer a child when I saw, on television, the following 'paranormal' performance by an alleged strongman. He had a fishing hook inserted in the skin of his back and he appeared to be towing, with a fishing line, a large, heavy railway truck. The skin of his naked back was pulled dramatically outwards, and he did lots of straining and groaning acting. Slowly but surely, the truck moved. My confession–and I overcome my shame to tell it as an illustration of how we are all vulnerable–is that I did *not* immediately dismiss it as a trick because the laws of physics simply cannot be violated in that way. Rather, my reaction was: 'Well, what a remarkable man. There are more things in Heaven and Earth, Horatio . . .' There. Got that confession out of the way and feel a complete and utter fool. But I know the gullible me of that time is lamentably far from alone.

The honesty of conjurors like Penn and Teller, James Randi and others is not in their commercial interests, by the way. Quite the contrary. The fakes and fraudsters who perform the same tricks (or more usually inferior tricks) but claim on television that they are supernatural, and then write bestselling books about their 'powers', must laugh all the way to the bank (or all the way to the oil or mining companies whose foolish executives pay them handsomely to 'divine' by 'psychic powers' where to drill for oil or precious minerals).

The philosophical interest goes further. Scientists of a rationalist bent are often challenged to say what might in principle cause them to change their minds and come to regard naturalism as falsified. What would it take to convince you of something supernatural? I used to pay lip service to the promise that I would become a super-naturalist overnight, the moment somebody showed me some convincing evidence. I presumed it obvious that such evidence would be easy for a god, say, to fur-nish. But now, stimulated by a thoughtful discussion by Steve Zara, one of the regular contributors to my web-site RichardDawkins.net, I am less sure. What would convincing evidence for supernaturalism look like? What could it possibly look like? A 'close magic' card trick by Jamy Ian Swiss appears to be as supernatural as almost any miracle I can imagine, and in this case I am assured by the honest conjuror that it really is only a trick, an illusion. If Jesus appeared to me in clouds of glory, or I saw the stars move into a new constellation spelling out the names of Zeus and the entire Olympian pantheon, why would I reject the hypothesis that I was dreaming, or hallucinating, or the victim of a cunning illusion, perhaps manufactured by extraterrestrial physicists or an alien David Copperfield-style conjuror, rather than succumb to the cop-out theory that the laws of nature had been overthrown by a 'supernatural' event? Super*human*, yes, why not? I'd be surprised if the vast universe was not home to superhuman intelligences. But 'supernatural'? What could supernatural even *mean*, other than falling outside our present, temporarily imperfect understanding of science?

Arthur C. Clarke, the famously prophetic science-fiction writer, made a related point in his 'third law': 'Any sufficiently advanced technology is indistinguishable from magic.' If we could somehow fly back to the middle ages in a Boeing 747 and invite people aboard to show them a laptop computer, a colour television or a mobile telephone, even their greatest intellects would conclude that all four devices were supernatural and we were gods. Again, what could 'supernatural' ever mean, other than 'beyond our *present* understanding'? Clever tricks by expert conjurors are beyond my present understanding and probably yours too. We are tempted to call them supernatural, but we resist the temptation because we know—the conjurors themselves assure us—they are not. As David Hume advised us, we should exercise the same scepticism over all alleged miracles because the alternative to the miracle hypothesis, even though implausible, is nevertheless more plausible than the miracle.

The other half of *Break the Science Barrier*'s message, promoting the wonder of science, we covered by, among other things, filming Professor Jocelyn Bell Burnell, the discoverer of pulsars, in the evocative location of the giant radio telescope at Jodrell Bank, near Manchester. What a moving spectacle that is: the giant, Cyclopean parabola, staring out through deep space into deep time. We also interviewed David Attenborough and–another *coup*–Douglas Adams. The speech about novels and science books I quoted in the introductory chapter came from this interview with Douglas, which ended with my asking him this: 'What is it about science that really gets your blood running?' And here is what he said, impromptu, and with his infectious enthusiasm somehow enhanced rather than lessened by the twinkle in the eye that endears us to his perennial readiness to laugh at himself.

> The world is a thing of utter inordinate complexity and richness and strangeness that is absolutely awesome. I mean the idea that such complexity can arise not only out of such simplicity, but probably absolutely out of nothing, is the most fabulous, extraordinary idea. And once you get some kind of inkling of how that might have happened–it's just wonderful. And . . . the opportunity to spend sev-

enty or eighty years of your life in such a universe
is time well spent as far as I am concerned.

Alas, he—and we—only got forty-nine.

This is as good a moment as any to mention my
friendship with Douglas, and how I came to know him.
The first of his books that I read was not *The Hitch-
hiker's Guide to the Galaxy* but *Dirk Gently's Holistic
Detective Agency.* It is certainly the only book I've ever
read from cover to cover and then immediately turned
back to page one and read again from cover to cover.
I did that because, the first time through, it took me a
while to twig all the Coleridge references and I wanted
to read it again, this time on the alert for them.

It is also the only book that has prompted me to
write a fan letter to its author. It was an early email,
sent at a time when emails were rare. The Apple Computer company had its own internal email network
called Applelink. You could only send emails to other
members of the Applelink circle, and in the late 1980s
there were only a few hundred of us in the whole world.
Douglas and I were among them, through the good
offices of Alan Kay. Alan had earlier been at Xerox
Parc, where he was one of the founding geniuses of the
WIMP interface (Windows, Icons, Menus [or Mouse],
Pointer), which Apple and later Microsoft were to

adopt. In the great diaspora from the computational Athens that was Xerox Parc, Alan moved to Apple with the honorific title of Apple Fellow, founding his own unit for developing educational software and adopting a very fortunate Los Angeles junior school as his test bed. Alan was a fan of both Douglas's and my books, and both of us were elected as honorary advisers to his educational unit. One of the perks was early member-ship in Applelink; and, given that there were so few of us on the network, it was easy for me to look Douglas's name up and email my fan letter to him.

He replied promptly, saying that he was a fan of my books too, and inviting me to visit him the next time I was in London. I arrived at his tall Islington house and rang the bell. Douglas opened the door, already laugh-ing. I immediately had the sense that he was laughing not at me but at himself, or perhaps more precisely at my anticipated reaction—he must have seen it many times before—to his spectacular height.[1] Or perhaps he was just laughing ironically at some absurdity of life which he presumed I would find equally amusing. I went in with him and he showed me round his house, bristling with guitars, Midi music equipment, futuris-tically giant loudspeakers and—as it seemed—dozens of

---

[1] When he was a boy, school expeditions were told to meet not under the clock but 'under Adams'.

retired Macintosh computers fallen afoul of Moore's Law and languishing in the shade of their state-of-the-art successors. It became obvious that we did indeed laugh at exactly the same things, and revelled in companionable recognition of the same comic absurdities. He would have guessed, for instance, that I must laugh delightedly at this:

> The fact that we live at the bottom of a deep gravity well, on the surface of a gas-covered planet going around a nuclear fireball 90 million miles away and think this to be normal is obviously some indication of how skewed our perspective tends to be . . .

And at the 'infinite improbability drive'. And at the Electric Monk, the labour-saving device that you buy to do your believing for you (the advanced version of which was capable of 'believing things they wouldn't believe in Salt Lake City'). And at the appetizingly suicidal, and morally sophisticated, 'Dish of the Day' in *The Restaurant at the End of the Universe* (introduced above in the chapter on my Christmas Lectures).

I've already explained how I met my wife at Douglas's fortieth birthday party. But 42 is a more significant number in the Adams canon, and he celebrated his forty-second birthday in characteristic style: a massive dinner for hundreds of guests. Although it was a sit-down dinner, it almost failed to consummate that promise–because of its remarkable seating plan. Shoving a card on each place mat bearing the guest's name was far too simple for Douglas. Douglas's place cards had two names, referring not to the person sitting there but to the neighbours on either side. 'The person on your left is Richard Dawkins. Ask him to say grace. The person on your right is Ed Victor. Turn to him and say, in an incredulous tone, "FIFTEEN?"' (Douglas's agent Ed Victor was then the only literary agent in London who took a commission as high as 15 per cent.) Sorting out this *placement* was a feat of such gratuitous complexity that it occupied Douglas (abetted, I suspect, by more than one of his fleet of Mac computers) most of the evening and we didn't finally sit down to dinner until nearly midnight. How I miss him, with his world-class sense of humour and–as has been said–world-class imagination.

Break the Science Barrier ended with a quintessentially Oxford scene: Lalla reclining in a punt while I poled her romantically up the Cherwell (plus the cameraman playing gooseberry, of course, but the audience is supposed not to work that out) with my voiceover extolling the beauty of scientific reality, as we both appreciated it.

Seven wonders

In the mid-nineties the BBC producer Christopher Sykes conceived the idea of a television series in which scientists were asked to name their own personal list of seven wonders of the world and talk impromptu about each one. Christopher illustrated their choices, presumably with footage from the BBC's vast library. My seven wonders were the spider's web, the bat's ear, the embryo, digital codes, the parabolic reflector, the pianist's fingers and Sir David Attenborough (which prompted a delightfully funny, handwritten letter from the great man). This half-hour of concentrated television was one of the few things I have done which apparently earned me no enemies (and lots of friends). Does that make it a good programme? It doesn't make it a bad one, notwithstanding Winston Churchill's 'You made enemies? Good, it means you were doing something right.' I have never gone out of my way to seek enemies, but they sometimes seem to loom up out of the darkness on the straight road ahead.

The seven wonders format threw up some wonderful candidates. Steven Pinker, for example, chose the bicycle, combinatorial systems, the language instinct, the camera, the eye, stereo vision and the mystery of consciousness. I don't think anyone chose 'the taxi driver's hippocampus' but perhaps they should have: London's black-cab drivers have to pass an exam testing their knowledge (it's even called 'The Knowledge') of every last little street and alleyway in one of the world's great cities, and it's been shown that in the driver's brain the part called the hippocampus is enlarged. There's a certain sadness in the thought that 'The Knowledge' may soon be made redundant by GPS navigation. Yet GPS systems have a way to go before they can rival The Knowledge of back-street shortcuts and how the best ones change with traffic conditions.

Other scientists in the series included my personal hero John Maynard Smith, Stephen Jay Gould, Danny Hillis (inventor of the parallel processing supercomputer), James Lovelock (the Gaia guru), and Miriam Rothschild. This remarkable old lady's seven wonders were ear mites, the monarch butterfly, the jump of the flea, dawn on the Jungfrau, the bizarrely complex life cycle of a parasitic worm, carotenoid pigments (such as the ones by which we see), and Jerusalem. Her delight in them was infectious–the enthusiasm of a child bubbling over in an 87-year-old–and her show was a type-specimen exemplar of Christopher Sykes's concept.

Dame Miriam

I didn't know Miriam well, but so remarkable a character demands a digression. She used to invite Lalla and me to her annual Dragonfly Party (so called because guests were encouraged to view the dragonfly conservation measures around her lake) at her country house at Ashton, near Oundle where I had earlier been at boarding school. Her garden was something to behold. There is a coffee-table book, *The New Englishwoman's Garden*, in which each double-page spread is devoted to the garden of some high-born or well-connected lady. The pages glow with immaculate lawns shaded by immemorial cedars, tastefully under-stated flower beds, herbaceous borders, shady arbours and ancient, brooding yew alleys. All is as expected, until you turn the page to the garden of the Honour-able Miriam Rothschild (they could have left off the Honourable and replaced it with FRS, but that would have been out of the book's character).[1] Her garden was stylishly her own. The plants were all such as the other ladies would have called weeds. They consisted entirely of wild English meadow flowers and unmown grasses. Waves of flower-decked long grass buffeted the walls of the house and crashed through the windows into the interior window boxes, which therefore looked like an indoor continuation of the garden. The large

---

[1] This little barb will make sense to non-British readers only if I ex-plain that in Britain 'the Honourable' is a title given to progeny of lords, and 'FRS' (Fellow of the Royal Society) is a genuine honour conferred on scientists.

house itself was so smothered in creepers you almost needed a machete to find it, like a fairy-tale castle in an enchanted forest. Under faded family photographs (including one of the bowler-hatted and full-bearded second Lord Rothschild driving through London in his coach pulled by four zebras) were the cases containing the celebrated Rothschild insect collections.

The luncheons themselves were sumptuous buffets. At one of these annual 'dragonfly bashes', she beckoned me over to her table: 'Come and sit by me, dear boy. But first, go and carve me a slice of venison: a very *small* slice, mind you, I'm a strict vegetarian.' To be fair, the deer had not been killed for food, but had died of an accident, so you could say that her vegetarian principles were being upheld in the spirit–if not the flesh . . . Miriam owned a herd of rare Père David's deer, brought back from China by her father with a view to conserving the species (they are extinct in the wild). One of these deer had unfortunately got itself entangled in a fence and died. Hence the venison on the ethical buffet table.

Miriam was once invited to give the prestigious annual Herbert Spencer Lecture in Oxford. The Vice-Chancellor and dignitaries were all seated in the front row of Christopher Wren's magnificent Sheldonian Theatre. They had probably processed in, begowned, mortar-boarded up and heralded by the Bedell with his mace, although I don't strictly remember that detail and may have embellished it. Miriam's lecture itself, I remember well. It turned out to be a heartfelt plea for animal rights and a passionate denunciation of meat-eating. I was seated immediately behind the Vice-Chancellor and noticed him begin to shift visibly in his seat with anxiety as the lecture progressed. Then I saw a note being passed discreetly along the row, and an aide hustled out, doubtless running hot-foot to the college kitchen where they were busy preparing the post-lecture dinner that the Vice-Chancellor was to host in Miriam's honour. You'd think she might have given his office a warning in advance, but I suspect her sense of mischief took over.

On another occasion, Lalla had been trying to raise money for Denville Hall, the wonderfully hospitable and sympathetic care home for retired actors of which she chairs the trustees. At that time her favoured art form was painting silk with beautiful animal designs. As well as ties (such as the warthog tie which failed to gain the seal of royal approval), she painted truly beautiful silk scarves, all with animal designs—butterflies, pigeons, chickens, whales, fish, shells, ducks, armadillos (Matt Ridley bought that one for his Texan wife, the armadillo being that state's mascot)—and offered them for sale in aid of her favourite charity. Knowing that Miriam habitually wore a headscarf, I encouraged Lalla to paint one for this wealthy and philanthropic old lady in the hope of securing a large donation. The obvious if unconventional subject, given Miriam's unmatched expertise in those acrobatic little suckers, was fleas: hugely magnified images of fleas, nine different species. Lalla painted the scarf beautifully, and I sent it off on her behalf, explaining the good cause. Finally, Miriam's reply came: 'Please thank your wife and tell her I shall keep the handkerchief [this 'handkerchief' was at least a metre square] but inform her that she has sadly underestimated the flea penis, which, as you doubtless know, is proportionately among the largest in the animal kingdom.' Miriam's letter was accompanied by a generous cheque for Denville Hall, and by the gift of her book on flea micro-anatomy, inscribed for Lalla with a note: 'See p. 112 for vagina of mole flea.'

Less happy television encounters

In addition to the science documentaries of which I have been the presenter, I have on many occasions found myself, one way or another, on the wrong side of a television camera. I won't list them all in detail here. Apart from the only two occasions (which I'll come on to) where I was the victim of deliberately deceitful editing, the series I remember with least affection is *The Brains Trust*. The title and format were inherited from a justly famous wartime radio series in which a panel of three people gave off-the-cuff responses to questions sent in by listeners and read out by a chairman. Panellists varied from week to week, but the renowned regulars were Julian Huxley, Commander A. B. Campbell, and C. E. M. Joad. At the time of the original broadcasts I was a toddler in Africa, but I've listened to recordings, redolent of a bygone era when friends called each other by their surnames and radio voices seemed to declaim rather than converse ('Thenk you, Cempbell. I say, Huxley, what is your cendid opinion?') The television version was never as successful as the radio original. I can't, now, imagine why I agreed to take part, but for some reason I did: three episodes, and I hated them all. I was not reassured when the woman in the chair greeted me with an expression of amazement that I was a scientist. She had apparently never met one before: 'We called them "grey men" at Oxford, and they used to go to 9 a.m. lectures while we were all still in bed.' She followed through by saying, when I mentioned Watson and Crick in one of my answers to a question, 'For the benefit of viewers, could you briefly explain who Watson and Crick are?' Would she have made a similar request if I had spoken of Wordsworth and Coleridge, or Aristotle and Plato? Or even Gilbert and Sullivan?

Famous name-pairings remind me of a nice story told by Francis Crick himself. He introduced Watson to somebody in Cambridge who said, 'Watson? But I thought *your* name was Watson-Crick.' Cue another digression. I feel privileged to have known both these men. The talents of both were essential to their remarkable achievement in stretching limited data to generate a conclusion of near-unlimited significance, and it's not obvious which of the two names should come first in the ubiquitous binary. The opening phrase of Watson's *The Double Helix* ('I have never seen Francis Crick in a modest mood') doesn't chime with my more limited experience of his senior partner, but it is true that both of them needed huge confidence to pull it off. In my jacket blurb for Crick's autobiography, *What Mad Pursuit*, I wrote of a

justified pride, almost arrogance, on behalf of a discipline–molecular biology–that earned the right to be arrogant by cutting the philosophical claptrap, getting its head down, and in short order solving many of the outstanding problems of life. Francis Crick seems to epitomise the ruthlessly successful science that he did so much to found.

He did much more than solve the structure of DNA. His demonstration, with Sydney Brenner and others, that the genetic code had to be a triplet code, must be one of the most ingenious experiments ever conceived.

Jim Watson too, if he is arrogant, earned the right to be. His ex-cathedra pronouncements can be ill-judged and his sense of humour can occasionally be cruel, but one gets the feeling he doesn't realize it out of a kind of naive innocence. His humour can also be baffling, as when he announced to me that, if he were to be portrayed on film, he wanted the actor to be the tennis player John McEnroe. What could that mean? How are we supposed to respond? But I treasure his answer to something I asked when I interviewed him in the garden of his old Cambridge college, Clare (for a BBC programme about Gregor Mendel, which culminated in the monastery where the great scientist monk did his pioneering work). I put it to Jim that many religious people wonder how atheists answer the question, 'What are we for?'

Well, I don't think we're for anything. We're just products of evolution. You can say, 'Gee, your life must be pretty bleak if you don't think there's a purpose.' But I'm looking forward to a good lunch.

Now that's vintage Jim—and the lunch was indeed good, made better by his company. Lalla and I got to know him and his wife Liz quite well when they bought a house in Oxford and spent several summers in our home city.

My fellow panellists on *The Brains Trust* varied from week to week. There was usually at least one philosopher, sometimes a historian, once a poetic novelist. I think I was the only scientist. Part of the conceit of the programme was that we were pointedly given no advance notice of the questions. The chair even made arch jokes about this, pretending to torment us with the secrecy and putting our limited reserves of spontaneous wit under pressure. The questions were things like 'What is the good life?' or 'What is happiness?' 'Happiness is a mountain stream . . .' was how one of my hapless fellow panellists began his answer. I'm sure mine was no better, if less pretentious, and it is a matter of some happiness to me that I have forgotten it.

I said that I'd mention the two occasions where I was stitched up by outright dishonest editing of television footage. I'm actually pleased that I can point to only two such examples, because the temptation for those with a losing agenda must be great. Creationists have ignominiously lost their argument, deception is their last recourse, and it is not surprising that my two stitch-ups were both at the hands of creationist organizations. In September 1997 I was approached by an Australian company, who said they were sending a team to Europe to make a film about the 'controversy' over evolution. Influenced, as I shall explain in the next chapter, by a conversation with Stephen Jay Gould, I had adopted a well-reasoned policy of never having debates with creationists, but this crew's pitch sounded like a bona fide attempt to document the argument without bias, so I agreed to talk to them.

The 'crew', when they arrived at my home, turned out to be amateurishly depauperate. The woman operating the camera also asked the questions. I answered them, despite my increasing doubts as to her competence to make a film at all, and my increasing regret that I had ever allowed her into my house. But then she asked a stock question which, as everyone involved in this so-called 'controversy' knows, is a dead-giveaway: only a dyed-in-the-wool creationist would ever say something like, 'Professor Dawkins, can you give an example of a genetic mutation, or an evolutionary process, which can be seen to increase the information in the genome?' It was now obvious that she had gained entry to my house under false pretences. She was, quite simply and obviously, a fundamentalist creationist, and I had been duped into granting her the attention such people crave and the opportunity to twist my words to her own barmy agenda.

What should I do? Should I summarily throw her out, or answer the question straight as if I hadn't rumbled her, or something in between? I paused, trying to decide what to do. Finally, after eleven seconds trying to make up my mind, I decided to throw her out because of the dishonesty of her original approach. I told her to stop the camera, and we repaired to my study where, in the presence of my assistant, I explained that I had detected her deception and she must leave forthwith. She pleaded with me, saying that she had travelled all the way from Australia to see me (an obvious lie, but let that pass). Finally, after much begging from her, I relented and agreed to resume filming. My intention was to give her a brief tutorial in some aspects of evolutionary theory of which she was obviously completely ignorant, rather than answer her silly questions–and certainly rather than attempt to explain information theory to someone incapable of understanding it. If you are interested in a full answer to her actual question, it is in *A Devil's Chaplain*, the chapter called 'The information challenge', which also gives a reference to Barry Williams's account, in the Australian *Skeptic* magazine, of the whole farcical episode.

Eventually she left, and I thought no more about the encounter until a year or so later when somebody called my attention to the film which had by then been released. It turned out that my eleven-second pause, while I was deciding whether to throw her out, was represented as me being 'stumped' by the question. She had edited the film so that the pause was followed by a cut to me talking about something completely different (from a different part of the interview), which made it look as though, in desperation after being 'stumped', I had wantonly changed the subject. As an amusing coda, she actually produced a second version of the film, in which the 'information' question was put not by her but by a male accomplice, in a bare, unfurnished room (presumably in Australia) very different from the one in which I was filmed. This was probably because the sound quality of her original question (she being behind the camera) was poor. It makes the deceptive editing even more obvious, but there apparently exists no level of obviousness sufficient to penetrate the intelligence of a certain type of creationist, and they have no doubt been triumphantly dining out ever since on how I was 'stumped'.

My second stitch-up was more serious because it was perpetrated by a proper film company with professional production values—albeit also with the same apparent level of dishonesty as the Australian amateurs. Again, the initial approach, in 2007, promised an objective look at the world of creationist apologetics, with no hint that the purpose was actually creationist propaganda. Indeed, so persuaded was I of the film-maker's intentions, I even went out of my way to help him find a venue for the filming, in London. Other evolutionists including Michael Ruse and P. Z. Myers confirm that they were misled in a similar way. I still had no inkling of the film's agenda, even right through the interview. The interviewer asked me whether I could conceive of any possible circumstances in which life on Earth *might* have been intelligently designed. My honest answer was a bending over backwards to try to imagine such possible circumstances. I said that the only way I could imagine it would be seeding by aliens from outer space, and that is something I do *not* believe. It was my way, in other words, of saying that I did *not* believe life on Earth was intelligently designed. With hindsight I should have guessed how easy it would be to twist! I still see frequent tweets and blogs saying things like 'Dawkins, the man who doesn't believe in God but believes in little green men'. However, the distortion of what I said is actually small beer compared to the rest of the film. My colleague Michael Ruse was stitched up in an analogous way, exploiting his own sincerity as an honest educator and twisting it to a dishonest agenda. The film even went so far as to blame Darwin for Hitler! (It's doubtful that Hitler ever read Darwin, whose name does not appear even once in *Mein Kampf*.)

Actually, my bending over backwards was even more generous than the interviewer or his duplicitous producer realized. Apologists for 'intelligent design' make no bones about who the 'designer' is when they are talking to the faithful: the God of the Jewish/Christian Bible, of course. However, there are times when they try to pretend that their case is a purely scientific one which would work just as well if the designer were an alien from outer space. In America, they need to put it that way in order not to fall foul of the US constitutional separation between church and state, when they try to argue for 'intelligent design' being taught in science classes. When the interviewer asked me if I could imagine any conceivable circumstances under which life on this planet might have been intelligently designed, my mentioning of aliens was a conscious and deliberate bending over backwards to be more than fair to the apologists whom–little did I know–he was supporting.

I've probably been fortunate in having experienced only two such episodes of outright dishonesty. And I don't want to make too much of what were, after all, rare occurrences among literally hundreds of television interviews over many years. Even so, such dishonesty has a disproportionately malign effect, in that it undermines one's natural impulse to trust people, a benign impulse the loss of which makes life the poorer. As a very different example of the same kind of thing, Lalla and I were once deceived by a young woman (a tutorial student of mine) into believing she was mortally ill with cancer. It eventually turned out that the only thing wrong with her was a version of Munchausen's Syndrome (the strange mental disorder whose sufferers feign illness), but before this was discovered Lalla had spent many hours sitting with her in hospital, holding her hand while she underwent painful tests. As soon as the doctors rumbled her, she instantly refused to see Lalla ever again, presumably out of embarrassment. We never discovered how many of her other stories were also lies–for instance, her claim to be a professional trumpet player. We both agreed that the worst aspect of this episode was the way it undermined our natural human kindness and desire to help disadvantaged people. Fortunately the undermining was only temporary, and Lalla continues, to this day, to devote a sizeable majority of her waking hours to unpaid, and highly skilled, charitable work.

Channel Four again

After *Break the Science Barrier* in 1996, I didn't get back into presenting full-length television documentaries until ten years later when I began my long and fruitful association with the independent producer/director Russell Barnes. Russell and I have worked together now on eleven hours of documentary television, spread over five different programmes on Channel Four. The first of these was about religion, broadcast in 2006 under the title *Root of All Evil?*. The question mark was Channel Four's sole concession to my distaste for the title. No single thing is the root of *all* evil, although religion, when it hits its stride, makes a pretty good go of it.

The budget of the film must have been quite generous, because our entire crew travelled to America, and also to Jerusalem and Lourdes. Lourdes served as a gently mocked monument to human gullibility, a gullibility born, perhaps, of desperation among the unwell. Lalla told me of her first visit to Lourdes many years earlier, in the company of the actor Malcolm McDowell (star of films such as *If* and *A Clockwork Orange*). They stopped their car at the top of the hill in Lourdes, and Malcolm ran wildly all the way down, shouting at the top of his voice, 'I can walk! I can walk! I can walk!' Did the pilgrims take it in their stride as yet another miracle such as their faith and hope had led them to expect?

Russell encouraged me, when interviewing the Lourdes pilgrims, to hide my scepticism and just let them talk. I also interviewed a resident Catholic priest. He appeared not to believe in miraculous cures himself but–and this is so typical of the religious mind–he didn't seem to *care* whether they were real or not. It was enough that the pilgrims *believed* they might be cured, and that this gave them comfort. For him, the real miracle was the pilgrims' faith. For me, a real miracle would have to include a cure (if not the regrowth of an amputated limb) and, as I pointed out–to his complete lack of consternation–the statistical cure rate at Lourdes is no more than would be expected by chance.

In all our films Russell encouraged me to remain quietly polite when interviewing creationists and the like. It amounts to giving them the rope to hang themselves. I tested the method almost to destruction in a later film made with Russell, *The Genius of Charles Darwin*, for which I interviewed Wendy Wright, president of 'Concerned Women for America', as an influential creationist. Her repeated refrain of 'Show me the evidence, show me the evidence, show me the evidence' in the face of–in the teeth of–clear and overwhelming evidence has become legendary on the internet, and so, it has to be said, has my patience in the (fake-smiling) face of her. I take no credit for it; I was simply following the director's instructions and fighting down my more natural–and less gentlemanly–impulses.

It was even harder to fight them down in some of the interviews for *Root of All Evil?*, which exposed me to actively unpleasant individuals: the snarl-smiling Ted Haggard, for example. We concentrated most of our American filming in Colorado Springs because it has become a hotbed of Christian revivalism, while the 'Garden of the Gods', in the foothills of the Rockies just outside the town, provided some magnificent backdrops for filming pieces to camera, for example on the metaphor of 'Mount Improbable' (see page 602). Whole areas of new (and, surprisingly for America, dull) housing in Colorado Springs have become de facto fundamentalist ghettos, and we went to one of them to film a decent but naive young family who were loyal regulars in 'Pastor Ted's' vast congregation.

Ted Haggard was a small man in a big church ('was' because he has since fallen from grace in a way that I won't spell out because I don't do *Schadenfreude*). We watched in amazement as his sheep flocked into the gigantic car park in their sedans and pickup trucks, clutching their bibles or prayer books. We listened in even greater amazement to the huge amplifiers booming out God-rock while the people jigged up and down the aisles with both arms raised to heaven and beatific expressions on their faith-doped faces. Finally, Pastor Ted himself strutted his entrance on to the stage, grinning wolfishly and encouraging the 14,000-strong congregation to intone the word 'obedience' in docile chorus. *'OBEDIENCE.'* After the service he gave me the full arm-around-the-shoulder greeting as we started our interview. He seemed mildly flattered when I compared his service to 'a Nuremberg Rally of which Dr Goebbels might have been proud' but, to be fair to him, it seems possible he had never heard of Nuremberg or indeed Joseph Goebbels. Things didn't turn nasty until I questioned his understanding of evolution. But however nasty they turned, nothing could shake the carnassial grin.

Later, our gifted cameraman Tim Cragg and I were packing up the kit after Tim had taken a few final shots in the car park when a pickup truck drove up fast and slammed to a halt just short of hitting us. Pastor Ted was at the wheel and he was furious, far more so than during the interview. With hindsight, we guessed that he probably went straight from the interview to Google my name and discovered who I was. At any rate, he berated us for abusing his hospitality, calling particular attention to his generosity in giving us tea *with milk.* He stressed the milk twice. And then, weirdest of all, he said to me, in accusatory tones: 'You called my children animals.' I was too baffled to reply. Afterwards the crew and I discussed what it might possibly mean. The consensus was that, although I hadn't explicitly mentioned either animals or the Haggard children, it would be implicit in the mind of a creationist that any evolutionist must regard all humans as animals. Correctly, as it happens, although why Pastor Ted chose to mention his own children rather than the entire human race was as puzzling as his homing in on milk in the tea. Perhaps he didn't mean his own biological children but his churchgoing flock, doped up on childlike 'OBEDIENCE'. Who can tell?

While ordering us off his land, Haggard threatened (among other things) to seize our film footage, a threat which our crew took seriously enough to carry it with us when we went out to dinner that evening, rather than leave it in Tim's hotel room. That sounds paranoid now, but Colorado Springs is such a hotbed of fundamentalist religion, and Pastor Ted's 'obedient' congregation so enormous, it was perhaps not totally unrealistic to think there was a risk.

Also in Colorado, I interviewed Michael Bray, another clergyman (although I'm not sure how much that means in America: the title 'Reverend' seems to be something you can acquire with the bare minimum of effort, complete with tax breaks and unearned prestige, without a qualification in theology or indeed anything else).[1] Bray had been in jail for violent attacks on doctors who carried out abortions, and I questioned him about his attitude and that of his friend Paul Hill, another 'Reverend', who had been executed in Florida for murdering an abortion doctor. I got the impression both men were sincere, honestly believing in the righteousness of their cause. Indeed, among Hill's last words were that he expected a 'great reward in Heaven': a chilling example of Steven Weinberg's much-quoted dictum: 'With or without religion, good people can behave well and bad people can do evil; but for good people to do evil–that takes religion.' And indeed, I suppose if you really do think a fetus is a 'baby' (as these people sincerely seem to do) you can put together some kind of moral case for taking the law into your own hands. At any rate, I couldn't find it in me to dislike Michael Bray the way I disliked Ted Haggard. I wished I could have found a way to talk some sense into

---

[1] I myself am a Minister of the Universal Life Church. My certificate of ordination, which hangs in the downstairs lavatory, was bought for me as a joke birthday present by Yan Wong. Lawrence Krauss is a minister in the same church and he has actually used the qualification to perform a marriage ceremony which he, and the couple concerned, are assured is legal.

him, but there wasn't time. Weirdly, he wanted to have a photograph taken of himself with me. I didn't know for what purpose, and I'm afraid I declined.

One can make a similarly sympathetic case for 'Pastor' Keenan Roberts, another of my Colorado interviewees, although he was a less appealing character. He ran an institution called Hell House, devoted to performing short plays designed to scare children out of their wits with threats of being barbecued for all eternity. We filmed rehearsals of two of these playlets. The lead character of both was a sadistically roaring Satan, noisily gloating, in the 'Ha-Haaar' manner of a Victorian melodrama baronet, over the eternal torments prepared for various sinners—a woman having an abortion in one play, a pair of lesbian lovers in the other. Afterwards, I interviewed Pastor Roberts. He told me his target audience was twelve-year-olds. I bridled at this and questioned the morality of threatening children with everlasting torture. His defence was robust: hell is such a terrible place that any measures to dissuade people, even or maybe especially children, from going there are justified. He had no answer to my question as to why he worshipped a God capable of sending children to hell, or why he believed in hell at all. It was simply his faith, and I had no entitlement to question his faith.

As with Michael Bray, I could kind of see where he was coming from. If you really do believe literally in hell, if you really do think abortion is murder, and if you really do think people will roast in hell for ever if they fall in love with a member of their own sex, I suppose you could say that any preventive measures, however illegal or however cruel, are the lesser of evils. Indeed, from that point of view, it's hard to see how any sincere believer could do anything *but* evangelize to try to save people from such a terrible fate. A bit like pulling people back from falling over a cliff. You'd feel bound to do it, even if you had to be pretty rough about it: another example of Weinberg's dictum.

I could find no such justification, however—not even a partial justification—for Joseph Cohen, alias Yousef al-Khattab. Russell and I and the crew were in Jerusalem, trying to get to grips with the religious enmities that plague this ancient city. We spoke to a pleasantly cultivated, educated Jewish spokesman, and we spoke to the Grand Mufti of Jerusalem, who used our local 'fixer' as interpreter. Seeking a middle way, somebody who could see both points of view, what could seem a more natural choice than a Jewish settler who had converted to Islam: Yousef al-Khattab, the former Joseph Cohen from New York? Surely he was in the best possible position to see both sides? How wrong we were. We found him in his small shop in a Jerusalem back street, selling perfumes. He greeted me cordially enough, but as soon as the camera was switched on so was the vitriol, a vitriol heated up by the authentic zeal of the convert. Having been a Jew, his most passionate hatred was reserved for the Jews. He openly expressed his admiration for Hitler. He longed for world domination secured by the victorious Soldiers of Allah. He refused to condemn the 9/11 raids. He attacked me as, in some warped way, responsible for western decadence, re-serving his especial disgust for 'the way you dress your women'. I momentarily let my anger show through as I made the obvious retort: 'I don't dress women, they dress themselves.'

In most of my films with Russell Barnes we have worked with the same cameraman, Tim Cragg, and the same sound man, Adam Prescod. Tim and Adam have worked together as a team on many more films all around the world, often with Russell. I have come to value friendship with all three men, and the kind of camaraderie that comes from working together day after day, travelling together, eating together, laughing together at the same absurdities, even getting thrown out of the same megachurch car park together. Tim is a handsome, smiling fellow, so dedicated to his craft that he never really stops looking at the world through an imaginary viewfinder or a real one, constantly seeking interesting, rewarding camera angles. Russell would happily send Tim off to get useful background footage on his own, knowing that he had no need of a director. Adam is similarly dedicated to, and similarly good at, his craft of sound recordist. He and Tim make a great team, knowing each other's game like doubles tennis partners. Someone we interviewed took one look at Adam's dreadlocks and dark skin and started asking him about reggae music. A classic case of judging a book by its cover, as Adam himself jovially observed to me (whenever I heard him humming to himself it was more likely to be J. S. Bach's suites for unaccompanied cello). As for Russell himself, he has the same virtues as a documentary director that I first identified in Jeremy Taylor. The best directors, such as Jeremy and Russell, are like academic scholars in that they become true experts in the subject of their current documentary, reading up the original research literature and visiting and talking to experts. Then, having planned their film, shot it and edited it, they switch to another subject and start reading up all over again. Does this chameleon-like switching make for a more satisfyingly varied life than that of the academic which it superficially resembles? I could readily imagine so.

For later films, I also enjoyed working with Russell's

business partner and fellow director Molly Milton. Preternaturally cheerful and friendly, she would charm her way through any barriers and breeze our whole crew through any red tape, come what might. Her Pollyanna-ish ways beguiled me too, but occasionally with mixed feelings. For the film *Sex, Death and the Meaning of Life*, she telephoned to ask me to go to India to interview the Dalai Lama. I was convinced (as it turned out, rightly) that the great spiritual leader would be much too busy to talk to me, and I used this as my de facto method of saying no to Molly: 'Har har har, well, if you har har har succeed in booking the Dalai Lama har har har, I'll go to India with you har har har.' I assumed that my laughing challenge was equivalent to a no, put the phone down and thought no more of it.

About three weeks later, Molly telephoned in high excitement: 'He's agreed, he's agreed, he's agreed, we can go to India, you promised you'd go if I booked the Dalai Lama, he's said yes, he's said yes, he's said yes, we're off to India, we're off to see the Dalai Lama.'

Well, I could not but deliver on my earlier promise. We went to India—and then, when we got there, it turned out that, exactly as I had guessed from the outset, the Dalai Lama was much too busy to see us. The full story then emerged. His office had said, 'Well, if you were to turn up on such and such a day, he *might* possibly be able to see you, but we can't guarantee it.' I believe that Molly, with her Pollyannitus ear and conviction that there's no obstacle she can't get round, literally *hears* 'Well, maybe' as 'Yes, definitely.' I forgave her: you couldn't fail to forgive somebody so winsomely charming, and we did end up filming some amazing scenes in India while we were there.

Molly and I share an embarrassing secret (embarrassing for me, not her), and I hereby confess it. Again it was *Sex, Death and the Meaning of Life*, and we were filming on the top of Beachy Head on the south coast of England. Its dizzying 500-foot chalk cliffs have made it a notorious suicide spot, and there are little low crosses, only knee high, lining the cliff path to commemorate the poor agonized souls who had launched despairingly into the void. I was to take a poignant walk along the path, while the camera focused in close-up on my feet as they passed the little crosses, each in their sad turn. I couldn't understand why my feet felt so uncomfortable, but I soldiered on while we shot several takes. When we finally had enough footage, I was able to sit down on the grass and take my shoes off–blessed relief. Molly came and sat with me to plan the next scene. It was then that we noticed why my feet had been so uncomfortable. I had somehow managed to put my left and right shoes on the wrong feet. Molly giggled delightedly and we agreed not to tell Russell and the rest of the crew. But my *faux pas* is preserved for posterity in close-up. I suppose I should be thankful my *pas* wasn't even more *faux*, given our proximity to the cliff edge.

I'm proud of all the films I made with Russell and his crew. Between *Root of All Evil?* (the first one) and *Sex, Death and the Meaning of Life* (the most recent) we did *Enemies of Reason* (about astrology, homoeopathy, dowsing, angels and other superstitious nonsense excluding religion), *The Genius of Charles Darwin* and *Faith Schools Menace*. The last of these included a memorable trip to Belfast to examine the educational roots of the tribal wars there and we took in an Orange parade among other disquieting sights including the huge, starkly realistic murals depicting masked men with guns.

Enemies of Reason contained a telling sequence on dowsing, coordinated by the London University psychologist Dr Chris French. Professional and amateur water diviners converged from far and wide to show their prowess, confident in their ability, proved to their own satisfaction over years. Alas, they had never before been subjected to a double-blind trial. Inside a big tent, Chris French laid out a rectangular array of buckets. Some of the buckets contained water, some contained sand. In a preliminary trial, the lids were removed from the buckets and the dowsers all had no difficulty: their divining rods, hazel twigs or pieces of bent wire all obediently twitched when they could see water, and didn't when they couldn't. But then came the real test, with lids on the buckets. Because it was a double-blind trial, neither the dowser nor Dr French (who was keeping score) knew which buckets contained water. The accomplice who set them up did so with the tent sealed, and he then disappeared so that he couldn't give the game away by any subtle cues. Under these double-blind conditions, not a single one of the dowsers scored above chance level. They were flabbergasted, desperately–in one case tearfully–disappointed, and obviously sincere. Such failure had never happened to them before. But they had never done a double-blind trial before.

I don't know who invented the double-blind trial, but it is a brilliantly effective yet simple technique. There's a telling story in John Diamond's courageous book *Snake Oil*, written when he was dying of cancer and beset by well-meaning quacks. The sceptical investigator Ray Hyman once did a double-blind trial of an 'alternative' diagnostic technique called applied kinesiology. As it happens, I have experienced kinesiology myself. I'd ricked my neck and was in pain. It was the weekend and I couldn't go to my normal doctor, so I decided to be open-minded and try an 'alternative' practitioner. Before beginning her manipulation she did a diagnostic test which consisted of pushing against my arm to test my strength while I was lying on my back–kinesiology. She demonstrated to her own satisfaction that my arm was stronger when I had a small vial of Vitamin C resting on my chest. The vial was sealed, there was no way for the vitamin to enter my body, so it was obvious that she was really–though probably subconsciously–pressing harder against my arm when the vial was not there than when it was. When I expressed my scepticism she gushed her enthusiasm: 'Yes, C is a *marvellous* vitamin, isn't it?'

Self-deception of that kind is precisely what the double-blind technique was invented to eliminate. In testing the efficacy of any medicine, not only must it be compared with a placebo control, it is vitally important that neither the patient, nor the experimenter, nor the nurse administering the dose should know which is experimental, which control. Ray Hyman did a double-blind trial of a slightly less far-fetched claim of kinesiology than used by my quack: that a drop of fructose placed on the tongue would strengthen a patient's arm, when compared with a drop of glucose. Under double-blind conditions there was no difference in strength. Whereat the chief kinesiologist delivered himself of this immortally indignant remark:

'You *see*? That is why we never do double-blind testing any more. It never works!'

In addition to the superseding of costly film stock by digital recording, other things have changed since my early films with Jeremy Taylor. In the 1980s, film crews were heavily unionized. There were statutory times for tea breaks, lunch breaks and the easeful 'It's a wrap' moment at the end of a day's work. If Jeremy wanted his camera crews to go on a bit late in the evening, because the filming was going so well and the light was so good, he had to ask them as a special favour. By the 2000s things had changed. There somehow seemed to be a greater sense of personal involvement in the film by the entire crew, and everybody was happy to go on as long as necessary. I suspect, too, that there was a certain amount of overmanning in the 1980s. The crews back then consisted of not only a cameraman, sound man and production assistant, but also an assistant cameraman (or 'focus puller') and at least one 'sparks' (electrician) to deal with the lights. I recall going up to Leeds around then for an ITV television show produced by Duncan Dallas, who, incidentally and irrelevantly, had been my exact contemporary at Balliol College,

Oxford, though we hardly knew each other. Duncan and I were alone in the studio (the crew had gone off for tea) and there was a large box obstructing the space where we were trying to work. Thinking I was being helpful, I was just about to pick it up when Duncan shouted in a panic: 'Don't touch it!' I recoiled as if he'd said it was a bomb, and he explained. Moving boxes was strictly the job of the scene-shifters and he couldn't answer for the consequences if I were seen picking it up. Duncan hesitated for a moment, looked nervously over his shoulder and then whispered: 'Dammit, let's risk it.' And we hastily moved the box before the crew came back from their tea break.[1]

Manchester television conference

In November 2006 I was invited to give a guest lecture in Manchester to a conference of science documentary makers. The title they gave me was 'Can television rescue science in an age of unreason?' My lecture was illustrated with clips from recent television documentaries, put together with the help of Simon Berthon, who also advised on the content of the lecture. I began by

---

[1] While working on this book I was sad to read in the newspaper that Duncan Dallas had died. In addition to his television work, he was the founder of the network of *Cafés Scientifiques*, an excellent grassroots organization for bringing science to a wide public which has spread from his home town of Leeds to the whole country and beyond.

apologizing for presuming to lecture professionals on how to do their job: my only excuse was that I had been invited to do so. I structured my lecture around a list of ten difficult choices–or ten sliding scales along which a film might be situated: choices that face anyone who makes a science documentary.

The first of the ten was the question of 'dumbing down'.

> The television producer rightly lives in dread of the remote control, knowing that, within any one second of his precious broadcast, literally thousands of viewers may be tempted to flip idly to another channel. There is a powerful temptation to pile on the 'fun', to lace it with gimmicks (speeding up laboratory procedures like Charlie Chaplin, for example), to shrink the science to soundbites whose real scientific nourishment is about as empty as a bucket of popcorn.

I sympathized with the need to chase ratings, but made an unfashionable plea for elitism–elitism as a mark of respect for the audience, rather than the patronizing, indeed insulting assumption that they need science to be dumbed down to render it accessible. The worst example of this patronizing attitude that

I ever came across was expressed by a participant in another conference on public understanding of science. He suggested that dumbing down might be necessary to bring 'minorities and women' to science. Seriously, that is what he said, and no doubt it brought a warm, cosy glow to his condescending little liberal breast. In my Manchester lecture I said:

> Elitism has become a dirty word, and it is a pity. Elitism is reprehensible only when it is snobbish and exclusive. The best sort of elitism tries to expand the élite by encouraging more and more people to join it . . . Science is inherently interesting, and the interest will shine through without the need for soundbites, gimmicks or dumbing down.

Another of my ten difficult choices concerned the perceived need to provide 'balance', something that especially afflicts the BBC because of its charter. I quoted a favourite maxim, which I think I first heard from Alan Grafen: 'When two opposing points of view are advocated with equal vigour, the truth doesn't necessarily lie halfway between. It's possible for one side to be simply wrong.'

The error shows itself in extreme form in the tendency for broadcasters to champion mavericks, who have nothing going for them except that they buck the orthodox trend. The most egregious example I know was a televised hagiography of a medical researcher who claimed that the triple MMR vaccine caused autism. His evidence was thin and is widely discounted within the medical profession. Yet his story unfortunately had what journalists call legs, giving full vent to the facile trope of the virile young rebel, played by a handsome and personable actor, fighting the stuffy old guard.

'Terry the pterodactyl' was another of my headings. The wonderful computer graphics techniques first given prominence by *Jurassic Park* were soon exploited by documentary makers. But instead of letting the wonders of the reconstructions speak for themselves, the documentaries succumbed to the same temptation as spoiled *Jurassic Park* itself: the perceived need to provide human interest. Not content with a computer-animated discussion of pterodactyls and their probable lifestyle, we are treated to a sob story about a particular, individually named pterodactyl (I don't think he was actually called Terry, but the point stands) getting lost and trying to find his family, or some such sentimental guff. Personified drama is not only superfluous: it perniciously blurs the distinction between speculation and real evidence:

Speculation about the habits and social life of pterosaurs or sabretooths or australopithecines is absolutely fine. But it needs to be presented as speculation. Sabretooths might have had a social and sex life similar to lions. Or similar to tigers. The trouble with telling stories about individual sabretooths called Half Tooth and The Brothers, is that it *forces* you to plump for one theory, say the lion theory, rather than another.

I quoted another film to illustrate the same tendency to let dramatic 'human interest' override scientific truth. The BBC conceived the interesting idea of tracing the mitochondrial and Y-chromosome DNA of three particular West Indian individuals back to their roots in Africa or Europe. The point about mitochondria and Y-chromosomes is that, unlike all other chromosomes, they are not subject to the comprehensive scrambling of genetic history that is caused by chromosomal crossing-over in the rest of the genome. You could travel to any particular moment in history, say 14 January 30,000 BC, and could theoretically locate the one individual female from whom your mitochondrial DNA comes. Your mitochondria come from only her and literally nobody else at that time except one and only one of her daughters (granddaughters etc.), plus her mother, her maternal grandmother etc. If you are male, your Y-chromosome comes from only one male alive in 30,000 BC (plus his father, paternal grandfather etc., and only one of his sons, grandsons etc.). All your other DNA comes from thousands of individuals, probably scattered all over the world.

So, great idea to take three people and trace the origins of the only two unmixed portions of their genomes, their mitochondria and Y-chromosomes. But the producers were not content with the scientific fascination of this quest. No, they had to ham it up. And in doing so they sadly misled those people into baseless sentimentality when they transported them back to their 'homeland'.

When Mark, later given the tribal name Kaigama, visited the Kanuri tribe in Niger, he believed he was 'returning' to the land of 'his people'. Beaula was welcomed as a long-lost daughter by eight women of the Bubi tribe on an island off the coast of Guinea, whose mitochondria matched hers. Beaula said, 'It was like blood touching blood . . . It was like family . . . I was just crying, my eyes were just filled with tears, my heart was pounding . . .'

She should never have been deceived into thinking this. All that she, or Mark, were really visiting—at least as far as they were given any reason to suppose–was individuals who shared their mito-chondria. As a matter of fact, Mark had already been told that his Y-chromosome came from Europe (which upset him and he was later palpably relieved to discover respectable African roots for his mitochondria!).

All the rest of their genes came from a wide variety of places, probably throughout the world.

A personal anecdote about Y-chromosomes at this point. In 2013, I was delighted to receive an email from James Dawkins, a young historian doing a PhD at University College, London, whose father's family came from Jamaica. His doctoral thesis is about the estates of a particular family of landed gentry, in England and Jamaica. The family concerned is the Dawkins family, who were sugar planters in Jamaica in the seventeenth and eighteenth centuries and, I'm sorry to say, slave owners. My regrettable family history means that Dawkins is a common family name in Jamaica, not just because of *droit de seigneur* but because the family gave its name to various places in 'our' region of Jamaica. My six-greats-uncle James Dawkins (1696–1766) actually had the nickname 'Jamaica Dawkins', as I learned from Boswell's *Life of Johnson*:

I have not observed (said he) that men of very large fortunes enjoy any thing extraordinary that makes happiness. What has the Duke of Bedford? What has the Duke of Devonshire? The only great instance that I have ever known of the enjoyment of wealth was that of Jamaica Dawkins, who, going to visit Palmyra, and hearing that the way was infested by robbers, hired a troop of Turkish horse to guard him.

Uncle James's family wealth was squandered long ago on futile lawsuits by the paranoid Colonel William Dawkins (1825–1914), who eventually died in penurious bankruptcy, and the once substantial family estates are now reduced to a small working farm in Oxfordshire. The modern James Dawkins has come several times to stay there as a welcome guest of my sister's family, while researching old tin boxes filled with dusty documents in my mother's attic. We were all hopeful that he might turn out to be a long-lost cousin, and the obvious way to find out was to look at our Y-chromosomes. The Oxford geneticist Bryan Sykes, author of *The Seven Daughters of Eve*,[1] kindly agreed to do the analysis, and

---

[1] Every European has mitochondria belonging to one of only seven types. Each one of us is descended from only one of seven mitochondrial matri-

both James and I sent cheek swabs to his company, Oxford Ancestors. When my result arrived, I wrote as a biologist to James as a historian, telling him what to look for in his.

> We each have a Y-chromosome which is nearly identical to our father and brothers. But, as the generations go by, occasional mutations arise. So, although your Y-chromosome is nearly identical to your paternal grandfather's, there is a slightly greater chance of a difference than in the case of your father. If we are both descended, in the paternal line from a sixteenth century Dawkins in Jamaica, our Y-chromosomes will be nearly, but not quite, identical . . .
>
> It's logically necessary that, if you go back sufficiently far, every human Y-chromosome in the world is descended from one ancestor, who is whimsically named Y-Chromosome Adam. He almost certainly lived in Africa, probably between 100,000 and 200,000 years ago. If you look at all the Y-

---

archs (themselves descended, much further back, from the African 'mitochondrial Eve'). Sykes dramatizes the story by giving names to each of the seven European matriarchs, telling us where they lived and inventing a fictional short story about each of them. Nice book. Recommended. Sykes has done the same kind of thing for Y-chromosomes, tracing all our Y-chromosomes back to only 17 Y-chromosomal patriarchs, all in turn descended from 'Y-chromosome Adam'.

chromosomes in the world, they are all descended from Y-Adam, but because of geographical separation and migrations etc., they can be classified into a dozen or so major 'clans'. Each of these clans can be traced to a hypothetical ancestor, a particular man who lived in a particular place. Bryan Sykes has given them all fanciful names. For example my Y-chromosome is derived from Oisin who lived in West Eurasia. That would be true of most Englishmen, and doesn't mean we are close cousins. However, if it should turn out that YOUR Y-chromosome is also descended from Oisin, that would be extremely interesting. Of course it might only mean that it came from a west European. But then the coincidence of surnames kicks in, and it would mean it would be worthwhile looking in more detail to see if our Y-chromosomes are CLOSER than any two west European Y-chromosomes. If, on the other hand, your Y-chromosome came from one of the three African progenitors coloured red on the rather pretty tree diagram, that would mean there's no point in our pursuing genetic cousinship any further. Which would be a pity!

When James's results came in, it turned out that we are not cousins descended from the same Dawkins ancestor. A pity indeed. James's Y-chromosome is derived not from the male tagged Oisin by Bryan Sykes, but from Eshu, who lived in Africa.

I've reproduced in the picture section Bryan's pedigree of all the human Y-chromosomes. James's and my faces are superimposed alongside our respective ancestors, (African) 'Eshu' and (west European) 'Oisin', the names given them by Bryan Sykes. You can see that our cousinship is actually rather distant. Well, strictly speaking, what you can see is only that our Y-cousinship is not close. We could share a more recent ancestor down a female line. But it does mean that our shared surname has no direct genetic significance of the kind that J. B. S. Haldane was referring to when he said 'I was born with a historically labelled Y-chromosome'–meaning an ancient surname. It's an interesting thought that aristocratic and royal families, who can trace their male lines back over centuries, are now in a position to question the legitimacy of each link in the chain, simply by looking at the Y-chromosomes of their alleged cousins in the male line. Will courts of law soon be asked, by long-forgotten distant cousins, to examine DNA challenges by pretenders to thrones or ducal houses?

Back to my presumptuous lecture to the science documentary makers. I included a brief section called

'Science as poetry or science as useful'. Nobody could deny the usefulness of science–frequently cited in terms of the myth that the space programme was justified by the spin-off of the non-stick frying pan–but I was keen to advocate the Carl Sagan, 'visionary' or 'poetic' end of the spectrum rather than the 'non-stick frying pan' end. As I put it on an earlier occasion: 'Concentrating only on the usefulness of science is a bit like celebrating music because it is good exercise for the violinist's right arm.'

One of my final sallies was an attack on the conventional wisdom among TV professionals that 'people don't want to watch talking heads'. I had no data to support my scepticism. But I remember the huge success of John Freeman's *Face to Face* series on BBC TV, in which even the interviewer's face was not seen–only the back of his head and one shoulder from behind. All the concentration was on the face–and, of course, the words–of the interviewee. The series has become legendary. Subjects included Bertrand Russell, Edith Sitwell, Adlai Stevenson, C. G. Jung, Tony Hancock, Henry Moore, Evelyn Waugh, Otto Klemperer, Augustus John, Simone Signoret and Jomo Kenyatta. I was interviewed by the wittily articulate sociologist Laurie Taylor, in a recent revival of the show. On a smaller scale, my 'mutual tutorial' videos (see pages 359–71) have been well received, and they are nothing but talking heads.

Nearly a decade before that Manchester conference, in fact, I had been lucky enough to take part in a project that used the 'talking heads' format to maximum effect. In the spring of 1997, I was approached by Graham Massey, sometime head of BBC Science and a previous BBC *Horizon* producer. He had had a lovely idea, inspired by his friend Christopher Sykes's celebrated interview with the great physicist Richard Feynman. He wanted to make a video archive of distinguished scientists talking at length about their careers. The format was to be an interview of each individual by a younger scientist who knew the field well enough to draw them out. The point was not to make programmes for immediate broadcasting, but to compile a record for the future: something that would survive perhaps for a very long time and be of interest to historians of science in later generations. I loved the idea and consequently felt hugely honoured to be invited to be the interviewer of John Maynard Smith.

The interview took place over two days, in John's house in Lewes in Sussex. John and his wife Sheila invited me to stay the night, and we all, including Graham and the crew, had lunch, on both days, in the local pub. The two days of conversation were divided into 102 'stories', each lasting a few minutes.[1] Each story has its own title and can be separately viewed, although they form a definite sequence and, if watched sequentially, form a wonderfully absorbing picture of the great man's scientific life. They include autobiographical accounts of his early life, education at Eton, wartime employment as an engineer designing aeroplanes, Marxist politics at Cambridge, and returning to university as a mature student after the war to read biology.

Some of the later 'stories' uncovered John's occasionally strained relations with Bill Hamilton. John told the story with disarming frankness. Neither he nor his famously idiosyncratic mentor, the great J. B. S. Haldane, recognized that the shy young man working in another department of their university was a consummate genius. John quoted Huxley on closing the *Origin of Species*: 'How extremely stupid of me not to have thought of that.' He reproached himself for not having offered support to Hamilton when the younger man needed it. In subsequent 'stories' he compared Bill Hamilton's concept of 'inclusive fitness' (a measure of what an individual organism is expected to maximize) with the 'gene's-eye view' adopted by Bill in other papers–the view which I, along with John, prefer (see

---

[1] They can be seen at a website called Web of Stories: http://www.webofstories.com/play/john.maynard.smith/1.

page 449), and which eventually arrives at the same answer as the inclusive fitness approach.

It was at University College London after the war that John came under the sway of J. B. S. Haldane, and the interview is peppered with lovely stories of that formidable eccentric. I'll quote one, to give the flavour:

> He and his wife, Helen, had the rather nice habit that on the night when we'd all finished our finals examination . . . of taking the class after the last exam, to The Marlborough, which is the pub just down the other side of the road, and buying us all a drink until closing time. It was very pleasant. And I went there the night I took finals. And when the pub shut, he said to me and to Pamela Robinson, who was also going to become a graduate student in, in fact, in palaeontology, would we like to go back to his flat and continue drinking, because we clearly hadn't had enough. And rather foolishly we said we would. So we went back to Prof's flat and we continued to drink and talk about the world, until about two o'clock in the morning Pamela said: 'Look, Prof, John and I really have to go home, but the tubes have stopped running and you're going to have to drive us home.' So Haldane said, 'All right, I'll drive you home.' So we piled into Prof's car . . .

It was a typical Haldane possession, it was extremely old and ramshackle and decrepit. And we started driving up [Parliament] hill. And about halfway up the hill, the car started filling with smoke. And I didn't like to say anything, I thought this was normal. But Pamela said, 'Prof, I think the car's on fire.'

'Oh? Oh well.' So we drew up . . . And, as an engineer, I was told to find out what was wrong. And it was clear nothing very serious was wrong. What had happened was that the carpeting had fallen down on the transmission and was burning, underneath the front seat. So, we looked at this for a bit. And Haldane said, 'The ladies will go and stand behind yonder lamp post.' I thought: What next? And he then turned to me and said, 'Smith, the method of Pantagruel. You have had more beer than I have. Put it out.' Now, part of this is, of course, you had to know the classical quotation, you had to know that Pantagruel had, in fact, put out Paris on fire by peeing on it. So I did. And I don't know, you know, if you've had a lot of beer and you start peeing, it's kind of hard to stop. He said, 'That's enough, boy, that's enough.'

But my point is that if you were going to work and live with Haldane, you had to be prepared to live in this slightly unpredictable environment, and I . . . I was . . . the other thing about Haldane was that if he said something you disagreed with, you could tell him to shut up and stop being such a silly old fart, he didn't mind. But you had to treat him like that, it was no good being polite, you had to fight back if he said things you didn't agree with.

The transcript is nice, but you really have to listen to John, he was such a wonderful raconteur.

The 'story' straight after that one reduced John to tears, as he recalled the moment when Haldane, about to depart to end his days in India, had confessed his affection for John's wife Sheila, and asked John to tell her because he was not able to do so himself. The emotion, actual and recollected, is powerful and moving. And all this was done by the method of 'talking heads'.

Debates and encounters

I am not a fan of the debate format, certainly not the rigidly structured and timed debate ending in a vote. As an undergraduate I regularly attended the Thursday evening debates at the Oxford Union, and I heard guest speeches, some of them extremely good, from leading politicians and orators of the day: Michael Foot, Hugh Gaitskell, Robert Kennedy, Edward Heath, Jeremy Thorpe, Harold Macmillan, Orson Welles, Brian Walden–even Oswald Mosley proved to be a mesmerizing speaker, however unpleasant his politics. Some undergraduate speakers, too, were extremely accomplished, for example Paul Foot, Michael's nephew, who later became a penetrating investigative journalist. But I have become disillusioned with the lawyer-style adversarial format of formal debates. Universities enter debating teams in competitions where the speakers are told, on the toss of a coin, which side they are to advocate. Good training for lawyers, no doubt, but I find something akin to whoring in young people learning to hone rhetorical skills in the service of an arbitrarily allocated cause in which they don't believe–maybe, indeed, advocating the opposite of what they believe. If I am moved by oratory, I want it to be sincerely meant.

But wait. Does a stage speech by an accomplished actor give the lie to my stricture? Does a rousing Henry V at the breach, or a Mark Antony 'come to bury Caesar', fail to convince because we're listening to an actor and not the real thing? I'd like to think not. I'd like to think that a great Portia transports herself into the skin of her character so deeply that her 'quality of mercy' speech really feels sincere in a way that a lawyer advocating a defence in which he doesn't believe cannot–indeed, *should* not. Lalla tells me crying on stage comes easily if you really live your character and embrace her pathos.

English law (and, I think, Scottish and American law too) is founded on the 'tug-of-war' principle: in any disagreement, pay somebody to put the strongest possible case for a proposition, whether they believe it or not, pay somebody else to put the strongest possible case against it, and see which way the tug-of-war moves. That's as opposed to the 'court of inquiry' principle more characteristic of west European law, which seems to me, in my naivety perhaps, more honest and humane: let's all sit down together, look at the evidence and try to work out what really happened here. English and American lawyers speak with unabashed admiration of legendary advocates from the past, so good that they even managed to get (fill in name of obviously guilty party) off. So much the better for the lawyer's reputation if any fool could see that his client was guilty but the great advocate *still* managed to win over the jury.

I was deeply shocked by a conversation with a bright young American defence lawyer who was exultant because the private detective she'd hired had found evidence that proved, beyond any doubt, the innocence of her client. 'Congratulations,' I said. 'What would you have done if your detective had found evidence that conclusively proved your client guilty?'

'I would have ignored it,' was her unabashed reply. It's up to the prosecution to find their own evidence: I'm not paid to give help to *the other side*' (my italics).

This was a murder case, and she was blithely happy to contemplate suppressing evidence, thereby letting a murderer go free perhaps to kill again, rather than lose a tug-of-war with the prosecuting lawyer 'on the other side'. How could any decent person fail to be shocked by that story? But I have yet to find a lawyer prepared to condemn it. They've inhaled the 'our side' versus 'the other side' smoke so deeply, they don't even notice it any more. I choke on it.

Incidentally, a version of the tug-of-war approach to getting at the truth has been adopted by a school of television interviewers, beginning (in Britain at least) with Robin Day. Only yesterday, as I write, I was in a BBC television studio waiting for my turn in the hot seat. As it turned out, I was not treated in this way, but in the minutes while I waited the interviewer was getting through a series of politicians, representatives of all three main political parties, on the topic at hand. His style of questioning was truculent from the outset. The presumption seemed to be that all three of them were lying, or at best incompetent. Maybe he believed they were. But I suspect the real reason was his training in a school of journalism which holds that the best way to get at the truth when interviewing somebody is to provoke him as hard as you can in order to see where the tug-of-war ends up. Perhaps that is indeed the best way, but it's not obviously so and it needs justifying.

Anyway, although I have occasionally accepted invitations to speak at both the Oxford and Cambridge Unions, I don't like the adversarial style of debating. My first experience of it was in 1986 at the Oxford Union, when John Maynard Smith and I took on a pair of creationists, Edgar Andrews and A. E. Wilder-Smith. The motion was 'that the doctrine of creation is more valid than the theory of evolution'. I certainly wouldn't agree to debate such a motion today, and even in 1986 I did it only to support a valued pupil from New College, Daniela Sieff, who had agreed to be the lead undergraduate speaker on the science side. Neither of the two guest speakers on the other side had any qualification in biology. Wilder-Smith, a chemist, turned out to be a harmlessly genial buffoon. Andrews, a physicist (and less genial), had written a number of books advocating fundamentalist creationism (including 'flood geology': yes, that's Noah's flood!), which I took the precaution of reading before the debate. Of course, naive creationism would be an instant debate-loser at the Oxford Union, so Andrews pretended to a more sophisticated philosophy-of-science approach. Nobody would have guessed that he—a professor of physics—could seriously be a naive creationist . . . until I began reading out passages from his books. Pathetically, and repeatedly, he stood up and tried to persuade the president to stop me reading from his own writings. Very properly she overruled him, and he sat with his head in his hands as I read out the significant passages, giving the lie to his philosophical pretensions. At the drinks after the debate, he had an altercation with John

Maynard Smith, the only time I ever saw that beloved good man flush red with anger.

The more specific reason why I now refuse to take part in formal debates with creationists is that every time a scientist agrees to such a debate it creates an illusion of equal standing. An audience is fooled by the presence of two chairs side by side on the platform, the allocation of equal time to 'both sides': fooled into thinking there really *are* two 'sides', fooled into thinking there is an issue of real substance to debate. It was Stephen Jay Gould who first opened my eyes to this 'two chairs effect'. I had been invited to debate with a creationist in America and I telephoned Steve to ask his opinion. 'Don't do it,' was his friendly advice. The moment a real scientist agrees to such a debate, the creationist has won his main objective, no matter what actually happens in the debate itself. 'They need the publicity,' Steve pointed out to me; 'you don't.' Robert May put the same thought with characteristically blunt Australian wit. When invited to participate in such a debate, his favoured reply is: 'That would look great on your CV, not so good on mine!' I've told that story so often that quite a few people think the witticism is mine. I wish it were!

So powerful is the 'two chairs effect' that it has actually, with petty malice, been turned around and used against me. I was once invited to have a debate, in Oxford, with an American Christian apologist called Craig, who had been badgering me for years to have a second debate with him (the first was at a big event in Mexico, where he was the least impressive of the three speakers on his side). As it happened, I had another speaking engagement in London on the evening of his proposed Oxford debate, but I would have refused in any case, for reasons I'll mention in a moment. So his supporters placed an empty chair on the stage in Oxford and pretended I'd been too cowardly to turn up!

In his case, I had already published in the *Guardian* a very particular reason for not wishing to share a platform with this individual ever again: my disgust at his justification of the biblical slaughter of the Canaanites. I wasn't complaining about the alleged massacre itself (like most of Old Testament 'history' it never happened). My point was that Craig, *believing* that it happened, *justified* it on the grotesquely immoral grounds that the Canaanites were all sinners, so they deserved what was coming to them. Moreover, all they had had to do was hand over their land to the invading 'Israelis' (*sic*) and their lives would have been spared.

> I have come to appreciate as a result of a closer reading of the biblical text that God's command to Israel was not primarily to exterminate the Canaanites but to drive them out of the land. It was the land that

was (and remains today!) paramount in the minds of these Ancient Near Eastern peoples. The Canaanite tribal kingdoms which occupied the land were to be destroyed as nation states, not as individuals. The judgment of God upon these tribal groups, which had become so incredibly debauched by that time, is that they were being divested of their land. Canaan was being given over to Israel, whom God had now brought out of Egypt. If the Canaanite tribes, seeing the armies of Israel, had simply chosen to flee, no one would have been killed at all.[1]

So it was the Canaanites' fault: they had it coming to them because God wanted their land for his pet tribe's *Lebensraum* and the incumbents just refused to get up and abandon their homes of their own accord. Craig even justified the slaughter of the children since they'd go to heaven anyway.

In passing, my *Guardian* article mentioned the 'empty chair' tactic (which had been well publicized in advance):

In an epitome of bullying presumption, Craig now proposes to place an empty chair on a stage in

---

[1] I quoted this passage in my *Guardian* article. For Craig's own 'justification', see http://www.reasonablefaith.org/the-slaughter-of-the-canaanites-re-visited.

Oxford next week to symbolise my absence. The idea of cashing in on another's name by conniving to share a stage with him is hardly new. But what are we to make of this attempt to turn my *non-appearance* into a self-promotion stunt? In the interests of transparency, I should point out that it isn't only Oxford that won't see me on the night Craig proposes to debate me in absentia: you can also see me not appear in Cambridge, Liverpool, Birmingham, Manchester, Edinburgh, Glasgow and, if time allows, Bristol.[1]

Craig reserved his special sympathy for the poor 'Israeli' soldiers who were obliged to carry out the unpleasant duty of massacring all those Canaanite women and children. The empty chair gambit, by the way, has since become known as 'Eastwooding' because it was used by the actor and director Clint Eastwood in an inept stunt aimed at President Obama during the 2012 presidential campaign.

My 'two chairs objection' to debating doesn't apply to scholarly theologians with real credentials. With them, I am happy to have debates (I'd prefer to say public conversations), and I have done so with two Archbishops of Canterbury, an Archbishop of York,

---

[1] You can read the full article at bit.ly/1fXPAGS.

several bishops, a cardinal, and two successive holders
of the office of British Chief Rabbi. In most cases these
have been amicable and civilized encounters. Some time
in 1993, for example, at the Royal Society, I was paired
with the distinguished cosmologist Sir Herman Bondi
against Hugh Montefiore, former Bishop of Birmingham, and Russell Stannard, a Christian physicist and
author of the excellent 'Uncle Albert' books explaining
modern physics to children. Stannard has written his
own account of the meeting:

> On being introduced to each other by one of the
> organizers, Dawkins straight away told me how
> much he had enjoyed my *Uncle Albert* books. He
> actually enjoyed them! I immediately thought that
> anyone who enjoyed Uncle Albert couldn't be all
> that bad, could they.
>
> But wait. Was this a trick to lull me into a false
> sense of security . . . As it turned out, I needn't
> have worried. The debate was conducted in a con-
> structive and courteous manner . . . This is not to
> say that the debate lacked tension. Far from it.
> There were cut and thrust exchanges, and out-and-
> out disagreement on various issues. But there was
> no acrimony, no cheap point-scoring.
>
> To underline just how good-humoured the

debate had been, the participants adjourned afterwards to a restaurant for a pleasant supper together!
I sat next to Dawkins and thoroughly enjoyed his
company.[1]

I have had four meetings with Rowan Williams, recently retired as Archbishop of Canterbury, and found
him to be one of the nicest men I have ever met: almost
impossible to argue with, he is so agreeable. And so
obligingly intelligent (in the literal sense of *intellego* = I
understand) that he actually finishes your sentences for
you, even when those sentences–in my understanding
of them–should have been devastating for his position
and he doesn't seem to have any comeback to them!
I first noticed this engaging habit when I interviewed
him for one of my Channel Four documentaries. He
later invited Lalla and me to a delightful party in Lam-
beth Palace (I think Lalla may have been the main draw
there, because his son Pip was a fan of her character in
Doctor Who). Then, a few years later, he and I had a
rather overpublicized 'debate' in the Sheldonian Theatre. I wanted this to be a friendly conversation without
a chairman, because I find (see below) that chairmen
often get in the way of the discussion: and so it proved in

---

[1] R. Stannard, *Doing Away with God* (London, Pickering, 1993).

this case. Afterwards, the archbishop and I sat together at dinner and I was again charmed by his company.

Our most recent encounter was as speakers on opposite sides in a debate at the Cambridge Union. This was after Dr Williams had stepped down from the archbishopric to become Master of Magdalene, and he told me at dinner of his sheer joy on waking every morning and remembering: 'I'm not Archbishop of Canterbury any more.' When it came to the debate itself, his side won, and the victory was widely credited to him. He did indeed make a decent speech, but the real victor, as was clear from audience reaction, was the last speaker on his side, the very personable journalist Douglas Murray. Murray proclaimed himself an atheist but thought—and this was really his only point—that religion was good for people: they'd be unhappy without it. I can't imagine Rowan Williams ever being so patronizingly condescending, but—surprisingly[1]—the Cambridge audience lapped it up.

I think the most revealing conversation I have had with a theologian was my filmed interview with the Jesuit Father George Coyne, erstwhile Director of the Vatican Observatory. We filmed it for the same Channel Four television documentary as my interview

---

[1] Until you recall the notoriously powerful lobbying influence of CICCU, the Cambridge Inter-Collegiate Christian Union.

with Archbishop Williams. Unfortunately the director felt he couldn't find the time to include both interviews and he dropped the one with Father Coyne.

This professional astronomer, a scientist to his fingertips, talked like an intelligent atheist for most of the interview. 'God', he said, 'is not an explanation. If I were seeking for a god of explanation . . . I'd probably be an atheist.' To which my inevitable reply was that that was exactly why I am an atheist. If an all-powerful creator God is really there, how could he not be an explanation for things? Or, if he is not the explanation for anything, what exactly does he do with his time that makes him worth worshipping?

Father Coyne also cheerfully agreed that his Catholic beliefs flowed from the accidental circumstance of his having been born into a Catholic family, and he accepted that he would have been an equally sincere Muslim if he had been born into a Muslim family. I was struck by his personal honesty, while at the same time marvelling at the professional dishonesty that his Catholic orders imposed upon him. He impressed me as a decent, humane, intelligent man.

As did the British Chief Rabbi Jonathan Sacks, who invited Lalla and me to dinner in his house with some of London's leading Jews. It was at that dinner that I learned the stunning fact that Jews, who constitute less than 1 per cent of the world's population, have won more than 20 per cent of all Nobel Prizes. This makes a poignant contrast with the derisorily low success rate of the world's Muslims, who are orders of magnitude more numerous in the world. I thought–still do–the comparison revealing. Whether you think of Judaism and Islam as religions or cultural systems (neither is a 'race', despite widespread misconceptions), how could it not be revealing that one of them has a success rate per head which is literally tens of thousands of times higher than the other, in the fields of intellectual endeavour celebrated by Nobel? Islamic scholars were notable in keeping the flame of Greek learning alive during the middle ages and dark ages of Christendom. What went wrong? Incidentally, Sir Harry Kroto has written to me of his belief that the great majority of Nobel laureates listed as Jews (including himself) are actually non-believers.

In a later encounter with Lord Sacks, in a television studio in Manchester, rather weirdly he publicly accused me of anti-Semitism. It turned out that the reason for this was my characterization, in *The God Delusion*, of the God of the Old Testament as 'arguably the most unpleasant character in all fiction'. I've quoted the rest of the sentence elsewhere in this book (see pages 616), and I agree that it does sound a bit polemical, although it can be justified in spades from the Bible. But my intention was directed less towards polemic than comedy. I had in the back of my mind Evelyn Waugh's rare purple set-pieces (and I glancingly signalled the allusion by recounting, in the very same paragraph, a story told by Waugh about Randolph Churchill). I couldn't, of course, deny that my sentence was anti-God. But anti-*Jewish*? It was not, by the way, the first time I was accused of anti-Semitism on similar grounds. I gave a lecture on a ship cruising the Galápagos archipelago and a fellow passenger objected. His sole reason was that I was against God, whom he apparently identified with his own Jewishness, and as a result he felt personally offended.

The Chief Rabbi was good enough to send me a gracious apology a few days later, and I take his remark in the studio to have been a temporary aberration: an anomalous mistake by a decent gentleman. I was not, to put it mildly, so impressed by the most senior Roman Catholic spokesman with whom I have debated, Cardinal George Pell, Archbishop of Sydney. We were pitted against each other in a television studio of the Australian Broadcasting Corporation. I had been warned in advance that he was a bully and a 'bruiser'—not a happy reputation, you would think, for a senior figure in a church purporting to be founded on more generous principles.

Pell played for cheap laughs from the gallery in a way that clerical gentlemen of the stature of Archbishop Williams, Chief Rabbi Sacks and Father George Coyne would never do. He was fortunate that a substantial fraction of the studio audience had obviously been hand-picked as partisans in his favour, because he had an almost endearing gift for putting his foot in his mouth, as when he spoiled his otherwise praiseworthy acceptance of evolution by adding the gratuitous error that humans were descended 'from Neanderthals'. Or when he told an anecdote about a time when he had been 'preparing some English boys . . .' and allowed an embarrassing pause to ensue before he completed the sentence '. . . .or first communion', a pause long enough to allow a minority of the audience to laugh suggestively. A less endearing *faux pas* was his apparent doubt as to the intelligence of Jews and his puzzlement at God's having chosen them. The chairman, Tony Jones, immediately jumped on that and the cardinal had to dig frantically to extricate himself. I let him dig, and resisted the temptation to quote the rhymed exchange between W. N. Ewer and Cecil Browne:

How odd
Of God
To choose
The Jews.

But not so odd
As those who choose
A Jewish God
Yet spurn the Jews.

He drew delighted applause from the partisan majority when he seemed to score a point by quoting Darwin's autobiography in evidence that he was a theist when he wrote it towards the end of his life. This is definitely false, and I said so. Pell, however, was briefed with notes and able to say that he was quoting 'page 92' of Darwin's memoir. It was this triumphant citation of 'page 92' that led his claque to applaud.

Here's a bit of a digression, but a televised misrepresentation of Darwin's religious beliefs by a prince of the church is surely important enough to make it necessary. Looking at the autobiography today, I'm inclined to think that Pell was not being deliberately dishonest in his triumphant citation of 'page 92'. Very probably an assistant furnished him with the quote, complete with page number, and failed to tell him what follows it. Judge for yourself. Here is what Pell quoted from Darwin's chapter on 'Religious belief', with the word that I think Pell should have emphasized picked out in bold type:

> Another source of conviction in the existence of God, connected with the reason and not with the feelings, impresses me as having much more weight. This follows from the extreme difficulty or rather impossibility of conceiving this immense and wonderful universe, including man with his capacity of looking far backwards and far into futurity, as the result of blind chance or necessity. **When** thus reflecting I feel compelled to look to a First Cause having an intelligent mind in some degree analogous to that of man; and I deserve to be called a Theist.

Pell could argue that the 'When' that begins the sentence did not have the conditional meaning that I see there, but was an absolute statement. The succeeding paragraph, however, which Pell did not read to us, leaves us in no doubt of Darwin's actual position when he wrote the chapter. Once again, I emphasize in bold type the key to interpreting his meaning:

This conclusion **was** strong in my mind about the time, as far as I can remember, when I wrote the *Origin of Species*; and it is since that time that it has very gradually with many fluctuations become weaker . . . The mystery of the beginning of all things is insoluble by us; and I for one must be content to remain an Agnostic.

I think we must exonerate Cardinal Pell of the charge of dishonesty. Let's allow that he (or his assistant) simply didn't read the second paragraph, and pardonably misunderstood the first. But I would like to hope that, if my book is seen by any of that Australian audience who cheered him on when he triumphantly said 'It's on page 92,' they will take the trouble to read the whole of Darwin's chapter on 'Religious belief' in his autobiography. In addition to the paragraph I quoted above, where Darwin concludes that he is content to remain an agnostic, much of the rest of the chapter consists of strong criticism of the Christian faith in which Darwin, as a young man, was a devout believer, destined for a career in the church. There's the famous sentence, for example, where Darwin says he could

hardly see how anyone ought to wish Christianity to be true; for if so the plain language of the text seems to show that the men who do not believe, and this would include my Father, Brother and almost all my best friends, will be everlastingly punished. And this is a damnable doctrine.

Darwin remained benevolently disposed towards his local parish church, supported it financially and wished to be buried there (a wish that was denied when his friends succeeded in getting him honoured in Westminster Abbey). And he ques-tioned what he perceived as the militant atheism of Edward Aveling (1849–98) and his German colleague Ludwig Büchner (1824–99). Aveling's account of their meeting at Darwin's lunch table in 1881 begins with a moving description of how the visitor fell 'under the spell of the frankest and the kindliest eyes that ever looked into mine', and then turns to their discussion of religion. Darwin asked: 'Why do you call yourselves atheists, and say there is no God?' Aveling and Büchner explained that they

were Atheists because there was no evidence of deity . . . that whilst we did not commit the folly of god-denial, we avoided with equal care the folly of god-assertion: that as god was not proven, we were without god ($\alpha'\vartheta\varepsilon o\iota$) and by consequence were with hope in this world, and in this world alone. As we spoke, it was evident from the change of light in the eyes that always met ours so frankly, that a new conception was arising in his mind. He had imagined until then that we were deniers of god, and he found the order of thought that was ours differing in no essential from his own. For with point after point of our argument he agreed; statement on statement that was made he endorsed, saying finally: 'I am with you in thought, but I should prefer the word Agnostic to the word Atheist.'[1]

To this day there is confusion over the word 'atheist', some taking it to mean a person who is positively convinced that there is no god (what the atheist Aveling called 'the folly of god-denial'), others a person who sees no reason to believe in any god and therefore lives their life in a god-free manner (i.e. what Darwin meant when he called himself an agnostic, and what Aveling meant by 'without god'). Probably rather few scientists would adopt the first sense, although they might add that the loophole they leave for a god is scarcely wider than that through which leprechauns or orbiting teapots or Easter Bunnies may jump. There's a spectrum between the two positions, and Darwin would admit-

---

[1] For the full account, see bit.ly/1rY74rY.

tedly have been less sceptical of gods than of flying teapots, as we can guess from a conversation with the Duke of Argyll, late in his life. By the Duke's account,

> I said to Mr. Darwin, with reference to some of his own remarkable works on *Fertilisation of Orchids*, and upon *The Earthworms*, and various other observations he had made of the wonderful contrivances for certain purposes of Nature–I said it was impossible to look at these without seeing that they were the effect of mind. I shall never forget Mr. Darwin's answer. He looked at me very hard, and said: 'Well, that often comes with overwhelming force; but at other times,' and he shook his head vaguely, adding, 'it seems to go away.'[1]

I too am marginally less sceptical of gods than of teapots in orbit around the sun, if only because the set of all imaginable things that might qualify as gods is larger than the set of orbiting projectiles that would qualify as teapots. But I think Darwin would agree with Aveling (and me) that the onus of proof is on the theist.

I hope I wasn't unfair in my assessment of Cardi-

---

[1] See http://www.electricscotland.com/history/glasgow/anec305.htm.

nal George Pell. Rather than take my word for it, you could listen to the debate itself.[1]

I don't think a recording has survived of my debate with another prelate, the then Archbishop of York, John Habgood, at the Edinburgh Science Festival in 1992. Perhaps it is just as well, as I'm not particularly proud of my performance, despite–or maybe even because of–the verdict of the *Observer* journalist (see below). If Archbishop Pell is regarded as a bully and a 'bruiser', I fear that may be how I (though physically of slighter build than they usually make bruisers) came over in my treatment of Dr Habgood. I wouldn't handle the encounter in the same way now–perhaps I have become more compassionate than I was, but I find I can't now bear hitting someone who is down. Twenty years ago, however, I seem to remember that I paxmanned him, repeating several times, and too unforgivingly, a question on what he really believed about the Virgin Birth (as opposed to what he professionally was expected to believe). And I'm afraid the audience joined in, barracking him with 'Answer the question! Answer the question!' The *Nullifidian*'s account of the evening seems to bear out my remembered misgivings:

---

[1] https://www.youtube.com/watch?v=tD1QHO_AVZA.

Richard Dawkins, well-known for his books on evolution, took part in a debate with the Archbishop of York, Dr John Habgood, on the existence of God at the Edinburgh science festival last Easter. The science correspondent of *The Observer* reported that the 'withering' Richard Dawkins clearly believed that 'God should be spoken of in the same way as Father Christmas or the Tooth Fairy'. He [the correspondent] overheard a gloomy cleric comment on the debate: 'That was easy to sum up. Lions 10, Christians nil'.[1]

The verb 'to paxman', for the benefit of non-British readers, stems from a notorious interview by the formidable Jeremy Paxman, Britain's most feared television journalist, of the then Home Secretary Michael Howard. Paxman relentlessly asked Howard the identical question no fewer than twelve times, while the poor man equally persistently evaded answering. I have just listened to the interview again,[2] and I know I would now be incapable of being so ruthless. Even then, I think that with Dr Habgood my limit was three repetitions of my awkward question about the Virgin Birth. Jeremy Paxman, by the way, has interviewed me

---

[1] http://bit.ly/1AUT0GJ.
[2] bit.ly/1iGJRVQ.

on BBC television on two occasions, and has chaired an on-stage encounter between me and the then Bishop of Oxford Dr Richard Harries. On all three occasions he was warm and sympathetic, and I have found him so when I've met him socially–for example at a summer dinner in his garden, or at the Hay-on-Wye Festival during the week in which I write this, when he joined me as I was having a solitary hotel breakfast. Maybe it's only politicians who have reason to fear him. I treasure the opening words of his interview with a notorious American political propagandist, when she was promoting her book in England: 'Your publishers gave us Chapter 1, Ann Coulter, and I've read it. Does it get any better?' I earlier referred to the school of pugnacious television journalists founded by Robin Day. Jeremy Paxman is an even more ruthless exponent. I prefer what I call the 'mutual tutorial' technique of interview or public conversation.

Mutual tutorials

More agreeable than debates have been on-stage conversations where the aim has been mutual enlightenment rather than scoring victories ('owning' or 'pwning'[1] as the net generation says). I think

---

[1] This spelling seems to have arisen as a serendipitous misprint, a mutant meme subsequently favoured. Gillian Somerscales suggests to me that it is

the phrase 'mutual tutorial' first occurred to me in February 1999 when I shared the stage at Central Hall, Westminster with the psychologist and linguist Steven Pinker. Billed as a 'debate' sponsored by the *Guardian* and chaired by their science editor Tim Radford, the event attracted an audience of 2,300, with many turned away outside. It wasn't a debate: there was no 'motion', nothing to vote on, and we agreed about most things anyway. And, as I said, it paved the way for what I was later to call the 'mutual tutorial'–a genre of on-stage conversation that I am increasingly pushing as a superior alternative to both the interview and the debate. Tim Radford, as it happened, did a good and unintrusive job. But it was this encounter that suggested to me the idea of the mutual tutorial without a chairman or 'host'.

The 'chairman interference effect' was especially noticeable in the encounter I mentioned above in Oxford's Sheldonian Theatre with the then Archbishop of Canterbury, Rowan Williams. Dr Williams and I were all set to have a civilized conversation and I had been greatly looking forward to it. But unfortunately it was continually derailed by the chairman, a distinguished

---

only ever seen in written form and nobody ever has to pronounce it. 'Do you think', she asks, 'that a "non-spoken" form of language may be emerging?' If so, 'LOL' might be another candidate for the text-only dictionary.

philosopher and very nice man, whose strenuous efforts to 'clarify' matters by injecting philosophical jargon had–as often seems to happen with philosophers–precisely the opposite effect.

The large audience attending my London 'mutual tutorial' with Steve Pinker (notwithstanding that 'mutual', I have to say that I learned more from him than he did from me) attracted the attention of the BBC. Would we like to go on television that evening in their *Newsnight* programme and reprise our discussion for a wider audience? We would. A little later I was telephoned by the BBC producer, wanting to be briefed on what to expect:

'Could you summarize for me the nature of your disagreement with Dr Pinker?'

'Er, well, actually, I'm not sure that we have much to offer in the way of disagreements. We seem to agree over most things. Is that a problem?'

There was a long pause from the other end of the line. 'No disagreement? *No* disagreement? Oh dear.'

And she promptly cancelled the invitation! Mutually informative conversation, it seems, is not 'good television'. There has to be disagreement, sparks must fly. If 'good television' means good for ratings, that's depressing. I'd like to hope that she was mistaken and that disagreement is not in truth good for ratings, but

I can't summon up much conviction. In any case, my own value judgement, as I said in the previous chapter, would place ratings rather low in the scale of what 'good television' ought to mean. Especially for the BBC, which doesn't have to worry about advertising revenue since it is financed by central government through the licence fee.

The Pinker 'debate that wasn't a debate' encouraged me to promote the format in a new series under the auspices of my charitable foundation, the Richard Dawkins Foundation for Reason and Science (RDFRS). The first one we did was a conversation before a large audience at Stanford University in March 2008, between the theoretical physicist Lawrence Krauss and me. I began by introducing the format to the audience: 'Well, I suppose I must bear some responsibility for the fact that we don't have a chairman sitting between us. I'm trying to pioneer a new method of public conversation . . .' And I went on to expound the 'mutual tutorial' and to explain my objections to chairmen at such events. I acknowledged that it threw a burden on us to keep the conversation going, and I then passed the burden to Lawrence by inviting him to start.

He began by reminding me of our first meeting, which had been somewhat less amicable. It was at a conference in New York State in 2006, shortly after publication of *The God Delusion*. I was fielding questions after my talk. I've had so much practice at this that I seldom find the questions challenging, but this was different. A questioner, not particularly tall but every inch a confident inch, stood up in the middle of the audience, and his very first sentence rang with an articulate and fluent conviction which is understandably rare on such public occasions. He roundly–almost aggressively–berated me for being too aggressive and insufficiently conciliatory when arguing with the faithful. I can't remember how I responded, but afterwards we had a drink together and Lawrence, in more friendly vein, suggested that we should continue our discussion in print. So we did, and the exchange appeared in the pages of *Scientific American*,[1] as he told that Stanford audience in his opening remarks. Lawrence and I have had several further public discussions since then, and our original disagreement has abated as we have become friends and drawn closer to each other's point of view. And, as we have learned from each other, our mutual tutorials have increasingly deserved the name. Several of these conversations have formed the basis of *The Unbelievers*, a documentary feature film produced by Gus and Luke Holwerda, showing Lawrence and

---

[1] http://www.scientificamerican.com/article/should-science-speak-to-faith-extended/; see also e-appendix.

me in various venues around the world, the most notable being the Sydney Opera House.

Lawrence is quirky, funny and fun. I've never quite known what 'comic timing' means but I suspect he has it. If he added introspective melancholy to his repertoire you might call him (as I once did) the Woody Allen of physics. And he's provocative in the best and most constructive sense: 'Every atom in your body came from a star that exploded, and the atoms in your left hand probably came from a different star than your right hand . . . forget Jesus: the stars died so you could be here today.'

The *Unbelievers* crew were once filming in a hired limousine on a humid, hot day in London. Almost everything you can imagine was wrong with that car, and Lawrence on the telephone to the company is a treasured memory (far from 'introspective'; and 'melancholy' comes nowhere near doing justice to his tirade), climaxing in a threat to wreak physical damage along the whole stretched length of that absurd vehicle. It was a virtuoso performance in exhibition invective, just what we needed to make us laugh in a stifling hot car with both the air-conditioning and the window-opening mechanism broken.

The model for the 'mutual tutorial' pioneered with Pinker and Krauss has proved successful in other public conversations, using the same chairman-free format. My partners in these dialogues have included Professor Aubrey Manning and Bishop Richard Holloway (possibly the two nicest men in Scotland). Aubrey and I share a common heritage as students of Niko Tinbergen (Aubrey a decade ahead of me), so our conversation included some reminiscing, with many laughs, about the Athens of ethology that was the Tinbergen group, but we also talked science itself. Bishop Holloway describes himself as a 'recovering Christian'. He's probably about as close to being an atheist as a bishop can get away with. We've had more than one encounter, including an on-stage conversation in Edinburgh, which moved the Glasgow journalist Muriel Gray to write as follows:

Holloway, as we all know, is the church leader who questioned his faith and found it wanting, and Dawkins of course is not simply world famous for his pioneering and award-winning scientific work, but also for his aggressive views concerning organised religion. A couple of audience members before the session began admitted they were worried that the two men might come to blows, or that a fundamentalist audience member might use the event to launch an offensive verbal attack on Dawkins. Instead the hour that seemed like five minutes was one filled with two startlingly intelligent men, each brimming with humanity, drawing personal pictures of just how awesome, mysterious and wonderful existence is. The sheer joy of hearing Holloway still trying to draw poetry and meaning from a religion he is not quite ready to dismiss completely out of hand, as Dawkins listened eagerly, trying to assist him without dismissing his desires as ignorant, was breathtakingly inspiring. And all this was book-ended with Dawkins's views on birthing universes, black holes, and the future of the human species when we start to form ourselves from silicon and alloys instead of vulnerable flesh. Now that's what I call entertainment . . . However, the most dreadful, in fact utterly unbearable, part of the evening was that it stopped after an hour.[1]

I think you could safely call that a mutual tutorial. Incidentally, I've since had two intellectually rewarding on-stage conversations, also in Edinburgh, with Muriel Gray herself.

Another wonderful encounter was with Neil DeGrasse Tyson, Director of the Hayden Planetarium,

---

[1] *Glasgow Sunday Herald*, 5 Sept. 2004.

New York. Our conversation[1] took place in 2010 at a conference organized by the RDFRS on the campus of Howard University in Washington, described as a 'historically black' university. In front of a lively student audience (though a smaller one than both Neil and I are used to because, as we later learned, religious leaders had 'discouraged' attendance), Neil and I talked of 'The poetry of science'. The phrase immediately makes one think of Carl Sagan, and Neil Tyson has magnificently, but with becoming humility, accepted the challenge of stepping into Sagan's unfillable shoes to present a new version of *Cosmos*. What a superb spokesman for science he is, this warm, friendly, witty, clever man, whose great knowledge is properly served by his ability to expound it. The only other person I can think of who might have understudied Carl Sagan so well is Carolyn Porco (of whom much more in the next chapter). Perhaps it is not entirely surprising that, of all scientific subjects, astronomy should be so well endowed with stellar ambassadors.

That was not the first time I had met Neil Tyson. Our first encounter, in San Diego in 2006, was almost a carbon copy of my introduction to Lawrence Krauss. I had just given a talk in which I was critical of the

---

[1] https://www.youtube.com/watch?v=eUMI3_QLmoM.

religiously inclined ecologist Joan Roughgarden. At question time Neil delivered a polite but serious–and impeccably phrased–attack on my style:

> I was in the back row as you spoke . . . and so I could see sort of the whole room as the words came out of your mouth as beautifully as they always do and as articulately as they always do. Let me just say your commentary had a sharpness of teeth that I had not even projected for you . . . You're Professor of the Public Understanding of Science, not professor of delivering truth to the public, and these are two different exercises. One of them is, you put the truth out there and, like you said, they either buy your book or they don't. Well, that's not being an educator. That's just putting it out there. Being an educator is not only getting the truth right, but there's got to be an act of persuasion in there as well. Persuasion isn't always 'Here's the facts, you are either an idiot or you're not.' It's 'Here's the facts, and here is a sensitivity to your state of mind.' And it's the facts plus the sensitivity, when convolved together, creates impact. And I worry that your methods, and how articulately barbed you can be, ends up simply being ineffective, when you have much more

power of influence than what is currently reflected in your output.

I was conscious that the chairman, Roger Bingham, was anxious to wrap up the session, so I replied briefly:

I gratefully accept the rebuke. Just one anecdote to show that I am not the worst in this thing. A former and highly successful editor of *New Scientist* magazine–he actually built up *New Scientist* to great new heights–was asked: 'What is your philosophy at *New Scientist*?' He said, 'Our philosophy in *New Scientist* is this: science is interesting, and if you don't agree you can fuck off.'

To the jovial sound of Neil Tyson's bellow of laughter, Roger Bingham closed the session.[1] Neil's critique was a good one–it was pretty much the same as Lawrence Krauss's, although more gently expressed–and it is one that I take to heart. I'll return to the question later, when I come to discuss *The God Delusion*.

In some of my 'mutual tutorials' I have had so much more to learn than my fellow conversationalist that the word 'mutual' needs to be dropped. Most daunting for

---

[1] A video of the encounter is here: https://www.youtube.com/watch?v=-_2xGIwQfik. It has had more than two million hits.

me was the formidable intellect of Steven Weinberg, Nobel Prize-winning physicist and cultured polymath. I hope I concealed my nervousness adequately, both during our filmed conversation and at the very pleasant dinner party which he hosted for me at his club in Austin, a city I have heard described as an intellectual oasis in Texas. There are some Nobel Prize-winners who–you can't help feeling–must have got lucky, to pair an American idiom with a little British understatement. You don't get that feeling when you meet Professor Weinberg–and I hope the British understatement is still coming through loud and clear. Good casting for a world-class genius.

The chairman-free format might seem unlikely to work when more than two discussants are involved, but we managed to make it work well with four of us in the meeting of the so-called 'Four Horsemen', filmed in 2008 by my Foundation,[1] in the book-lined Washington apartment of Christopher Hitchens. Dan Dennett and Sam Harris joined Christopher and me, and we got on just fine without a chairman. We had invited Ayaan Hirsi Ali to make a fifth; unfortunately she had to dash off on a sudden emergency visit to the Netherlands, where she had been a Member of Parliament. So we were four after all, and the 'Horsemen' title stuck.

---

[1] https://www.youtube.com/watch?v=n7IHU28aR2E.

...ies. The only ties I wear now are the beautiful ones designed and hand-painted by Lalla with animal motifs. Here I am (*clockwise from top left*) with Lalla in my dugong tie; with Rowan Williams, Archbishop of Canterbury (praying mantis tie); with Robert Winston, dedicatee of *Climbing Mount Improbable* (chameleon tie); with Joan Bakewell at Hay-on-Wye (penguin tie); signing books (scarlet ibis tie) and receiving an honorary doctorate from the Open University (zebra tie).

Oxford. *Top, left and right*: Alan Grafen and Bill Hamilton in action in the Great Annual Punt Race. *Lef* John Krebs (in glasses, right on the river bank after the race, with friends from the Animal Behaviour Group. *Below*, Mark Ridley, having smuggled a joke into *Oxford Surveys in Evolutionary Biology*, of which he and I were the founding editors.

OXFORD SURVEYS IN
EVOLUTIONARY
BIOLOGY

VOLUME 2 · 1985

Edited by R. Dawkins and M. Ridley

OXFORD UNIVERSITY PRESS

Intellectual giants and good friends. *Top*: Lalla and I entertaining Francis Crick (second from left) and Richard Gregory (right) to dinner in our Oxford flat. *Right*: Richard Attenborough after presenting me with an honorary doctorate at Sussex University ('But why have you come as Liquorice Allsorts?') *Below, right*: 'poet of the planets' Carolyn Porco with Lalla in our Oxford garden; *far right*: adventurer *extraordinaire* Redmond O'Hanlon among just a few of his books.

Royal Institution Christmas Lectures and (*opposite page*) the Japanese summer reprise. *Right*: Giant child volunteer Douglas Adams; *below right*: eyeballing the cannonball that is just about to not quite break my nose; *bottom*: demonstrating the defensive chemistry of the Bombardier Beetle.

It was during this trip to Japan that Lalla and I first met Sir John Boyd (*far left*), then Ambassador to Japan and still our good friend today. Lalla joined me on stage for the lectures in Japan (*above*) and once with (*left*) the python we hired for the occasion.

Below: Some of the dogs gathered to illustrate the power of artificial selection.

The deep. *Clockwise from top*: On Ray Dalio's research vessel *Alucia*, about to enter the Triton submersible in search of the giant squid; the first photograph of a living specimen of this extraordinary creature; Edith Widder, whose 'electric jellyfish' lured it to the camera; inside the Triton with the pilot, Mark Taylor, and (right) Tsunemi Kubodera, the first scientist to see a live giant squid.

...ands in the sun. My second
...pedition on *Alucia* took me
...Raja Ampat (*top*), where
...ried kayaking. The book
...ur with *Climbing Mount
...probable* took Lalla and
...e to Heron Island on the
...reat Barrier Reef (*right
...d below*). Here I was taken
...orkelling among the sharks.
...dly I never caught sight of
...e lovely *Dawkinsia rohani
...bove*).

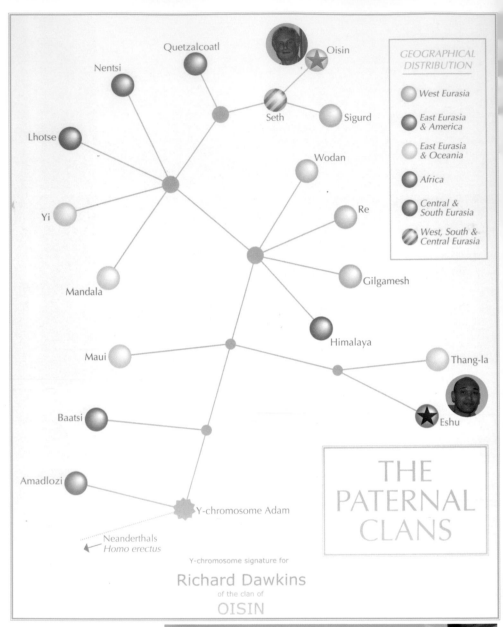

THE
PATERNAL
CLANS

Y-chromosome signature for

Richard Dawkins
of the clan of
OISIN

I was hoping James Dawkins and I were descended from a shared Dawkins ancestor in Jamaica, but DNA disproves this. As you can see from the geneticist Bryan Sykes's pedigree of human Y-chromosomes (*above*), we belong to different Y-chromosome 'clans', Eshu and Oisin respectively.

The hours of discussion around the table passed surprisingly quickly, with no one of us dominating, and I strongly suspect in this case that a chairman would almost certainly have spoiled the atmosphere.

Christopher

I need to say something about that hero of the mind, Christopher Hitchens. I didn't know him well. I was not one of his inner circle of friends from youth, but I got to know him when *God is Not Great* was published, and its natural bracketing with *The God Delusion* led to our sharing public platforms of various kinds. I first met him at a debate in London in March 2007 at Central Hall, Westminster, a large venue holding more than 2,000 people—the same venue as for my encounter with Steven Pinker, already mentioned. Christopher and I were teamed with A. C. Grayling, one of my favourite philosophers, to propose the motion: 'We'd be better off without religion.' It was the promised participation of these two admired colleagues that tempted me to relax my normal objection to formal debates. On the other side were an anthropologist, Nigel Spivey; Roger Scruton, a philosopher; and Rabbi Julia Neuberger, whom I've already mentioned. What I chiefly remember from the debate is a splendid 'How dare you? How DARE you?' from Christopher. But it was a false memory on my part, which I should document because false memory syndrome needs to be more widely known and because I once recounted this false memory in a public speech in Melbourne, eulogizing Christopher after his death.

I was convinced I remembered Christopher's 'How DARE you?' as an intervention during Rabbi Neuberger's speech. It wasn't. The videotape clearly shows that it was in response to a questioner from the audience who claimed (because he tried to be a good person) to be religious although he didn't believe in God. Christopher did intervene during Julia Neuberger's speech, but it was to say something quite different (not clearly audible on the video because his microphone wasn't switched on, but definitely not 'How dare you?'). False memory syndrome is real, interesting, and disturbing. I hope the evidence for it is taught to law students and all those involved in the use of evidence from witnesses, but I fear it is not. Eye-witness evidence is far less reliable than many people think, including jury members. Witnesses in court don't only lie. They are honestly self-deceived. I was first convinced of this when I had the pleasure of meeting Elizabeth Loftus, a brave and personable American psychologist who frequently testifies on behalf of people wrongly accused, for example, of child molestation. In some of the cases she has dealt with, the problem is exacerbated by unscrupulous practitioners deliberately implanting false memories in witnesses—something Elizabeth persuaded me is disquietingly easy to do, especially in children. Nobody went out of their way to implant my false memory about Christopher's intervention. My own brain did it, all by itself, subconsciously conflating two real memories.[1]

I should apologize for that, but at least I'm in good company. In 1982 the Nobel Prize-winning molecular biologist François Jacob wrote an excellent book called *The Possible and the Actual.* I read the English translation around the time it came out and came to a paragraph that seemed oddly familiar. I searched and found the reason. Jacob must have read *The Selfish Gene*, perhaps in the French translation; perhaps he had a photographic memory, or perhaps he copied out a paragraph which he later found and falsely remembered composing it himself. I've reproduced the two paragraphs overleaf.

---

[1] While this book was in press, I happened to meet Professor A. C. Grayling, my fellow debater, at a luncheon. The subject of false memories came up and I recounted this story to him. To the amazement of both of us, he confessed that he had exactly the same false memory. He was incredulous when I told him the true story. But the film evidence is unequivocal. Both of us had concocted the same false memory. I wonder how often that happens? To me it seems to undermine eye-witness testimony even more gravely than I had thought. Imagine that the incident apparently witnessed was a serious crime instead of an intervention in a debate. Would any jury throw out the identically corroborated, independent evidence of two witnesses, both university professors, if a barrister tried to argue that both witnesses suffered from false memory syndrome?

The Selfish Gene by Richard Dawkins	*The Possible and the Actual* by Francois Jacob
1st edition OUP 1976, p. 49	1st edition, Pantheon Books, 1982, p. 20
Another branch. now known as animals, "discovered" how to exploit the chemical labours of the plants, either by eating them or by eating other animals. Both main branches of survival machines evolved more and more ingenious tricks to increase their efficiency in their various ways of life, and new ways of life were continually being opened up. Sub-branches and sub-sub-branches evolved, each one excelling in a particular specialized way of making a living, in the sea, on the ground, in the air, underground, up trees, inside other living bodies. This sub-branching has given rise to the immense diversity of animals and plants which so impresses us today.	Another branch called animals became able to use the biochemical capacity of the plants, either directly by eating them or indirectly by eating other animals that eat plants. Both branches found ever new ways of living under ever diversified environmental conditions. Subbranches appeared and sub-sub-branches, each one becoming able to live in a particular environment, in the sea, on the land, in the air, in the polar regions, in hot springs, inside other organisms, etc. This progressive ramification over billions of years has generated the tremendous diversity and adaptation that baffle us in the living world of today.

I don't for a moment think this is wanton plagiarism. Why ever would a distinguished Nobel laureate need to do that? I think it is a genuine case of a failure of memory–or maybe too *good* a memory for the text itself and a failure to remember its provenance.

Returning to the London debate, it was run by an outfit called Intelligence Squared, whose practice it is to hold a vote both before and after the debate, to see whether the speeches changed anybody's mind. In the case of our debate on the motion 'We'd be better off without religion', the table below shows how the voting went. I'm not quite sure what to make of the fact that the total votes counted at the end of the debate came to 112 more than those registered at the beginning, and the increment is probably even greater because at the end the Don't knows weren't counted. However, it's gratifying that our side both won the debate absolutely and increased our proportion.

	Before	After
Yes	826	1,205
No	681	778
Don't know	364	Not recorded
Total	1,871	1,983
Margin for Yes over No	145	427

At dinner after the debate, I sat opposite Roger Scruton, whom I hadn't met before, and found him quietly charming. We were joined by Martin Amis (among others), and it was lovely to see Martin and Christopher in a witty mock dispute over which one had been the bigger fan of Lalla (who sat between them) in her *Doctor Who* role.

I think I was probably the last person to interview Christopher Hitchens formally. I had been invited to guest-edit *New Statesman*'s Christmas double issue for

2011, and among the pieces I included was an abridged transcript of my own long interview with Christopher. It took place on 7 October 2011 in Houston, Texas, where he was receiving advanced treatment for his cancer. He and his wife had been lent a large and beautiful house while the owners were abroad, and it was there that they entertained me to dinner, together with the charismatic writer and film-maker (who also happens to be Darwin's great-grandson) Matthew Chapman. Christopher was a wonderful host at table, witty, charming and solicitous, although too ill to eat.

Before dinner, Christopher and I sat at a garden table and talked, for the benefit of *New Statesman*. I was so terrified of missing anything he said that I used no fewer than three recording devices. They all worked, and my account of the interview can be read in the e-appendix. Here, I'll just pick out one brief exchange, because it meant a great deal to me and still consoles me when I occasionally feel beleaguered:

RD: One of my main beefs with religion is the way they label children as a 'Catholic child' or a 'Muslim child'. I've become a bit of a bore about it.

CH: You must never be afraid of that charge, any more than stridency.

RD: I will remember that.

CH: If I was strident, it doesn't matter–I was a job-bing hack, I bang my drum. You have a discipline in which you are very distinguished. You've edu-cated a lot of people; nobody denies that, not even your worst enemies. You see your discipline being attacked and defamed and attempts made to drive it out. Stridency is the least you should muster . . . It's the shame of your colleagues that they don't form ranks and say, 'Listen, we're going to defend our colleagues from these appalling and obfuscating el-ements.'

He reiterated the point in his final column for *Free In-quiry*, published posthumously and entitled 'In defense of Richard Dawkins'.[1]

The day after the *New Statesman* interview, Chris-topher and I were both at the Texas Freethought Con-vention where, at the banquet in the evening, I was to present him with the Richard Dawkins Award of the Atheist Alliance of America. This annual award has now been given twelve times,[2] starting in 2003 when it went to James Randi, followed in successive years

[1] http://www.secularhumanism.org/index.php/articles/3136.
[2] Those before 2011 were given by the Atheist Alliance International.

by Ann Druyan, Penn & Teller (joint recipients), Julia Sweeney, Daniel Dennett, Ayaan Hirsi Ali, Bill Maher, Susan Jacoby, Christopher Hitchens, Eugenie Scott, Steven Pinker and, most recently, Rebecca Goldstein. Christopher was too ill to eat the dinner, and he made his entrance at the end, to a standing ovation which made me feel quite tearful. I then made a speech, after which Christopher mounted the platform to another standing ovation and I presented the award. His acceptance speech was a tour de force, made the more powerful by the poignant fact that his splendid voice was fading, along with his life. It was impromptu, but mine was written, and I reproduce here the opening and closing paragraphs in his memory:[1]

> Today I am called upon to honour a man whose name will be joined, in the history of our movement, with those of Bertrand Russell, Robert Ingersoll, Thomas Paine, David Hume.
>
> He is a writer and an orator with a matchless style, commanding a vocabulary and a range of literary and historical allusion far wider than anybody

---

[1] For my speech in full, see the e-appendix. You can hear both our speeches, followed by questions, at: https://www.youtube.com/watch?v=8UmdzqLE6wM.

I know. And I live in Oxford, his alma mater and mine.

He is a reader whose breadth of reading is simultaneously so deep and comprehensive as to deserve the slightly stuffy word 'learned'–except that Christopher is the least stuffy learned person you will ever meet.

He is a debater, who will kick the stuffing out of a hapless victim, yet he does it with a grace that disarms his opponent while simultaneously eviscerating him. He is emphatically not of the (all too common) school that thinks the winner of a debate is he who shouts loudest. His opponents may shout and shriek. Indeed they do. But Hitch doesn't need to shout . . .

Though not a scientist and with no pretensions in that direction, he understands the importance of science in the advancement of our species and the destruction of religion and superstition: 'One must state it plainly. Religion comes from the period of human prehistory where nobody–not even the mighty Democritus who concluded that all matter was made from atoms–had the smallest idea what was going on. It comes from the bawling and fearful infancy of our species, and is a babyish attempt to meet our inescapable demand for knowledge (as well as for comfort, reassurance and other infantile needs). Today the least educated of my children knows much more about the natural order than any of the founders of religion . . .'

He has inspired and energised and encouraged us. He has us cheering him on almost daily. He's even begotten a new word–the hitchslap. We don't just admire his intellect, we admire his pugnacity, his spirit, his refusal to countenance ignoble compromise, his forthrightness, his indomitable spirit, his brutal honesty.

And in the very way he is looking his illness in the eye, he is embodying one part of the case against religion. Leave it to the religious to mewl and whimper at the feet of an imaginary deity in their fear of death; leave it to them to spend their lives in denial of its reality. Hitch is looking it squarely in the eye: not denying it, not giving in to it, but facing up to it squarely and honestly and with a courage that inspires us all.

Before his illness, it was as an erudite author and essayist, a sparkling, devastating speaker that this valiant horseman led the charge against the follies and lies of religion. Since his illness he has added another weapon to his armoury and ours–perhaps the most formidable and powerful weapon of all:

his very character has become an outstanding and unmistakable symbol of the honesty and dignity of atheism, as well as of the worth and dignity of the human being when not debased by the infantile babblings of religion.

Every day he is demonstrating the falsehood of that most squalid of Christian lies: that there are no atheists in foxholes. Hitch is in a foxhole, and he is dealing with it with a courage, an honesty and a dignity that any of us would be, and should be, proud to be able to muster. And in the process, he is showing himself to be even more deserving of our admiration, respect, and love.

I was asked to honour Christopher Hitchens today. I need hardly say that he does me the far greater honour, by accepting this award in my name. Ladies and gentlemen, comrades, I give you Christopher Hitchens.

Simonyi Professor

I enjoyed tutoring in the early days, and I think I was adequately good at it. The Senior Tutor of New College, in a piece of ingenious statistical research during my time, unearthed the fact that New College biology students were significantly more likely to win first-class degrees than biology students in the university as a whole. (The same was true of New College mathematics students, but was not clearly shown for any other subject.) Can my teaching take some of the credit for that? I can't be sure, though few things would give me greater pleasure than to think so.

In those early days I still had the enthusiasm of youth and I really cared about imparting understanding to my students: not just knowledge but understanding. I enjoy explaining things, and the experience of tutoring perhaps honed in me certain skills in the art

of explanation–to students of less as well as more ability–which later helped me in writing my books. But I
cannot deny that, by the time I reached my fifties and
had clocked up more than six hundred hours of one-
on-one tutoring, I had begun to feel a little jaded. I
probably wasn't as good as I should have been, probably
not as good as I had been earlier. I was doing my best,
but I still had some fifteen years to run until I reached
retirement age, and I increasingly wondered whether
New College biology tutoring might benefit from new
blood. At the same time, I had the positive feeling I
might leave the world a better place if I devoted my
remaining working life to explaining things to a wider
public outside the walls of Oxford. How could I achieve
that? I started to think along the following lines.

My books were best-sellers. Whether or not I was
any good as a lecturer to Oxford students, I was in
demand as a lecturer around the world. I'd had some
experience of television and journalism. Various people
made me aware that I had enterprising, entrepreneur-
ial–well, rich–readers, some of whom were enthusiastic
enough to be called fans. Oxford, like all universities,
was by then heavily involved in fundraising, and it had
set up a branch Development Office in New York. It
was suggested to me that Oxford's professional fund-
raisers, perhaps especially the American office, might

do well to go out in search of a benefactor to endow a new Professorship of Public Understanding of Science, with me as the incumbent. With the support of Sir Richard Southwood, Oxford's Vice-Chancellor, whom I knew because he was also the Linacre Professor of Zoology, I attended various planning meetings to discuss the possibility with officials in the Oxford Development Office. They handed the assignment over to their New York branch and I temporarily forgot about it and carried on with my duties.

The initiative was now with Michael Cunningham of the New York office. I told him that, as a fellow guest at the Connecticut farm of my literary agent John Brockman, I had met Nathan Myhrvold (see page 6 for a later meeting with him in Oxford), who had since become Microsoft's Chief Technology Officer. Michael got in touch with Nathan and arranged a meeting between the three of us in New York. Nathan took on board the idea of finding a benefactor for Oxford's proposed Chair of Public Understanding of Science, and went away to discuss it with some of his friends at Microsoft. Among these was Charles Simonyi.

Charles

Charles Simonyi is a Hungarian American software pioneer. A brilliant software designer, he had been one of that charmed circle that came together at Xerox Parc where the modern personal computer with its 'WIMP' interface was conceived. He was recruited to Microsoft as early as 1981, where he advocated the object-oriented programming developed at Xerox Parc, and his own 'Hungarian Notation' for programmers, whose ingenuity intrigues me although I haven't used it myself. He was the supervising architect of the original Microsoft Office software suite. As an early investor in Microsoft, he grew wealthy from the growth of his shares in the company over a long period. Nathan reported to Michael that Charles was tentatively interested in the Oxford idea and wanted to meet me to discuss it.

So it was that in the spring of 1995 Lalla and I flew to Seattle, where we were joined by Michael Cunningham who came in from New York. Charles put us up in a nice waterfront hotel, and we prepared for the ordeal of the evening: a dinner in a Seattle restaurant for about fifty of Charles's guests—I say 'ordeal' only because it was clearly going to be my 'audition' (to quote Lalla's thespian simile) 'for the part'. Charles arranged the dinner's seating plan with great care, including a remove halfway through the evening to an equally carefully crafted second configuration (Oxford colleges occasionally do the same thing, but only at very formal dinners with a dessert course after dinner). I sat on in the same place: all the other guests moved. For the first half I was placed next to Bill Gates. No surprise that he turned out to be highly intelligent and very interesting—but the same seemed to be true of most of the other guests. This became alarmingly evident when Charles called on me to speak, and then to answer questions from anybody in the room. I have fielded questions from university audiences all over the world, including Cambridge, Oxford, Harvard, Yale, Princeton, Berkeley and Stanford, and I declare I've never been grilled so penetratingly as by that mostly quite young audience from Seattle and Silicon Valley: high-tech digerati, entrepreneurs, venture capitalists, computer pioneers and biotechnologists. Somehow I managed to answer all their questions—even those from one guest who seemed critical to the point of heckling—and I ended the evening feeling it had gone reasonably well.

The next day we were to spend with Charles, getting to know each other. Lalla and I met him and his friend Angela Siddall, and Charles drove us to one of Seattle's airfields where we boarded his helicopter, together with a professional pilot who, with Charles under his instruction, flew us north up Puget Sound towards (but not into) Canada. We landed on an island for lunch, and had the rare treat of seeing a Bald Eagle through the restaurant window. On the return journey the dreamlike atmosphere continued as we dodged and danced among the skyscrapers of downtown Seattle. From the airfield Charles drove us back to the hotel, where he had a ten-minute meeting closeted with Michael Cunningham. Charles and Angela then left, and Michael emerged to tell Lalla and me that the deal was done: the Charles Simonyi Professorship of Public Understanding of Science was really going to happen, give or take some details which would be sorted out with Oxford.

One of the details was that, although Charles had endowed a full professorship, I would have to be appointed, initially, at the same rank as I already was, namely reader. This was because Oxford has a strict rule against benefactions buying promotion for named individuals (a scrupulous and sensible safeguard against rich uncles buying preferment–'simony', as Charles himself later punned on his own name). So I was not initially promoted: I stayed at the rank of reader and actually took a slight cut in salary. A year later I was promoted to professor on merit, my qualifications being scrutinized on the same objective basis as anybody else's. So I actually became the Simonyi Professor a year after my original appointment as Simonyi Reader. My successors will all be appointed Simonyi Professor from the outset.

'Successors'? Yes, because Charles graciously agreed to endow the professorship in perpetuity. That is to say, instead of giving only enough money to last until I reached retirement age (which was all that the original proposal had dared to suggest), he would donate a capital sum to be invested by Oxford, the annual income from which would pay not only my salary and expenses but those of a series of successors into the indefinite future. That itself was splendidly generous, but Charles added to his magnanimity an imaginative vision which I dare say is quite rare in major benefactors. He wrote what amounted to a far-sighted manifesto to accompany his gift. The gist was that he was looking to the distant future, and therefore he would avowedly *not* try to specify exactly how the terms of his benefaction should be interpreted in centuries to come. He explicitly eschewed legal red tape, saying, in effect: 'Future centuries will inevitably be different and we can't predict how. I trust you, future generations at Oxford, to interpret the *spirit* of what I am trying to achieve in Public Understanding of Science, in the light of your own times.' In Charles's own words, 'this is where we were in 1995, this was the kernel of agreement between me, between the University, and between Prof. Richard Dawkins, the first occupant of the chair. Deviate from this point if you must, but do it knowingly. Return to it if you can.'

Here, in full, is the text of Dr Simonyi's trusting, and blessedly un-lawyered-up, missive to the future. And if future generations at Oxford betray his trust, may my ghost come back to haunt them. Or, to put the same thought in more practical terms, I hope that printing his manifesto in (what I hope and intend will be) a permanent book will make it very hard for anyone to betray it.

Là, tout n'est qu'ordre et beauté
Luxe, calme, et volupté (Baudelaire)

Since I am a computer scientist, it seems appropriate that the present description of my intentions to create a chair in 'Public Understanding of Science' at Oxford University should be called a 'program'! Just as a computer program sets the processor on an inexorable future course, shouldn't this program guide the chair's Appointing Committee for generations to come? Quite evidently, the metaphor is weak. Administrative affairs being as they are I can only vainly hope that the distinguished committee members will take my comments to heart before deciding on a new appointment. Yet I begrudge by no means the uncertainty and flexibility that is built into the appointment process so that the University can adapt, evolve, and flourish.

This flexibility can be used for experimentation and exploration of new arrangements but over time it can also result in an accumulating creep or drift in direction that may not even be noticed. The purpose of this program is then to be a fixed navigation point on the sea of possibilities. It says: this is where we were in 1995, this was the kernel of agreement between me, between the University, and between Prof. Richard Dawkins, the first occupant of the chair. Deviate from this point if you must, but do it knowingly. Return to it if you can.

The chair is for 'Public Understanding of Science', that is the holder will be expected to make important contributions to the public's understanding of some scientific field rather than study the public's perception of the same. By 'public' we mean the largest possible audience, provided, however, that people who have the power and ability to propagate or oppose the ideas (especially scholars in other sciences and in humanities, engineers, business people, journalists, politicians, professionals, and artists) are not lost in the process. Here it is useful to distinguish between the roles of scholars and popularizers. The university chair is intended for accomplished scholars who have made original contributions to their field, and who are able to grasp the subject, when necessary, at the highest levels of abstraction. A popularizer, on the other hand, focuses mainly on the size of the audience and frequently gets separated from the world of scholarship. Popularizers often write on immediate concerns or even fads. In some cases they seduced less educated audiences by offering a patronizingly oversimplified or exaggerated view of the state of the art or the scientific process itself. This is best seen in hindsight, as we remember the 'giant brains' computer books of yesteryear but I suspect many current science books will in time be recognized as having fallen into this category. While the role of populariser may still be valuable, nevertheless it is not one supported by this chair. The public's expectation of scholars is high, and it is only fitting that we have a high expectation of the public.

'Understanding' in this instance should be taken a little poetically as well as literally. The goal is for the public to appreciate the order and beauty of the abstract and natural worlds which is there, hidden, layer-upon-layer. To share the excitement and awe that scientists feel when confronting the greatest of riddles. To have empathy for the scientists who are humbled by the grandeur of it all. Those in the audience who reach the understanding sufficient to *reveal the order and beauty in science will also gain greater insight into the connectedness of science and their everyday life.*

Finally, 'science' here means not only the natural and mathematical sciences but also the history of science and the philosophy of science as well. However, preference should be given to specialities which express or achieve their results mainly by symbolic manipulation, such as Particle physics, Molecular biology, Cosmology, Genetics, Computer science, Linguistics, Brain research, and, of course, Mathematics. The reason for this is more than a personal predilection. Symbolic expression enables the highest degree of abstraction and thence the utilization of powerful mathematical and data processing tools ensures tremendous progress. At the same time the very means of success tends to isolate the scientists from the lay audience and prevents the communication of the results. Considering the profoundly vital interdependence between the society at large and the scientific world, the dearth of effective information flow is positively dangerous.

In order to accomplish the above goals, the appointees to the chair must have a pedagogical range that goes beyond the traditional university setting. They should be able to communicate effectively with audiences of all kinds and in different media. Above all, they must approach the public with the utmost candor. Naturally, they will interact with political, religious, and other societal forces, but they must not, under any circumstances, let these forces affect the scientific validity of what they say. Conversely, they should be also candid about the limits of scientific knowledge at any given time, and communicate the uncertainties, frustrations, scientifically perplexing phenomena, and even the failures in their area of expertise.

Scientific speculation, when so labeled, and when the concept of speculation and its place in the scientific method has been made clear to the audience, can be very exciting. It is a very effective communication tool, and it is by no means discouraged.

We recognize that persons with these combined qualifications are rare. Therefore, the preferences listed above for particular scientific specialities should be taken secondary to the appointees' pedagogical and communication talents.

The appointees should have the opportunity to continue their scientific work. This is best accomplished if their appointment in the Department closest to their field would be held jointly with the

Department of Continuing Education. While being firmly based in Oxford, the appointees should receive every possible support from the University for travel and for visiting professorships. In accordance with this, their teaching and administrative responsibilities within Oxford should be correspondingly limited, and should be directed primarily towards the education of nonspecialists. They would be expected to write books and magazine articles in any medium for the popular as well as scientific audiences, participate in public lectures, whether through the University or otherwise, and generally participate in the expression of the 'Public Understanding of Science'.

There is always a potential danger that a benefaction can prove counterproductive if the first holder's previous post is not filled when he or she vacates it. I make this gift on the assumption that Richard Dawkins's existing post in the Department of Zoology will be filled, in a similar field, as a matter of routine when he vacates it.

I gratefully acknowledge the contribution from Prof. Dawkins, who provided me with a framework for the present program.

Charles Simonyi

Bellevue, 15 May 1995

Obviously the members of the appointment boards for all future Simonyi professors must read his letter in full, and it should be in front of each one of them around the committee table. But I would call attention especially to a number of points. He makes the distinction between popularizers of science and scientists (with original scientific contributions to their credit) who also popularize. He interprets 'understanding' of science 'a little poetically'. He wrote the letter three years before I published *Unweaving the Rainbow*, and I like to think that when that book finally appeared he found it resonant with his wish. My preface to that book contains a tribute to him as a Renaissance man with an 'imaginative vision of science and how it should be communicated'. I explained how we had talked over these matters since becoming friends and I offered *Unweaving the Rainbow* as my written contribution to the conversation 'and as my inaugural statement as Simonyi Professor'.

In an especially telling passage of his manifesto, Charles urges future Simonyi professors to be candid about the limitations of science, while never allowing religious or political forces to affect the scientific validity of what they say.

Finally, a more short-term but important point, Charles realized that his gift might backfire if I were simply moved sideways and my lecturing position in zoology lost. One of my motivations in seeking the move was specifically so that I might be replaced by new blood bringing fresh enthusiasm to Oxford zoology, even as I carried my own renewed enthusiasm to the world outside. I was indeed replaced, by a succession of excellent younger zoologists: David Goldstein, Eddie Holmes, Oliver Pybus–each of whom soon went on to a prestigious professorship–and now the wonderful Ashleigh Griffin (who, I hope, will be with us a long time before the same happens to her).

Simonyi Lectures

One of the first things I did as Simonyi Professor was to endow, in a much smaller way with my own money from book royalties, an annual Charles Simonyi Lecture in Oxford. In accordance with Charles's manifesto, the lecturers I invited have all been distinguished scholars in their own right, and have all been successful in increasing the public understanding of science. I'm proud to say it is a pretty starry list. Here they are, with the titles of their lectures:

| 1999 | Daniel Dennett | The evolution of culture |
| 2000 | Richard Gregory | Shaking hands with the universe |

2001	Jared Diamond	Why did human history unfold differently on different continents?
2002	Steven Pinker	The blank slate
2003	Martin Rees	The mystery of our complex cosmos
2004	Richard Leakey	Why our origins matter
2005	Carolyn Porco	In orbit! Cassini explores the Saturn system
2006	Harry Kroto	Can the internet save the Enlightenment?
2007	Paul Nurse	The great ideas of biology

Finally, in 2008, the year of my retirement, I gave the tenth Simonyi Lecture myself, my valedictory swansong, with the title 'The purpose of purpose'.

A high point of that same year, incidentally, was the wonderful retirement dinner that the Vice-Chancellor, John Hood, arranged for me in the University Museum, with a guest list every bit as distinguished as that for my seventieth birthday dinner three years later.

Except for the first two, which took place in the Zoology Department, all the Simonyi Lectures were in the comfortable and stylish surroundings of the Oxford Playhouse. The enlightened managers of the Playhouse were keen to promote science as well as drama. I've already mentioned that they put on Michael Frayn's important play *Copenhagen*, about the enigma of Werner Heisenberg's wartime visit to Niels Bohr, and that afterwards they invited Oxford physicists to a question and answer session with Michael Frayn himself. Michael told Lalla and me afterwards that he found this quite a testing experience, but I thought he handled it extremely well, and so did distinguished physicists that I spoke to, for example Sir Roger Penrose and Sir Roger Elliott.

The Heisenberg–Bohr meeting, if I may digress again, is of historic importance because of the enigma of Germany's failure to develop an atomic bomb. If anyone could have led such a project, it would have been Heisenberg. When he miscalculated that it couldn't realistically be done, was it a deliberate mistake? To think so would be a tribute to his memory, but unfortunately the answer is probably no, as I first learned from Roger Elliott's predecessor as Wykeham Professor of Physics, Sir Rudolf Peierls, my senior colleague at New College. Peierls was one of the two British physicists (both Jewish refugees from Hitler) who first correctly calculated that an atomic super-bomb *was* possible, and alerted the Allies to that fact (the 'Frisch–Peierls memorandum'). In his aged widowerhood, Sir Rudolf invited Lalla and me to a large dinner party in his Oxford flat, for which he did the cooking entirely by himself. When all the other guests had left we stayed to help wash the dishes, and he told us the story of Heisenberg's apparently sincere (secretly recorded) astonishment on first learning the news of Hiroshima. Also while washing the dishes, we were fascinated to hear of the ingenuity by which Sir Rudolf had earlier guessed that the Germans were not putting any serious effort into an atom bomb project. Knowing the world of German physics intimately, he carefully examined university lecture lists and noted that Professor This, Professor von That and Doctor the Other were all still giving their lectures in their respective universities, at a time when they would surely have been seconded to an equivalent of the Manhattan Project if such existed. Lovely piece of detective work! And he was a lovely man who, after the war, struggled like Robert Oppenheimer to reduce the dangers of the terrible weapons they had helped to create, becoming a prominent member of the Pugwash movement for world peace. I attended his funeral in 1995, and was sorry he was not around to give a Simonyi Lecture, for he had a great interest in public understanding of science and presented me with a signed copy of his book *The Laws of Nature*, which explains physics to people like me.

Each of the Simonyi Lectures was followed by a dinner for about sixteen people, usually in New College but on two occasions in the timelessly beautiful Wytham Abbey, just outside Oxford, through the kindness of its owners, Michael and Martine Stewart, who also graced the table with their own lively company. Charles himself flew in (piloted his jet into Oxford's tiny airport) for several of the lectures. It was at one of these post-lecture dinners that Charles presented me with one of my most treasured possessions, a first edition *Origin of Species*, one of the original print run of only 1,250. I was tongue-tied with emotion when he stood up and made a gracious speech as he presented it to me.

It is a privilege to have known all nine of 'my' Simonyi Lecturers. I first became aware of Dan Dennett when he and his colleague Douglas Hofstadter invited a chapter of *The Selfish Gene* (the meme chapter) into their thought-provoking anthology *The Mind's I*. The anthology also includes the text of Dan's own 'Where am I?', a tour de force of a lecture in which he pretended that his brain ('Yorick') was secured on a life-support system in a vat, communicating with his body by radio and running in perfect synchrony with an exact copy ('Hubert') downloaded into a computer. It made no difference which of the two 'brains' controlled his body. So confident was he of their interchangeability that, as his climax to the lecture, he switched from one to the other–with dramatically hammed-up results that fully justified the standing ovation I have no doubt the lecture received.

The 'Where am I?' lecture is one of those philosophical works–indeed, Dan is one of those philosophers (along with A. C. Grayling, Jonathan Glover and Rebecca Goldstein)–that enable me (and I think many scientists) to 'get' what philosophers can be good for. His thinking has a high-spirited, teasing quality as well as great depth, and he is one of that new breed of philosophers of science who are knowledgeable *about* science and able to talk, on equal terms, to leading scientists about their own field. He is a warmly sympathetic friend, and the sort of conversationalist who 'raises the game' of whoever he is talking to. When I have a conversation with Dan, I can almost feel my intelligence quotient rising towards (though never reaching) his.

This 'raising the game' is a curious ability, rare but not unknown in others (for example Steven Pinker, to pick a further name from my list of Simonyi Lecturers) and it might repay research by education theorists. The late Bernard Williams (another distinguished philosopher who became a friend, together with his sweet wife Patricia) had a similar effect, but in his case he seemed to make his companion become wittier and more amusing. So does the literary scholar and biographer Hermione Lee, another New College colleague, now

President of Wolfson College, Oxford and still a good, though now less often seen, friend. I don't know where the phrase 'raise one's game' comes from, but it is apt to all these people.

As I shall mention in the 'Memes' section of the next chapter, Dan Dennett is one of those (another is the briskly intelligent psychologist Susan Blackmore, author of *The Meme Machine*) who has taken the idea of memes and run with it. Memes play a significant role in several of Dan's books, including *Darwin's Dangerous Idea*, *Consciousness Explained* and *Breaking the Spell*, among others. He is an evocative phrase-maker, with a bulging quiverful[1] of *Intuition Pumps* (to quote the title of another of his books, itself an intuition pump in its own right): 'cranes' and 'skyhooks' are among my favourites. He is also a devastating deflater of obscurantism and pretentious 'deepities' (his excellent coining, which could be defined ostensively as 'almost anything ever said by Deepak Chopra, Karen Armstrong or Teilhard de Chardin').

Years after Dan's Simonyi Lecture, I was with John Brockman in New York and John confided to a group of us that Dan had suddenly been taken dangerously ill. The prognosis was dire and we, his friends, had

---

[1] Yes, I know it's a mixed metaphor but I can't help liking it.

already prepared ourselves to mourn when the news began to get slightly better. Dan was saved by heroic American medicine and state-of-the-art heart surgery. While recuperating in hospital, he wrote a deeply moving article called 'Thank goodness'. The contrast with the conventional 'Thank God' was calculated. He was thanking the goodness of the team of surgeons and doctors, the nurses, the inventors of the advanced scientific equipment that enabled them to diagnose and treat him, even the launderers of his bloody sheets. With gentle sarcasm he mocked those who had written to say they were praying for him: 'And did you also sacrifice a goat?' Read his paper: it is an exultant heart-cry of gratitude to recipients who really deserve gratitude (and who really exist).[1]

Richard Gregory died in 2010, a great loss to our shared enterprise of raising public appreciation of science. He was a psychologist specializing in visual illusions as a window shining light into the workings of the mind, but he combined his psychology with the skills and intuitions of an inventive engineer and he also had a deep knowledge of the history of science. He pioneered the style of 'hands-on' science museum that became well-known through his own

---

[1] http://edge.org/conversation/thank-goodness

Exploratory at Bristol and the Exploratorium in San Francisco.

His personal manner was one of joyous, hopping-about enthusiasm. When explaining a favourite piece of science he almost seemed to dance on the spot, chortling with joy, like a big schoolboy bursting with excitement as he unwraps a new toy for Christmas. He greatly appreciated his election as a Fellow of the Royal Society, and he made a point of writing me a kindly glowing letter when I myself received the same honour much later: 'It's much more fun being "in" than "out"!'

I first met Richard when he came to give a talk in Oxford while I was a graduate student. As an afterthought to his psychology lecture, in response to a question he described his wickedly ingenious invention of an attachment to an astronomical telescope. The idea (now superseded by the computer equivalent of the same trick) was a cunning technique of taking photographs through previously exposed photographic negatives, in order to average out the random 'noise' of interference in the upper atmosphere.

I next met him when he was again visiting Oxford, and Lalla and I invited him to dinner in our flat, together with Francis Crick and his wife Odile (who took the photo that appears in the picture section). It was a huge privilege for Lalla and me to have these two intellectual giants at our table and hear them spark off each other–a kind of forerunner of what I would later call a mutual tutorial.

I just mentioned Sue Blackmore, and I am reminded of her affectionate obituary of Richard Gregory, which captures the man beautifully. She describes her own first meeting with him in his Bristol lab in 1978, and her memory of

> a whirlwind tour of an early flight simulator made of plaster and bits of wood, a 3D drawing machine with metal arms and joints, and a spinning bowl of mercury which he was hoping to use as some kind of reflecting telescope (imagine that being allowed today).
>
> 'Isn't this fun?' Gregory would gasp as he went from one quirky and interesting question to another . . .
>
> There will never be anyone quite as wacky, inventive, eclectic, brilliant, or engaging as Gregory, but I hope there will be many more scientists who have his playful curiosity, his delight in science–and whose enthusiasm can survive our current culture of targets, measurement, and obsession with usefulness.

The title of his Simonyi Lecture, 'Shaking hands with the universe', was, of course, a reference to his 'hands-on' approach, and his lecture was a feast of vivid demonstrations.

I first met Jared Diamond in 1987 in Los Angeles. I was spending two weeks there, working intensively as a guest in Alan Kay's research offshoot of the Apple Computer company, writing the colour version of my Blind Watchmaker 'biomorphs' program (see pages 554-55). The working atmosphere was ideal. I shared an open-plan office with clever young programmers and could turn to them at any time for advice on the recondite inner workings of the Mac Toolbox. Even better was my living environment as a house guest of the delightful Gwen Roberts–mathematics teacher and puzzler *extraordinaire*–as one of her motley but intriguing population of migrant visitors. A quirkily entertaining companion, if she were an author she would have a most exotic pulverbatch.[1] Every morning I commuted by bus from Gwen's house to the office, and usually had lunch with the geeks–sandwiches, for which they sent out to a neighbouring deli. On one of these days, however, I was invited out to lunch by a UCLA professor, whose name was prominent in my

---

[1] Google it together with 'Douglas Adams'.

field of biology, but whom I had not so far met: Jared Diamond.

We agreed that he would pick me up in his car, outside the Apple office. His books were best-sellers, and so as I stood on the corner I was looking out for something exuding quiet luxury: prosperous if not flash. I didn't give a second glance at the ancient Volkswagen Beetle that chugged and lurched slowly and erratically towards me from far along the dead straight road–until it creaked to a halt, and there was the smiling Dr Diamond. I got in, dodging the curtain of upholstery that had come unglued and was hanging loose from the car's ceiling. I hadn't known quite what kind of restaurant he would take me to. Perhaps the charming Volkswagen should have given me a clue. We parked the Beetle on the UCLA campus and walked to a cool, grassy bank by a stream, gently shaded by trees. Here we sat on the grass, and Jared produced lunch, wrapped in a large cloth: a hunk of cheese and some crusty bread, which he cut with a Swiss army knife. Perfect! So much more conducive to interesting conversation than a noisy restaurant with waiters coyly informing us 'I'm Jason and I'll be your server for today' and reeling off lists of specials, then later interrupting the conversation to ask 'How's everything tasting?' And Jared's bread and cheese, in those bucolic surroundings, really did taste good.

In English pubs, by the way, bread and cheese is called 'Ploughman's Lunch'. Not an ancient name, it was presumably coined by some whizz-kid in Marketing. It was the subject of an amusing anachronism on *The Archers*[1] when an old farmhand complained nostalgically that the Ploughman's Lunch he had been given in the village inn was not a patch on the Plewman's Lunches of his youth in the good old days.

My next encounter with Jared was in 1990 when Jim Watson, Director of the Cold Spring Harbor Laboratory on Long Island, invited Jared and me to organize a conference there to celebrate the centenary of that prestigious institution. The conference was titled 'Evolution: molecules to culture', but what I remember most vividly was the presence of a rather outspoken school of linguists from Russia. Jared had taken the initiative in inviting them, and I must have fondly imagined that linguists and evolutionary biologists would have much in common. Language changes gradually over historical time, in a way that bears a strong, if superficial, resemblance to the way living species change over geological time. Linguists have perfected techniques for reconstructing ancient dead languages, such as proto-Indo-European, by careful compara-

---

[1] For non-British readers: a popular soap opera on BBC Radio, chronicling the lives and feuds of farming folk in a fictitious rural village.

tive analysis of their descendants–techniques that look winningly familiar to evolutionary biologists, especially molecular taxonomists who deal in what can–in these post-Watson–Crick days–be called molecular texts. Moreover, the first stirrings of linguistic capability in our hominin ancestors is a topic of great curiosity to biologists, even though some linguists have treated it as a forbidden (because intractable) one. Notoriously, in 1866 the Linguistic Society of Paris banned discussion of the question, on the grounds that it was forever unanswerable.

Such a prohibition seems absurdly negative to me. However hard it might be to reconstruct, language clearly must have had an origin–or origins. There must have been a transitional period when language evolved from our ancestors' pre-linguistic state. The transition was a real phenomenon, it really happened whether the Paris Society liked it or not, and there is surely no harm in at least speculating about it. Did our ancestors pass through a stage resembling chimpanzee sign language–large vocabulary but no hierarchically nested syntax of the kind that is now unique to humans? Did the capacity for hierarchically embedded grammatical structures arise suddenly in one genius individual? If so, who did she talk to? Could it have arisen as a software tool for internal, unvoiced thoughts, becoming externalized in the form of audible language only later? Do fossils tell us anything about the range of sounds that our various ancestors were capable of uttering? These are all questions to which there has to be a definite answer, even if the answer is beyond our reach in practice, and I'll return to them in the next chapter (see pages 546–52).

Jared and I had an enjoyable back-and-forth correspondence, putting together the invitations to the conference, and it has to be said that most of the expertise came from him. The conference itself, when it finally happened, was the occasion for a certain bafflement on my part. I was impressed by the confidence with which the linguists claimed to reconstruct relatively recent ancestral languages such as proto-Indo-European (c.3500 BC). I could stomach similar reconstructions of other ur-languages such as proto-Uralic and proto-Altaic. By more strained analogy, I supposed that in principle you could feed those proto-languages back into the same reconstructive mill and come up with the ur-language to end (or rather begin) all ur-languages–'proto-Nostratic'–although I gathered that even many of the linguists themselves found that a bit of a stretch.

So far, so intriguing. But I came unstuck when I ventured what seemed like a pretty obvious suggestion, what is vulgarly called a no-brainer. Seeking to make a contribution as an evolutionary biologist, I proposed what I thought was a salient difference between linguistic and genetic evolution. Once a biological species has split in two–perhaps separated by an accident of geography–once the divergence has gone far enough to preclude interbreeding, it is for ever. The two gene pools, previously mixed by sexual reproduction, never merge again even if they meet.[1] Indeed, that's how we define the separation of species. Languages, by contrast, having diverged far apart, often come together again and form gloriously rich hybrids. This would mean that, although biologists can safely trace all the surviving mammals, say, back to one single matriarchal individual who lived and died about 180 million years ago, it would not be true that all Indo-European languages trace back to a single, unique ancestral language spoken by a particular tribe somewhere in eastern Europe about three-and-a-half millennia ago.

The Russian linguists were almost apoplectic in their indignation. Languages *never* merge. But, but, but, I stammered, what about English? Nonsense, they shot back: English is a purely Germanic language. 'What percentage of English words is of Romance origin?' I asked. 'Oh, about 80 per cent,' was their unabashed,

---

[1] There may be extremely rare exceptions–too rare to detain us here.

almost contemptuously paradoxical reply. I sank back into my biologist's shell, crushed but unsatisfied.

I think the conference was a success, and Jared and I both were pleased with it. When he came to Oxford to give his Simonyi Lecture, he was a most gracious guest. Belying–or no, maybe in accordance with–his bread-and-cheese lunch, he showed an appreciation for the finer things in life when he presented Lalla and me with a good vintage Napa Valley Cabernet Sauvignon, and carefully wrote on the bottle that it should be drunk between 2005 and 2017. We shall open it to celebrate publication of this book in 2015. In addition to being a distinguished physiologist, ornithologist and ecologist, he is a widely cultivated man, a polyglot, deeply knowledgeable about anthropology and world history, and we benefited from this in his Simonyi Lecture, which was centred on his book *Guns, Germs, and Steel.* It is a tour de force, and one cannot help wondering why no historian ever got around to writing it before him. Why did it take a scientist to develop its fascinating historical thesis? One could make the same point, perhaps even more forcefully, about *The Better Angels of our Nature*, by the next of my Simonyi Lecturers, Steven Pinker.

Apart from reading Chomsky himself and one or two other books (when I was teaching myself to write a grammar-generating computer program–see *An Appetite for Wonder* and pages 381–2 below), most of what I know about linguistics comes from Steven Pinker. The same is true for much of what I know about modern cognitive psychology. And the history of human violence.

Steve Pinker and I are among the handful of scientists in the world (along with Jim Watson and Craig Venter) who have had our genomes sequenced in their entirety. Steve's genes suggest that he should have high intelligence (no surprise there) but also, amusingly (look at any photo of him on the internet), that he ought to be bald. It's an important lesson to learn: the known effects of genes in many cases only slightly change the statistical probability of a particular outcome. With conspicuous exceptions like Huntington's disease, they don't determine it with high probability, but interact with many other factors including lots of other genes. It's especially important to remember this when dealing with genes 'for' diseases. People are sometimes frightened to look at their genome for fear it will tell them exactly when and how they will die–a kind of death sentence. If that were a realistic fear, identical twins would die simultaneously!

As psychologists go, Steve has a reputation for veering slightly towards the nativist wing, but that really means only that he is not part of the extreme environmentalism that characterized some schools of academic psychology and social science through much of the twentieth century. This comes through in his book *The Blank Slate*, which was also the title of his Simonyi Lecture in 2002. He is a leader of the growing, but still somewhat beleaguered school of evolutionary psychologists: a stance that has made him strangely unpopular among some psychologists and philosophers including, even more strangely, the late Bernard Williams who was, in other respects, extremely reasonable.

As I mentioned in the previous chapter, the Atheist Alliance International honoured me in 2003 by instituting the Richard Dawkins Award, given each year to an individual for raising public consciousness of atheism. Since 2011, after the AAI spun off two daughter societies, the award has been given by the Atheist Alliance of America. The recipient is chosen by a committee in which I play no part, but I usually make an effort to present the prize personally during the annual conference of the Alliance. In those years when I haven't been able to travel I have videotaped a speech. The full text of my speech for Steve Pinker, the 2013 winner, is in the e-appendix; here I will confine myself to reproducing the opening and closing paragraphs:

Newspapers and magazines frequently publish ranked lists of Public Intellectuals throughout the world. Steven Pinker is always near the top of such lists and rightly so. I think he'd probably be top of my world list. And I'm truly delighted that he is being given this award in my name.

Wonderfully readable, he introduces non-specialist readers to his own specialist subject. He's not the only person to do that, although he does it superlatively well. But what is truly remarkable is that he does it for several different subjects, and, unlike a science journalist, he is a genuine world-class expert in all the subjects he writes about. His scholarship is as deep as his writing style is gripping.

I went on to give a brief account of his various books, and then concluded:

After achieving so much, you might have expected him to rest on his very considerable laurels. And, now that I think about it, a laurel wreath would look quite becoming on the famous coiffure. But resting on his laurels was exactly what Steve didn't do. He produced what can only be described as a *magnum opus* and he did it by moving into a completely new field, namely history. *The Better Angels of our Nature* is a magisterial work of history, but it is unmistakably the work of a scientist. And a scientist at the height of his powers.

The Better Angels of our Nature is not just a scholarly *tour de force*. It is a document of hope and optimism. Hope and optimism are much needed today, and that very fact should make us suspicious of anyone who steps up to offer them. But our suspicion is battered into submission by the sheer weight of scholarship. And if 'weight of scholarship' suggests heavy going that is exactly wrong. The book is light and easy reading. Good company, witty and amusing like its author.

I am humbled and honoured that the Atheist Alliance has chosen so illustrious a scholar, and my personal hero, for the award in my name.

Where British science is concerned, Martin Rees pretty much *is* the great and the good: Astronomer Royal, President of the Royal Society, Master of the largest, richest and arguably most distinguished (certainly most distinguished in science) of all the colleges of Cambridge or Oxford, knighted, raised to the peerage. And . . . Templeton Prize-winner: ay, there's the rub, for in that dream of 'spiritual dimension' what corruption of true science may come?

In its early years the Templeton Prize, pegged by its naively benevolent founder to exceed the Nobel in monetary (though of course no other) value, was awarded to frankly religious figures such as Mother Teresa and Billy Graham. A little later, the black spot moved on to scientists of no great distinction but who happened to be openly devout. As an exact reversal, yet more recent prizewinners have included scientists of genuine and enormous distinction, not really religious at all but willing to utter the occasional 'spiritual' deepity and therefore sprinkle on religion some of the gold dust of true science. Freeman Dyson and Martin Rees are the prime examples. What is the next Faustian progression: notorious atheists prepared to stage a Damascene conversion? Dan Dennett, begetter of the excellent 'deepity' coinage itself, might seem to be a prime candidate; or, as he himself said to me, 'Richard, if ever you fall on hard times . . .'

The greater you are as a scientist, the greater the danger that Templeton can exploit you. Martin Rees is a truly great scientist, as well as a good and exceptionally nice man, and I want to apologize if any of my negativity towards Templeton, either here or in the past, has seemed to be aimed at him personally. I have the very highest regard for him, and I can see exactly why Templeton would seek to recruit such a bright star to burnish its shabby image.

Martin Rees is not just a great scientist, he's also a great communicator of his science–no easy task where cosmology is concerned. Cosmologists really do have to grapple with some of the deepest questions any scientist faces, and Martin manages to make it clear without dumbing down, fascinating without selling out to demotic populism. His Simonyi Lecture was a model of how to deal with the deep problems of existence simply but not simplistically. His title, 'The mystery of our complex cosmos', led him to elucidate what he meant by complexity, illustrating it with a lovely image: stars are vast, but 'a star is much simpler than a butterfly'. He was rightly firm about the entitlement of science, as opposed to metaphysics, to ask speculative questions about, for instance, the likelihood of finding bio-friendly planets in the universe, and even bio-friendly universes in a multiverse of billions of universes (an idea he had beautifully explored in *Just Six Numbers*). To quote his lecture, 'This is not metaphysics but science, albeit speculative science.'

I first met Richard Leakey when he wrote me a somewhat unusual letter. He had a charitable interest in a London college of which he was a trustee, and he was trying to persuade a rich man from America to come through with a big donation. The would-be benefactor had read my books and expressed a wish to meet me. Richard wrote to ask whether I would have lunch with the two of them, in an Oxford restaurant.

I said I would, mostly because I wanted to meet Richard Leakey. Both men were, in their different ways, larger-than-life figures. Our host turned out to be a polymathic and prolific talker, of strong and decisive will, who went some way towards living up to his preferred nickname of 'Philosopher King'. When we had ordered our meals, he handed the wine list to Richard and unsuspectingly invited him to choose the wine. Did a wicked smile cross Richard's lips as he scanned the list, exchanged a quiet word with the sommelier and handed it back? If it did, I didn't notice. The meal was convivial and the wine was excellent. As indeed it had every right to be—and thereby hangs this tale. But I knew nothing of it until the waiter gave the Philosopher King the bill. His face went white and his jaw dropped, but he paid without a word. At the time I didn't know what the problem was, but Richard told me afterwards, in high merriment. His whispered instruction to the wine waiter had been to fetch a bottle costing more than £200. Not, you might think, the best way to endear yourself to a man from whom you are hoping to raise a large charitable benefaction. I believe the *mot juste* is chutzpah, and that's Richard all over, as I later came to appreciate. For all I know, he may even have got away with it.

The next time I met him was at another lunch: this

time a celebratory launch of the Science Masters series initiated by John Brockman and Anthony Cheetham (see page 204), in which he and I both had slim volumes: *River Out of Eden* in my case, the excellent *The Origin of Humankind* in his. Lalla happened to sit next to him at the lunch, and they got on so well that he invited her (and incidentally me) to spend Christmas with his family in Kenya, at their house on the Indian Ocean coast. We went, and the encounter reminded us again of his darkly humorous indomitability. Here is what I wrote of him after that Christmas visit in *The Sunday Times* (reprinted in *A Devil's Chaplain*, in the section called 'All Africa and her prodigies'):

Richard Leakey is a robust hero of a man, who actually lives up to the cliché, 'a big man in every sense of the word'. Like other big men, he is loved by many, feared by some, and not over-preoccupied with the judgments of any. He lost both legs in a near-fatal air crash in 1994, at the end of his rampantly successful years crusading against poachers. As director of the Kenya Wildlife Service, he transformed the previously demoralised rangers into a crack fighting army with modern weapons to match those of the poachers and, more importantly, with an *esprit de corps* and a will to hit back at them. In

1989 he persuaded President Moi to light a bonfire of more than 2,000 seized tusks, a uniquely Leakeyan masterstroke of public relations that did much to destroy the ivory trade and save the elephant. But jealousies were aroused by his international prestige, which helped raise funds for his department, money that other officials coveted. Hardest to forgive, he conspicuously proved it possible to run a big department in Kenya efficiently and without corruption. Leakey had to go, and he did. Coincidentally, his plane had unexplained engine failure and now he swings along on two artificial legs (with a spare pair with flippers specially made for swimming). He again races his sailing boat with his wife and daughters for crew, he lost no time in regaining his pilot's licence, and his spirit will not be crushed.

Should that 'coincidentally' have been in quotation marks? I suppose we shall never know, but it seems odd that the near-fatal engine failure happened soon after take-off on the very first flight after his plane had been serviced.

Richard tells a lovely, if slightly macabre, story about his legs. After they were amputated in Cambridge he wanted, for sentimental reasons, to bury them in his beloved Kenya. He had to get permission to transport them, and bureaucracy insisted that this was possible only if he could produce a death certificate. He very reasonably argued that he wasn't dead, and eventually the dundridges saw the justice of this and agreed. They stipulated, however, that he must take them in his hand luggage. Legs may not be checked in. Richard hilariously describes the double-take of the previously bored official watching the X-ray screen as the bag containing the legs went through, and his facial expression as he frantically beckoned his colleagues to come over and have a look.

Richard was a natural choice for Simonyi Lecturer, and he gave a stellar performance. As usual it was spontaneous and without notes. In Hitchensian vein, his fluent performance was the more impressive because he arrived at the theatre directly from yet another large lunch–at the very same restaurant as the Philosopher King's (and accompanied, for all I know, by equally good wine)–with another potential benefactor, this time from Holland.

I first met Carolyn Porco in 1998 in Los Angeles, when we were both invited by the Alfred P. Sloan Foundation to a meeting of scientists and film-makers, to try to persuade Hollywood to frame science in a more sympathetic light. We were reminded that scientists in fiction, from Dr Frankenstein to Dr Strangelove, were typically portrayed as heartless eccentrics, gradgrinds, psychopaths or worse. A 1943 film about Marie Curie represented her as indifferent to her husband's death, whereas in fact, as one delegate said, 'We know from a letter that when they brought her husband's corpse in she hurled herself at it and kissed it and cried.' The film directors at our Hollywood conference included one infuriating contrarian who seemed hell-bent on wrecking the whole meeting and all it stood for. This was unfortunate as he was powerful and influential, a well-known name in the world of television. Jim Watson lost patience and hissed a wonderfully Watsonian put-down: 'Are you for real? You sound like an escapee from the Yale English Department.' But I was equally impressed by the insouciant disdain shown towards him by the bright, articulate, courageous and beautiful astronomer sitting next to him on the same panel: Carolyn Porco. At one point she quietly whispered something to him, which prompted him to bellow to the whole audience, 'Oh, so now she's calling me an asshole.'

There was a lot of talk about trying to initiate a science soap opera, with sympathetically portrayed scientists supplying the human interest. Carolyn would have made an ideal role model for the heroine of such a drama. Indeed, one of two rumours holds that Carolyn was the model for Ellie, the heroine of Carl Sagan's science fiction book *Contact* (the other candidate being Jill Tarter, the admirable Director of SETI, the Search for Extraterrestrial Intelligence). My contribution to the discussion was the somewhat heretical suggestion that science is so interesting in its own right, it doesn't need the sort of human interest that a soap would provide. The *New York Times* report of the meeting quoted me as wondering why *Jurassic Park* had to have any people in it at all, when it had dinosaurs. I have just watched that film again and, even on a small airliner screen, was as enthralled as ever by the dinosaurs. But I had forgotten quite how anti-scientific its 'human interest' message was. The concluding negativity of even the scientist characters was so untrue to life. However terrifying their experience, which included watching a lawyer being swallowed whole by a tyrannosaur, how could any scientist not remain captivated by the very idea of recovering viable dinosaur DNA from the last blood meal of a mosquito embalmed in amber? Presumably the ludicrous shoehorning in of 'chaos theory' was in deference to the pop science flavour of the month at the time the film was being made. Nowadays the equivalent fad, enjoying its fifteen minutes of pop science voguery, would be 'epigenetics' (and no, there are some in-jokes that are best not spelled out).

After our panel discussion, I made a pretty shameless point of manoeuvring to sit next to Carolyn on the bus as we toured Hollywood and were shown around one of the big studios. In this legendary town of starry people, I was star-struck by a charismatic scientist—which I suppose sums up the purpose of the conference. Now that I think about it, Barbara Kingsolver's *Flight Behavior* is a novel that exactly fits that same purpose: a beautiful story of scientists as sympathetic human beings, and how they work and think. Hollywood please take note. It would make a lovely film.

Carolyn came to visit us in Oxford (see picture section), and has been friends with Lalla and me ever since. She is a planetary scientist, in charge of NASA's Cassini imaging team—the team that has brought us those stunning pictures sent back from Saturn and its many moons. But she is more than just a good scientist; she is inspired by the poetry of science, especially the romance of the spheres that share our sun. She is the nearest approach I know to a female Carl Sagan, a poet of the planets and singer of the stars. Whether or not the heroine of the book *Contact* was actually modelled on her, it is a fact that Carl Sagan invited her to be the character consultant on the film version. The scene where Ellie first hears the unmistakable communication from far space still gives me goose bumps when I think of it. The slender, clever young woman, woken up by the mind-shattering signal, bouncing back to base in her open car, exultantly yelling the celestial co-ordinates into the intercom for her dozing assistants: numbers, numbers, the spine-tingling poetry of those numbers and their arc-second precision. And how po-etically right that the hero of the numbers should have been a woman. A role model, just like Carolyn.

An anecdote displays the poetry of Carolyn, and I related it in the Oxford Playhouse when introducing her Simonyi Lecture. A beloved professor from her days at Caltech was the geologist Eugene Shoemaker, co-discoverer, with his wife and David Levy, of the famous Shoemaker–Levy comet. A pioneer of astro-geology, Shoemaker was part of the Apollo space pro-gramme. He was in the running to be the first geologist on the moon, but to his sad regret had to drop out for health reasons, and he turned to training astronauts in-stead of being one. In 1997 Shoemaker was killed in a car crash in Australia. Carolyn, in her grief, raced into action. She knew that NASA was about to launch an unmanned craft, which was programmed to crash-land on the moon after its mission was accomplished. She managed to persuade the mission manager, as well as the head of the planetary exploration programme at NASA, to add her teacher's ashes to the spacecraft's payload. Gene Shoemaker's ambition to be an astronaut was denied him in life, but his ashes now lie on the moon's surface where no wind stirs them (it is said that Neil Armstrong's footprints are almost certainly still intact), and with a photographic engraving bearing these words that Carolyn chose, from *Romeo and Juliet*:

> . . . and, when he shall die
> Take him and cut him out in little stars,
> And he will make the face of heaven so fine
> That all the world will be in love with night,
> And pay no worship to the garish sun.

I have dined out on that story from time to time, but I usually cannot manage to recite the Shakespeare, and turn to Lalla to rescue me. When she speaks the lines from memory in her beautiful voice, I think I am not the only one around the table to choke up.

Carolyn's Simonyi Lecture was, as you would expect, gorgeously illustrated, the beauty of her images matching the poetry of her words. The ovation that the Oxford audience gave her made me proud to have initiated the series, and delighted that this was one that Charles himself was able to attend. I placed Carolyn next to him at the dinner and I believe they have stayed in touch since. It is thanks to Carolyn, by the way, that Asteroid 8331, a main belt asteroid discovered on 27 May 1982 by Shoemaker and Bus, is named Dawkins.

I ended my run of Simonyi Lectures on a high, with two Nobel Prize-winners, Sir Harry Kroto in 2006 and Sir Paul Nurse in 2007. Immensely distinguished as they are–and in spite of the fact that Paul Nurse is now President of the Royal Society–neither of them fits the establishment model of 'great and good'. Harry Kroto, especially, probably wouldn't mind being called a maverick. He won his Nobel with two other chemists for their discovery of the remarkable molecule buckminsterfullerene ('buckyballs'), consisting of sixty carbon atoms C60. It has long been known that you can assemble an elegant sphere-like shape from twenty hexagons and twelve pentagons (it's the 'truncated icosahedron' of classical geometers; soccer balls are often constructed in this way). It is also well known that carbon atoms link up with each other in a 'tinkertoy'-like way to form structures of indefinite size, of which the best known are graphite and diamond crystals. It was therefore a theoretical possibility that sixty carbon atoms might link arms to make a 'soccer ball', a truncated icosahedron. It was almost too good to be true when that possibility was realized in the lab by Harry Kroto and his colleagues. Harry named the molecule 'buckminsterfullerene' after the visionary architect Buckminster Fuller (whom I met in his nineties, by the way, as a fellow speaker at a strange conference in France, where he held the audience spellbound for three hours). 'Bucky' invented the geodesic dome, a stable structure whose resemblance to C60 was spotted by Harry Kroto. Amazingly, buckyballs have turned up in meteorites. Even more amazingly, buckyballs, though gigantic compared with quanta, behave like quanta in the famously counter-intuitive two-slit experiment. (Presumably nobody has been quixotic enough to try it with golf balls?—but that aside is surely straying absurdly too far aside.)

Harry Kroto's Simonyi Lecture was a passionate plea to save the Enlightenment and rescue rational thinking, and he unexpectedly launched a thundering broadside against the Templeton Foundation. This was music to my ears—he went further in his denunciation than I would ever have dared. He illustrated the lecture with examples of his wonderful series of educational resources, small films that science teachers can use. I met him again at the second Starmus conference (see page 143), where he was as stimulating as ever and well deserved his standing ovation (I think he was the only lecturer at the conference to get one).

Incidentally, Harry's Starmus lecture, like his Simonyi Lecture, was a tour de force of PowerPoint virtuosity, using a technique that anyone might do well to emulate. Like most lecturers, I find that my lectures often draw from the same modular groups of slides, but different modules in each lecture. It is wasteful to duplicate the same slides every time you put together a presentation. The sensible strategy, one that would occur to any computer programmer, is to have only one copy of each slide, or modular group of slides, and 'call' it each time you need it in different lectures. Harry is the only person I know who actually does this, and does it properly, so that each lecture is simply a collection of *pointers* to units which are stored elsewhere on his hard disk. Annoyingly, it is not possible to do this in Keynote, which is Apple's otherwise superior rival to PowerPoint. I have repeatedly tried to persuade Apple to implement 'sub-routine jump' hyperlinks instead of absolute jumps. The point about sub-routine jumps is that they remember where they came from and return there. This is essential for the Kroto ploy. I cannot see that sub-routine jumps should be any more difficult to implement than the absolute jumps that are already there (and shouldn't be: absolute jumps are notoriously bad programming practice anyway).

I met Paul Nurse a couple of times when he was still in Oxford, running in Port Meadow, for instance, but I didn't have a long conversation with him until April

2007, when I won the Lewis Thomas Prize given by Rockefeller University in New York. Paul, as President of the University, hosted me when I travelled over to receive it. I was especially pleased to win this prize, as Lewis Thomas was a much admired lyrical stylist among biologists, a prose poet. Paul was a delightfully informal and friendly president, the sort of man you can't help instantly liking and continuing to like. He told me the strange story of his birth, which has now become well known but which he had then only just uncovered. The woman whom he had thought to be his mother was actually his grandmother. And the woman whom he had thought to be his big sister was in fact his mother. Both had died while still maintaining the pretence. Paul seemed more amused than shocked by his recent discovery of his true origins, although he said it did take a bit of getting used to. What strange quirks of fate, I wondered, lead to the uncovering of genius from unexpected beginnings? How many geniuses remain undiscovered for lack of opportunity? How many Ramanujans have died unrecognized? How many talented women in Islamic theocracies are reduced to uneducated serfdom?

Paul Nurse, like Harry Kroto, is far from an 'establishment' figure, and I suggested to him that he would consequently make an ideal President of the Royal Society in succession to Martin Rees. He hinted discreetly that it might be a possibility. I am delighted that it did indeed come to pass, in 2010. Three years earlier, his Simonyi Lecture in 2007, 'The great ideas of biology', was already the kind of magisterial survey you would expect a President of the Royal Society to give, a little reminiscent (but of course up to date, which makes a big difference in modern biology) of Peter Medawar's 1963 presidential address to the British Association.

At the end of *Other Men's Flowers*, a charming and surprising (for a field marshal) anthology of poems, most of which he had had in memory at one time or another, Lord Wavell inserted his own 'little wayside dandelion', his 'Sonnet for the Madonna of the Cherries', whose final rhyming couplet, coming in the wake of the three sensitive quatrains, moves me deeply despite its Christian tenor:

> For all that loveliness, that warmth, that light,
> Blessed Madonna, I go back to fight.

I quote Field Marshal Wavell here only for the becoming modesty of his apology for including his own poem in such company. I felt the same diffidence when I decided that I myself should give the final Simonyi Lecture of my tenure. I was conscious that I had never given an inaugural lecture, as new professors are supposed to do. This was technically because, as explained above, I was initially appointed Simonyi Reader and only later became Simonyi Professor. In practice I had thought of my Dimbleby Lecture (see page 218) as filling the role of inaugural, but that was given on national television rather than in an Oxford hall. So I decided that I would make good the omission by giving a valedictory lecture in the Oxford Playhouse; and that it would be the final Simonyi Lecture of my run, my 'wayside dandelion' outside the garden of the distinguished nine. As part of my presentation, I showed pictures of all the Simonyi Lecturers with the titles of their lectures.

In my own lecture, called 'The purpose of purpose', I made a distinction between two meanings of purpose. I defined 'neo-purpose' as true, deliberate, human purpose, as in creative design: purpose as in goal and ambition. And I defined 'archi-purpose' as its ancient predecessor, the pseudo-purpose mocked up by Darwinian natural selection. My thesis was that neo-purpose is itself a Darwinian adaptation with its own archi-purpose. Like other Darwinian adaptations, it has its limitations—and I illustrated its dark side—but also its huge virtues and breathtaking possibilities.

I hope Charles Simonyi appreciated my instituting this series of lectures in his honour, and I'm glad that my successor, Marcus du Sautoy, is continuing the tradition. I was gratified that Charles made every effort to fly in each year, including 1999 when Dan Dennett was the first lecturer of the series.

On that occasion, at the post-lecture dinner, I proposed the health of both Dan and Charles. The full text of what I said can be found in the e-appendix, but I'd like to close this chapter as I drew to the close of my speech:

It is incredible to think that I am now in my fourth year as the Simonyi Professor. I can't tell you how fortunate I feel in this position, and therefore how grateful to Charles for his generosity. Not just on behalf of myself, but on behalf of the university for it is, I don't need to remind you, an endowment in perpetuity to a university with which Charles had no previous connection. In perpetuity means that we only have to put up with me for another ten years before we get a new Simonyi Professor.

But Charles has also, during that time, become a really good personal friend to Lalla and me. And a good *colleague*, for we talk a great deal about science and the world of the mind, and I find myself continually learning from him, and sharpening my arguments in discussion with him.

I think of Charles as a sort of intellectual James Bond. He lives life to the full, and he won't mind my saying that he lives it in the fast lane. He loves gadgets and fast cars, pilots his own helicopter and his own jet planes, both the supersonic and the ordinary kind. But the *conversation* you are likely to have with him in the helicopter or the speedboat is not at all what you would expect from James Bond. It is much more likely to be about the Nature of Consciousness or the Singular Beginning of Time; the Principle of Free Speech, or the hope for a Grand Unified Theory of Everything.

Charles has now stayed with us in our home four or five times, and it is always a delight to have him. We have visited Seattle slightly less often, mainly because of our relative lack of Lear Jets and Falcons. But we did attend his memorable housewarming party in his unprecedentedly memorable house. Villa Simonyi is one of the most imaginatively planned buildings I have ever seen, the glass walls abutting at unbelievable angles, the ultramodern architecture the perfect backdrop to the Vasarely paintings and the wall to wall computer screen inside.[1]

---

[1] I had occasion to mention this in a poem that I wrote for Charles's pyjama party, which I have unfortunately lost and of which I can recall only this couplet:

There's the finest champagne, and the best from the deli
(The walls are of glass, when they're not Vasarely).

We unfortunately had to miss last year's party for his fiftieth birthday, but it was possible to imagine what it was like, and attend it in spirit, in the form of a little verse that I composed in honour of the occasion. I should explain that this happened to coincide with the publication of my book *Unweaving the Rainbow,* which is about Keats and Newton, science and poets.

Never mind about John Keats,
Or Newton's scientific feats.
Forget your William Butler Yeats,
William Wordsworth, William Gates.
Never mind about unweaving:
Here's a man beyond believing.
Here's a man so smart and swift he
Penetrates Mach 2 at fifty!
And that's not all he'll penetrate . . .
(Even Windows 98
Is not beyond his understanding.)
Happy take-off. Happy landing.
See his supersonic plane go–
Vanishing right *through* the rainbow!

Unweaving the threads from a scientist's loom

My twelve books have punctuated my decades, and their research, composition and revision have dominated my waking thoughts. But, since they are all available to be read, it seems pointless in an autobiography to plod through them one by one, summarizing each and then turning to the next. I've already mentioned the titles, more or less chronologically, in the context of my relationships with agents and publishers. If I now descry various themes that recur throughout my books, I don't mean it to sound grandiose when I hope that these themes taken together might add up to a kind of biologist's world-view, with an aspiration at least to coherence. Chronology will be only loosely in evidence as I trace each theme through the books in which it is serially developed, and try to look back to its original entry into my life.

The Taxicab Theory of Evolution

In *An Appetite for Wonder* I recalled a visit to Oxford by a Japanese television crew. They turned up, bristling with tripods, lights, umbrella reflectors and camera gear, in a London taxi, and the director was keen to conduct the interview in the moving vehicle. This proved difficult, partly because the official interpreter's English was unintelligible to me, so 'interview' perforce became 'impromptu monologue' while the unfortunate interpreter was banished to walk the streets for an hour; and partly because the bemused taxi driver from London didn't know Oxford, so I had to interrupt my discourse at frequent intervals to bark 'turn left' or 'turn right' over my shoulder. When we returned to New College, I was curious to know why it had all had to be done in the taxi, so I asked the director, and received the puzzled reply: 'Hoh! Are you not author of Taxicab Theory of Evolution?' It was my turn to be puzzled, until I later worked out the probable origin of his phrase. In my writings I have made frequent reference to the body as the 'survival machine' or 'vehicle' for the genes that 'ride inside it'. My guess–although I have never checked it–is that a Japanese translator of one of my writings must have rendered 'vehicle', with a little poetic licence, as 'taxicab'. Television being television, that would have been a sufficient reason for conducting the interview in a moving taxi. But never mind about taxicabs in particular: I need to explain the theoretical importance of the 'vehicle'.

One of the most persistent–and annoying–criticisms of *The Selfish Gene* is that it mistakes the level at which natural selection acts. Characteristically, the error was most articulately expressed by Stephen Gould, whose genius for getting things wrong matched the eloquence with which he did so:

Challenges to Darwin's focus on individuals have sparked some lively debates among evolutionists. These challenges have come from above and from below. From above, Scottish biologist V. C. Wynne-Edwards raised orthodox hackles fifteen years ago by arguing that groups, not individuals, are units of selection, at least for the evolution of social behavior. From below, English biologist Richard Dawkins has recently raised my hackles with his claim that genes themselves are units of selection, and individuals merely their temporary receptacles.[1]

Gould was right that Darwin focused on the individual organism as the unit of natural selection, and right

[1] S. J. Gould, 'Caring groups and selfish genes', ch. 8 in *The Panda's Thumb* (New York, Norton, 1980).

about Wynne-Edwards arguing for group selection as an alternative. Right, too, that I regard individuals as temporary receptacles for genes. But wrong wrong wrong to see that as a challenge to Darwin's focus on the individual. That whole 'above/below' rhetoric is as misconceived as it is seductive. The gene, the individual and the group are not rungs on the same ladder. If we must talk ladders, the gene is off to one side, more like a single rung all on its own. The gene and the individual are both units of natural selection, but in two different senses of 'unit': as replicator and as vehicle. Replicators (on this planet they are usually stretches of DNA code, occasionally RNA) are the units that actually survive–potentially for millions of years–or fail to survive. The world becomes full of successful replicators and empty of unsuccessful ones, where 'successful' means literally good at surviving as copies through very many generations, even through geological deep time.

What makes a replicator successful is its talent for influencing the world to promote its own survival (exactly how it does so varies massively from species to species, but it typically involves influencing the development of vehicles to make them good at reproducing). And if it succeeds in surviving it potentially survives again, and again, and again . . . into the indefinite future. So the difference between success and failure really matters.

Really matters for a replicator, that is. The same is not true of vehicles: no matter how successful or unsuccessful an organism may be, it will last only a single generation. Success, for an organism, means passing genes on to the distant future before inevitably dying in the comparatively near future. Not even asexually reproducing animals like aphids or stick insects are replicators, as you can tell if you pull one leg off (there's no need to do anything so sadistic: you know what the result would be). That kind of 'mutation' is not inherited. Remove, or change, a bit of DNA, however, and the change–a true mutation–may survive for a million generations.

The word 'phenotype' denotes the outward and physical levers used by replicators to promote their own survival (whether successfully or not). In practice, phenotypes normally consist of features of individual organisms. And organisms are built by embryological processes influenced by the replicators that ride inside them. Organisms, especially animals (plants less so), are coherent, unified bodies that either survive as a whole or die as a whole. And when an animal dies, all its replicators die with it, except those that have previously been handed on to another organism in the process of reproduction. Do you begin to see how apt the word 'vehicle' is? And 'throwaway survival machine'?

Most animals reproduce sexually, which means that the replicators inside them are continually changing partners, sharing new bodies with new combinations of replicators–serving yet again to emphasize the temporary nature of the individual 'survival machine', the mortal vehicle for immortal genes. This is not a way of thinking that would have occurred to most biologists a few decades ago. Genes would have been seen as tools used by organisms, rather than the other way around, as we now see things.

Do you see, too, how persuasive–yet in very different ways–are the *unitary* qualities of both the gene (replicator) *and* the individual (vehicle)? And do you see that *both* are units of natural selection but in their two different senses? I tried, and signally failed, to explain this to Steve Gould when we had a much publicized debate in Oxford's Sheldonian Theatre in the late 1980s. The event was sponsored by Gould's publishers, W. W. Norton, and chaired by John Durant, then of Oxford's Department of Continuing Education. John hosted dinner for the three of us beforehand, in the Randolph Hotel where Steve was staying. I remember it as a rather frosty occasion, perhaps because Steve was not particularly friendly, perhaps because I was intimidated by the looming thought of Oxford's largest and most hallowed theatre, despite rehearsing and expert prepping with my closest friend at the time, Helena Cronin. My nervousness persisted into the debate itself, although I think I performed well enough, especially in the public conversation that followed our two prepared speeches. An audiotape of the two formal speeches was preserved, and later broadcast by Robyn Williams, star science journalist of the Australian Broadcasting Corporation. Unfortunately, no recording seems to have survived of the cut and thrust after the speeches, when most of the interesting points were made. The loss of this part of the tape is a matter of great regret to me because I believe it would show—well, I would say that, wouldn't I, and Steve, alas, is no longer here to dissent—that I was right and he simply couldn't get it.

Two images add colour to 'this view of life' (the Darwinian phrase borrowed by Steve Gould to head his column in *Natural History*, but here I am re-borrowing it from Darwin for my own view of life). The first is from *The Blind Watchmaker*: a willow tree at the bottom of my garden, pumping out downy seeds to cover the ground in all directions, and down the Oxford Canal as far as my binoculars could reach.

It is raining DNA outside . . . The whole performance, cotton wool, catkins, tree and all is in aid of one thing and one thing only, the spreading of DNA around the countryside . . . Those fluffy specks are, literally, spreading instructions for making themselves. They are there because their ancestors succeeded in doing the same. It is raining instructions out there; it's raining programs; it's raining tree-growing, fluff-spreading algorithms. That is not a metaphor, it is the plain truth. It couldn't be any plainer if it were raining floppy discs.

Floppy discs–that dates it. But the 'plain truth' is timeless and deep, its truth undiminished by Moore's Law chipping away at the superficial imagery. Here, from January 2015, is a heartwarming illustration of what Twitter does well (there's a lot that it does badly). A woman quoted the above passage and added her own delightful reaction:

It's winter outside, but spring inside. Suddenly, I'm lying on the grass under a willow tree.[1]

The second image is from *Climbing Mount Improbable*, ten years later. I made a strong analogy between computer viruses and biological viruses. Both are programs that say 'Duplicate me' and little else

---

[1] My thanks to Natalie Batalha for her permission to reproduce this message.

besides. What, then, of a large animal like an elephant? Elephant DNA instructions

> are also saying 'Duplicate me', but they are saying it in a much more roundabout way. The DNA of an elephant constitutes a gigantic program, analogous to a computer program. Like the virus DNA it is fundamentally a Duplicate Me program but it contains an almost fantastically large digression as an essential part of the efficient execution of its fundamental message. That digression is an elephant. The program says: 'Duplicate me by the roundabout route of building an elephant first.'

It is because an individual organism like an elephant is such a unitary and coherent entity, such a plausible and persuasive vehicle, that the great majority of evolutionary biologists have followed Darwin in treating the organism as the main *agent* of biological adaptation. Ethologists follow Darwin in seeing animal behaviour as a striving, by individual animals, to survive and reproduce. This is correct, but you have to take a sophisticated view of the quantity that this agent is striving to maximize. Population geneticists call it 'fitness', which is (or is proportional to) a kind of weighted sum of children, grandchildren and other descendants.

Parental care, and self-sacrifice in the interests of offspring, are obviously easily accommodated in this formulation, as is 'Darwin's other theory' of sexual selection. But R. A. Fisher, J. B. S. Haldane and above all W. D. Hamilton realized that natural selection could also favour individuals who care for collateral kin who have a statistical probability of sharing the genes that mediate the caring.

One way to approach the argument is a thought experiment which, in *The Selfish Gene*, I dubbed the 'Green Beard'. Most new mutations have more than one effect on bodies (the phenomenon is called pleiotropy). Imagine a gene which gives individuals a conspicuous label like a green beard and simultaneously confers benign feelings towards green beards, and a tendency to help green-bearded individuals to survive and reproduce. The contingency is a remote one, Sir, as Jeeves would say, but it makes the point. Such a gene would spread through the population. The idea caught on (a Google search for 'Green Beard Effect' achieves lots of hits and even some photos), but my purpose was only to pave the way for an explanation of kin selection. Pleiotropic coincidences of bodily peculiarities coupled with altruistic tendencies towards those peculiarities are improbable (although several suggestive examples have since appeared in the scientific literature). What is not at all improbable, however, is a statistical equivalent of the Green Beard Effect–a statistical watering down of it. If you 'know' who your brother is, you don't have to specify a particular gene like a green beard gene. You can calculate the probability that he will share *any* particular gene with you.[1] Such a gene could plausibly be a hypothetical gene for being nice to brothers. Or, more practically, a hypothetical gene for being nice to individuals with whom you shared a childhood nest; or individuals who smell like you–a practical rule of thumb for identifying siblings–could easily be favoured in practice, for the same reason as a green-beard gene would be favoured in theory. Kinship–or in practice nest-sharing, or smelling like me–is a realistic statistical proxy for the unrealistic green beard.

Hamilton in 1964 published a mathematical way to redefine 'fitness' to take account of kinship from the individual organism's point of view. He came up with the concept of *inclusive fitness*. I informally (perhaps

---

[1] It's actually a bit more complicated than that. Genes that are common in the population are perforce shared by most of us anyway (and indeed by most individuals of some other species). The particular sense of 'probability of sharing' that is relevant to the theory of kin selection is something more like 'probability over and above the baseline set by the population as a whole'. The best way to visualize this subtle idea is with a geometric model developed by Alan Grafen: see his chapter in R. Dawkins and M. Ridley, eds, *Oxford Surveys in Evolutionary Biology*, vol. 2 (Oxford, Oxford University Press, 1985), pp. 28–9.

too informally, but Hamilton himself gave his blessing) redefined inclusive fitness as 'that quantity which an individual will appear to be maximizing, when what is really being maximized is gene survival'. The table below summarizes the two ideas of 'replicator' and 'vehicle' and explains how both are units of selection but in different senses.

Unit of selection	Role	Quantity maximized
Gene	Replicator	Survival
Individual organism	Vehicle	Inclusive fitness

In *The Extended Phenotype*, I used the analogy of the Necker Cube (see below) to argue that both these

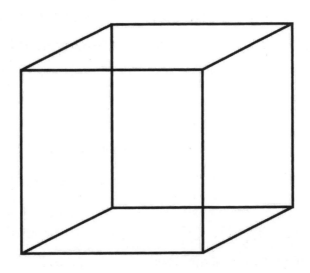

ways of looking at natural selection amount to the same thing, just as both views of the Necker Cube are equally compatible with the information flowing in from the eyes. I shall return to the Necker Cube in a later section of this chapter.

In *The Selfish Gene* I claimed to follow Hamilton, but Hamilton himself switched between two ways of expressing himself, the gene-centred way and the individual-centred, inclusive fitness way. Here's how he put the gene-centred view:

> A gene is being favoured in natural selection if the aggregate of its replicas form an increasing fraction of the total gene pool. We are going to be concerned with genes supposed to affect the social behaviour of their bearers, so let us try to make the argument more vivid by attributing to the genes, temporarily, intelligence, and a certain freedom of choice. Imagine that a gene is considering the problem of increasing the number of its replicas . . .

He wrote those words some eight years after his 'inclusive fitness' paper, but it's entirely clear that the same gene-centred view undergirded the epic 1964 tour de force as well. In *The Selfish Gene* I too switched pretty freely between the gene itself as metaphorical agent or decision-maker, and a kind of informal inclusive fitness approach in which I allowed individual organisms to soliloquize about what would be best for their genes. Neither of these kinds of subjective soliloquy, needless to say, is intended to be taken literally. The 'agents' in both cases should be thought of as behaving *as if* calculating optimal courses of action. But only 'as if'.

Although Hamilton based his idea on the gene's-eye view, inclusive fitness is nevertheless a way of preserving the traditional focus on the individual organism, the vehicle. Indeed, I see it as a regrettably cumbersome bending over backwards to rescue the individual as the focus of our Darwinian attention instead of the gene. But why is the organism such a salient and discrete vehicle at all? Why do we take the organism for granted? Just as we ask why wings and eyes, antlers and penises exist, and expect to get a gene-centred answer, shouldn't the gene's-eye view lead us to ask why organisms themselves exist? Genes survive by pulling phenotypic levers of power. But why are those phenotypic levers bundled into discrete vehicles which we call organisms, and why do the immortal replicators 'choose' to gang up with other genes to share vehicles? This was the point at which I went beyond Hamilton, while never really going against anything he said. I did so in my second book, *The Extended Phenotype*.

After interviewing me for his own book *A Reason for Everything*, subtitled *Natural Selection and the English Imagination*, Marek Kohn put the point well.

> Having based his first book on the assumption that 'if adaptations are to be treated as "for the good of" something, that something is the gene', Dawkins was now mounting an attempt 'to free the selfish gene from the individual organism which has been its conceptual prison'.
>
> One of the jailers of this Bastille was Bill Hamilton, who found himself cast in the role of the revolutionary deemed insufficiently radical by his disciple. Although Dawkins never wavered in his admiration for Hamilton, he felt that the idea of inclusive fitness was an obstacle to seeing biological facts through the gene's eye. Inclusive fitness was about the selection of genes, but it complicated the issue by trying to fit into biology's existing frame of reference. 'Before Hamilton's revolution, our world was peopled by individual organisms working single-mindedly to keep themselves alive and to have children. In those days it was natural to measure success in this undertaking at the level of the individual organism. Hamilton changed all that but unfortunately, instead of following his ideas

through to their logical conclusion and sweeping the organism from its pedestal . . . he exerted his genius in devising a means of rescuing the individual.'

John Maynard Smith, in my 'Web of Stories' interview with him in 1997 (see page 331),[1] said very much the same thing.

Extending the phenotype

If St Peter were to twist my arm at the Pearly Gates and force me to come up with an answer to the question of how, if at all, I might justify having occupied a little space on this Earth and exchanged a fraction of its air, the best I could do would be to point to *The Extended Phenotype*. It isn't really a new hypothesis which might be right or might be wrong, to be tested by experiment or observation. It's more like a new way of looking at what is already familiar: a way of looking at biology that helps things fall into place and make sense. I suppose it's a bit like 'Today is the first day of the rest of your life.' Trite, necessarily true, definitely not the sort of statement for which you would go out and seek supporting evidence: but nevertheless we recognize it as

---

[1] Story no. 40, 'W. D. Hamilton: inclusive fitness'. See http://www.webofstories.com/play/john.maynard.smith/40.

a truth that changes the way we see things. Obvious as it is, it can be worth bothering to say to somebody as a way of influencing the way they do things. That's how I see the extended phenotype. But instead of being summarized in a pithy aphorism it needs explaining. One way to say it is that it amounts to a questioning of the assumed centrality of the 'vehicle'.

The phenotypic influence of a gene had been thought to end at the body wall of the individual containing that gene. Genes influence bodies via the processes of embryonic development. A mutant version of a gene subtly changes some detail of the shape of a swift's wing. As a consequence the bird flies slightly faster for the same expenditure of energy, and this makes it slightly more likely to survive and therefore pass on the very same gene to future generations. Multiply the effect over many swifts and over many generations, and the result will be that the mutant gene comes to dominate the population, at the expense of alternative alleles.[1]

All genes exert their immediate effects in ways that are buried deep in the internal biochemistry of the individual body, and these are usually invisible to all but specialist scientists. But the phenotypic effect which we finally recognize as the adaptation or survival tool is

---

[1] Alternative versions of the gene, sitting at the same locus on the chromosomes of the population.

usually external and visible to the naked eye—as in the case of the swift's wing. There is a cascade of internally buried causes and effects, often beginning with the synthesis of a protein precisely coded by the DNA sequence. We could arbitrarily identify a 'phenotype' at any point along the cascade—the protein itself, its immediate effect as a catalyst of cellular biochemistry, the consequent effect on the behaviour of cells interacting in tissues, many more downstream consequences—before we hit something visible on the outside of the animal: a webbier foot on a duck, perhaps, or a larger wing on a wasp, or a more stilted courtship gesture in an albatross. All these are properly called phenotypic effects of a gene.

What I added in *The Extended Phenotype* was the thought that the sequence of causes and effects doesn't have to stop at the body wall. Take, for instance, the set of tubes built by Jane Brockmann's mud-daubing wasps, *Trypoxylon politum* (see pages 102–3 and picture section). Each tube is like an organ of the body: an external uterus for the nurturing of young. It has been shaped to a useful purpose by natural selection, in just the same kind of way as the wasp's wings or legs or antennae. The genes exerted their influence via building behaviour, before that via a carefully rigged nervous system, before that by cell growth programs in embryology, before that by biochemical influences on cell growth, before that by protein synthesis under the influence of genes in the nuclei of cells. As with legs and wings, genes that influence for the better the shape and size of the (mud) 'organ' have been favoured by natural selection. And, as with legs and wings, genes, in interaction with lots of other genes, influence the shape and size of the mud 'organ' by indirect processes, beginning with an effect on cellular chemistry and running through a cascade of intermediate causes to the final phenotype.

Yes, that's right, *phenotype*. That's the point I am making. This 'phenotype' is made of mud rather than living cells–hence *extended* phenotype–but is no less a true phenotype. The mud in the stream before the wasp fetches it is not phenotype. It becomes phenotype when it is shaped to a biological purpose, in this case the purpose of protecting a growing larva. It is phenotype because the shape and other properties of the tube have evolved, over many generations of increasing perfection. So there must be genes for tube length, genes for tube diameter, genes for thickness of tube wall, genes for distance between partitions down the length of the tube.

How do I know those genes exist? I don't. Not in the sense that anyone has ever done a genetic study of the phenotypic traits I have just listed. But I am confident that, if such a genetic study were undertaken–and it certainly could be–all those phenotypic traits would be found to vary under genetic control. Why so confident? Because the tubes constructed by the mud-dauber have obviously been shaped to their well-designed form by natural selection, and the logic of natural selection implies the involvement of genes. How else than by favouring certain genes over other genes could natural selection have shaped mud tubes towards increasing suitability to their function of protecting larvae? Of course, to repeat, the genes affect the tubes only indirectly, via the building behaviour of wasps. And before that in the causal chain via wasp nervous systems. And before that via cellular processes making wasp nervous systems. But all phenotypic effects are indirect, anyway. The influence of genes on mud tubes is indirect in exactly the same kind of way as the influence of genes on wings, legs and antennae. And the extended phenotype we are looking at may not be the last point on the cascade of causes and effects. Anything caused by it, 'downstream' in the cascade, could be seen as a further extended phenotype, provided only that genes responsible for it have been favoured as a result by natural selection.

The picture reproduced in this book shows variation

in tube colour. Are there, then, genes for tube colour? Maybe. Here I am less confident, but only because it is not obvious that tube colour has been favoured by natural selection. It is possible that some colours are better than others, and possible that genes make wasps fussy about the colour of mud they gather. On the other hand it could well be that the wasps are indifferent to colour of mud, simply gathering it from whatever is available in a local stream, which might happen to be light brown, dark brown or reddish brown. Why doesn't the same 'indifference' argument apply to length of tube or thickness of wall? It might, but it seems unlikely in those cases. It is easy to see that the tube wall could be too thin for purpose (offering inadequate protection for the larva, or even falling apart). And it could be too thick (using too much mud, requiring more time-consuming trips to the stream). It is hard to see how thickness of tube wall could *not* be subject to natural selection. I personally suspect that tube colour is also subject to natural selection (some colours might be easier for predators to see), but it is entirely plausible that the time-saving advantages of fetching mud from the nearest stream regardless of colour (rather than searching far and wide for a stream with better-coloured mud) could be paramount.

These are hypothetical details for illustration only.

The point is that the logic of natural selection (its choice of genes by virtue of their phenotypic effects) compels us to recognize that such functional phenotypes are not limited to the individual body, the 'vehicle'. Animal artefacts provide the clearest and simplest example, and here I benefited from my close friendship as a graduate student at Oxford, where we shared a flat, with Michael Hansell. Mike is now the world's leading authority on animal artefacts and author of several books on the subject, including the beautiful *Built by Animals*, which cleverly uses the subject of artefacts as a platform to talk about many aspects of animal behaviour more generally. *The Extended Phenotype* has a whole chapter on animal artefacts, from caddis larva houses to bird nests to termite mounds to beaver dams. Even the lake created by a beaver dam is properly regarded as (extended) phenotypic expression of beaver genes, probably the largest phenotype in the world.

If *The Extended Phenotype* had restricted its scope to artefacts such as Jane Brockmann's mud-dauber tubes or Mike Hansell's caddis mobile homes, I wouldn't have bothered to say (and the publishers wouldn't have bothered to print on the paperback cover) 'If you never read anything else of mine, please at least read *this*.' But the extension goes further. The chapter on animal artefacts softens the reader up for the more radical ideas of parasite manipulation of hosts and 'action at a distance'. A fluke lives inside its snail shell, as a caddis larva lives inside its stone house. The fluke doesn't 'build' its shell as the caddis builds its house. But if the fluke could find a way to modify the snail shell to its own advantage, and if we can be sure the modification was favoured by natural selection, neo-Darwinian logic forces us to recognize a fluke gene 'for' snail-shell characteristics. The logic of the extended phenotype, if you buy the analogy with the caddis house (and how could you not?), concludes that fluke genomes contain genes 'for' snail phenotypes, at least in the same sense as they contain genes 'for' fluke phenotypes.

The shell of a snail is its protective dwelling, just as the stone house is that of the caddis. We would not be surprised to find that parasitic infection debilitates a snail, for example causing the shell to be thinner than it should be and the snail more vulnerable. But what shall we say if a snail's shell is *thicker* when it has a parasite? Such is indeed the case for snails parasitized by some flukes. Is the snail better protected as a result of some parasitic influence? Is the fluke doing the snail an altruistic good turn? Is the snail actually better off for housing a parasite?

In one sense probably yes, but not a good Darwinian sense. Here's what I think. Everything about an animal is a compromise between conflicting pressures. Just as a shell can be too thin for the snail's own good, so it can also be too thick. How so? It's a question of economics, as so often in evolutionary theory. The wherewithal to make a shell, calcium for example, is expensive. As we saw in the chapter on digger wasp economics, too much investment in one part of the body's economy has to be paid for in the form of too little investment in some other part. A snail that invests too heavily in its shell must skimp on something else, and will be less successful than a rival snail that invests a bit less in the shell (and therefore more somewhere else). We may presume that the average shell thickness of unparasitized snails is an optimum. When a fluke forces its snail to thicken its shell, it is pushing that snail away from the snail optimum and in the direction of a different and more costly optimum, which is the fluke optimum.

Is it plausible that the fluke optimum should be thicker than the snail optimum? Yes, actually very plausible. Any animal has to balance the needs of individual survival against those of reproduction. Peacocks and peahens sit at different optima along the sex–survival continuum. Hens 'care' more about survival; cocks more about reproduction, even at the cost of a shorter life. This is because, not having to lay large, costly eggs, a cock can potentially pack far more reproduction into a short life than a hen can. Most cocks are less successful in terms of passing on genes than the average hen, but a few 'elite' cocks are far more successful than the average hen and are so even if they die young. Cocks tend to inherit their characteristics from ancestors belonging to the elite minority who died young after a bonanza of reproduction. So a tendency to shift the bodily economy away from the individual survival optimum and towards the reproduction optimum is favoured in cocks.

The snail 'cares' about its reproduction: the end towards which its survival is just the means. The fluke doesn't 'care' at all about the reproductive success of the particular snail which is its present domicile. As between snail survival and snail reproduction, the fluke genes reach a different compromise from the snail genes. The snail genes 'want' to save some resources for snail reproduction and so compromise on survival. The fluke genes 'want' the snail to put all its resources into preserving the protective house in which the fluke rides–snail survival, and to hell with snail reproduction. Things would be different if flukes were passed directly from parent snail to offspring snail: in that case the flukes too would 'care' about snail reproduction, not just snail survival. This is one of the most important lessons from *The Ex-*

tended Phenotype. Parasites become gentler to their hosts, more symbiotic, to the extent that their offspring infect the offspring of their particular hosts, rather than getting passed on to random members of the host species.

Parasite genes, then, can have 'extended' effects on host phenotypes. The parasitology literature is filled with fascinating, even macabre, accounts of hosts whose habits are manipulated to facilitate the lifecycles of the parasites that ride inside them, and my chapter 'Host phenotypes of parasite genes' lists lots of examples. It's almost as though the parasite is pulling the host's puppet strings, and the logic of natural selection compels us to carry that image down to the level of the parasite's genes. In 2012, David Hughes and his colleagues published a splendid book on *Host Manipulation by Parasites*,[1] which takes a very 'extended phenotype' view of the facts.

Action at a distance

But parasites don't necessarily ride in (or on) their hosts. Empty space supervenes between cuckoo and host, but the cuckoo is no less a parasite, and the distorted parental behaviour of the foster parent is no less

---

[1] D. P. Hughes, J. Brodeur and F. Thomas, eds, *Host Manipulation by Parasites* (Oxford: Oxford University Press, 2012).

a naturally selected adaptation of the cuckoo nestling. By what black arts of seduction does the monstrous cuckoo nestling cajole the nervous system of the tiny wren? We don't know, but it is surely the product of an evolutionary arms race (see page 483-4). In that arms race, natural selection of cuckoos implies selection of cuckoo genes 'for' manipulating hosts. And that's just another way of saying there are cuckoo genes 'for' host behaviour, cuckoo genes whose phenotypic effects are manifested as complaisant changes in host behaviour. So the extended phenotype extends beyond the body wall, beyond the stone house enclosing the caddis, beyond the snail shell enclosing the fluke, right outside the body and across the space between cuckoo and host–a space in which something is transmitted from the one and picked up by the other. This is the meaning of 'action at a distance', which is the title of the penultimate chapter of *The Extended Phenotype*. And it doesn't apply only to parasites and hosts.

If a physiologist wants to bring a female canary into reproductive condition, increase the size of her functional ovary and cause her to start nest-building and other reproductive behaviour patterns, there are various things he can do. He can inject her with gonadotropins or oestrogens. He can use electric light to increase the day-length that she experiences. Or, most interestingly from our point of view, he can play her a tape recording of male canary song. It apparently has to be canary song; budgerigar song will not do, although budgerigar song has a similar effect on female budgerigars.

That quotation is from an earlier chapter of *The Extended Phenotype*, the one on 'Arms races and manipulation', but it serves equally to illustrate action at a distance. Genes in male canaries have been naturally selected for their extended phenotypic effect–at a distance–on female canaries.

This theme was foreshadowed in a 1978 paper which I wrote jointly with my friend John Krebs (see page 482), called 'Animal signals: information or manipulation?' That paper could be credited with importing the 'selfish gene' revolution into the study of animal signals such as bird song. Hitherto, under the influence of Niko Tinbergen, Mike Cullen, Desmond Morris and other ethologists of the Tinbergen–Lorenz school, animal signals had been treated in a cooperative spirit: both parties to the communication would benefit from a flow of accurate information between them ('I am informing you, for our mutual benefit, that I am a male of your species, possessing a territory, and I am ready to mate'). John Krebs and I turned that on its head by regarding the sender of the signal as *manipulating* the receiver, as if flooding her nervous system with a drug, or as if stimulating her brain electrically with micro-electrodes. I put the point with calculated bathos in *The Extended Phenotype*:

> The snort of a pig-frog *Rana gryllio* may affect another pig-frog as the nightingale affected Keats, or the skylark Shelley.

And much later, in *Unweaving the Rainbow* (whose title was paraphrased from Keats) I said something like it again, after quoting the 'Ode to a Nightingale':

> Keats may not have intended it literally, but the idea of nightingale song working as a drug is not totally far fetched. Consider what it is doing in nature, and what natural selection has shaped it to do. Male nightingales need to influence the behaviour of female nightingales, and of other males. Some ornithologists have thought of song as conveying information: 'I am a male of the species *Luscinia megarhynchos*, in breeding condition, with a territory, hormonally primed to mate and

build a nest.' Yes, the song does contain that information, in the sense that a female who acted on the assumption that it was true could benefit thereby. But another way to look at it has always seemed to me more vivid. The song is not informing the female but *manipulating* her. It is not so much changing what the female knows as directly changing the internal physiological state of her brain. It is acting like a drug.

There is experimental evidence from measuring the hormone levels of female doves and canaries, as well as their behaviour, that the sexual state of females is directly influenced by the vocalisations of males, the effects being integrated over a period of days. The sounds from a male canary flood through the female's ears into her brain where they have an effect that is indistinguishable from one that an experimenter can procure with a hypodermic syringe. The male's 'drug' enters the female through the portals of her ears rather than through a hypodermic, but this difference does not seem particularly telling.

In my more grandiose moments I dreamed of encompassing the whole field of animal communication in extended phenotypic action at a distance. In theory, genetic action at a distance could include almost all interactions between individuals of the same or different species. The living world can be seen as a network of interlocking fields of replicator power.

Unfortunately, it is still

hard for me to imagine the kind of mathematics that the understanding of the details will eventually demand. I have a dim vision of phenotypic characters in an evolutionary space being tugged in different directions by replicators under selection.

And it is still true that

I have not the wings to fly in mathematical spaces. There must be a verbal message . . . Most serious field biologists now subscribe to the theorem, largely due to Hamilton, that animals are expected to behave as if maximising the survival chances of all the genes inside them. I have amended this to a new central theorem of the extended phenotype: An animal's behaviour tends to maximise the survival of the genes 'for' that behaviour, whether or not those genes happen to be in the body of the particular animal performing the behaviour. The two theorems would amount to the same thing if animal phenotypes were always under the unadulterated control of their own genotypes and uninfluenced by the genes of other organisms.

Rediscovering the organism: passengers and stowaways

What, then, of the organism as vehicle? There may be planets out there with life forms whose replicators (I conjecture that there must be replicators at the root, wherever life might be found) have no bounded vehicles; planets where the whole biosphere is a criss-crossing web of extended phenotypic influences radiating out from unbounded replicators. But on our planet it isn't like that. Organisms, discrete units shared by lots of cooperating replicators, dominate. Almost all replicators, instead of being free, ride together inside massive vehicles–'swarm inside great lumbering robots, sealed off from the outside world', as I put it in a controversially much quoted passage of *The Selfish Gene*. Why do our genes swarm together and work together to one end? Whence the organism?

In *The Extended Phenotype* I conjured a thought experiment with two hypothetical seaweeds, renamed in the second edition of *The Selfish Gene* as 'Splurge-

weed' (which simply grows at its margins and then fragments vegetatively) and 'Bottlewrack' (whose genes,
unlike those of Splurgeweed, are funnelled down to a
single-celled propagule, a genetic bottleneck in every
generation). Rather than repeating the argument here
I'll go straight to the more practical conclusion which,
in a sense, flows naturally from the very idea of the extended phenotype. The genes in a discretely 'vehicular'
organism work together to a common end because they
all share the same ('bottleneck') exit route to the future–
the sperms or eggs of the organism that they share.
If some genes have a different exit route, for instance
getting sneezed out of the present organism instead of
ejaculated, they don't cooperate and we use a name like
'virus'. The coherent unity of the organism depends on
the fact that its genes share an exit route and hence share
their expectations, even 'hopes', of the future.

Fluke genes and snail genes favour different optima
for snail-shell thickness. Snail genes are more 'interested' in snail reproduction and fluke genes more 'interested' in snail survival. Fluke genes would 'agree with'
snail genes only if their reproductive propagules made
the journey to the next generation in the sperms or eggs
of their shared snail. If a bacterium had no other way to
reach the future than to enter its host's eggs and hence
the bodies of only its host's offspring, its genes and the

host's genes would be subject to near-identical selection pressures. Both would 'want' not only that the host should survive but also that she should build a nest, attract a mate, ward off egg-stealers, feed the young, even care for grandchildren. Such a parasite would cease to deserve the name. Its genes would evolve to become so intimately wedded to those of the host that its identity would merge into that of the host, with only a Cheshire Cat grin left behind to betray its parasitic origin. Mitochondria (those vital little energy-releasing bodies that swarm inside all our cells) started out as bacterial stowaways, but they became proper passengers because they came to share an exit route–the vehicle's eggs–with all the other genes of the cooperative. So subtle was the Cheshire grin of the mitochondria (the image is borrowed from my sometime Oxford colleague Professor David C. Smith) that we've only just noticed that they originated as bacteria. The reason they cooperate with us rather than fight us is that they share not only the big vehicles that we call bodies (many virulent parasites do that) but also–crucially–the mini-vehicles, eggs, which, in this hypothetical case, transport them from body to body. The surreal-sounding conclusion, which follows from the logic of the extended phenotype, is that all our genes, all our 'own' genes, all our *own* genes, can be thought of as a gigantic colony of viruses: amicable viruses, distinguished from the malevolent ones only by the fact that their expected route to the future is to be not sneezed out, not coughed or breathed or excreted out, but pumped out through the 'legitimate' conduit of sperms or eggs directly into the offspring of the present host.

Our 'own' genes, 'amicable viruses', can be thought of as paid-up passengers in the vehicle, as opposed to 'stowaways' like the chickenpox virus or the various flu viruses. At its deepest level, the difference between the two lies in their exit route from the vehicle. This is perhaps the main message of *The Extended Phenotype*, and it would be my Exhibit A at the pearly court of St Peter. It's nearly obvious when you think it through, but I don't think anybody else ever put it in this way.

Aftermaths to *The Extended Phenotype*

Three aftermaths to *The Extended Phenotype* have given me particular pleasure. The first, in 1999, was the marvellously insightful afterword to a new paperback printing, written by the distinguished philosopher of science (and first Simonyi Lecturer, also in 1999) Daniel Dennett. The second was a special issue of the journal *Biology and Philosophy* devoted to a critically retrospective look at the first twenty years of *The Extended Phenotype*. And the third was a conference near

Copenhagen organized by David Hughes, convened to review the successes and failures of the idea of the extended phenotype.

Dan Dennett's afterword to the 1999 reprinting gave me especial joy because here was a philosopher arguing the case for *The Extended Phenotype* as a work of philosophy. I confess to a certain exasperation when I read people bending over backwards to be complimentary about my science, as a prelude to saying that I should stick to science and not stray into the territory of philosophy. But what is the territory of philosophy other than clear and logical thinking? Don't scientists have to think clearly and logically too? It is of course true that a professional biologist is typically not as well read in the philosophers of the past as he would be if his degree were in philosophy. This might make him neglect an apposite citation of Hume, Locke or Wittgenstein. But that doesn't, of itself, mean he can't present a clear and logical argument of a philosophical character. I hope I won't sound too defensive, therefore, if I quote Dennett on the subject:

Why is a philosopher writing an Afterword for this book? Is *The Extended Phenotype* science or philosophy? It is both; it is science, certainly, but it is also what philosophy should be, and only intermittently is: A scrupulously reasoned argument that opens our eyes to a new perspective, clarifying what had been murky and ill-understood, and giving us a new way of thinking about topics we thought we already understood. As Richard Dawkins says at the outset, 'The extended phenotype may not constitute a testable hypothesis in itself, but it so far changes the way we see animals and plants that it may cause us to think of testable hypotheses that we would otherwise never have dreamed of.' And what is this new way of thinking? It is not just the 'gene's-eye point of view' made famous in Dawkins's 1976 book, *The Selfish Gene*. Building here on that foundation, he shows how our traditional way of thinking about organisms should be replaced by a richer version in which the boundary between organism and environment first dissolves and then gets partially rebuilt on a deeper foundation . . .

For the professional philosopher, I cannot resist adding, there is a feast: some of the most masterful, sustained chains of reasoning I have ever encountered . . .

Forgive the self-indulgence in my quoting that last sentence. I am, perhaps oversensitively, trying to redress the balance after being described as philosophically naive. Dennett develops his theme, illustrating it with page citations from the book. His examples include some of my thought experiments, and this is especially interesting as he himself is a pre-eminent master of the thought experiment as an 'intuition pump'.

Continuing the theme of *The Extended Phenotype* as a work of philosophy, in 2002 the Australian philosopher Kim Sterelny, editor of *Biology and Philosophy*, decided to mark the twentieth anniversary of the book with a special issue of that interdisciplinary journal. In the event, owing to various delays, the commemorative issue didn't finally come out until 2004, but that didn't matter. Sterelny commissioned three scholars, Kevin Laland, J. Scott Turner and Eva Jablonka, each to write a retrospective evaluation and critique of the book, to be followed by a detailed response from me. We all four accepted the invitation, and I must say I enjoyed reading the papers and writing my response more than I expected to.

The title of my reply was 'Extended phenotype– but not too extended'. 'Not too extended' is a phrase I had used before, in response to audience questions about human artefacts. 'If a weaver bird's nest is an extended phenotype, would you say the same of the Sydney Opera House or the Chrysler Building?' No I wouldn't, and the answer is more interesting than the question. A bird's nest or a caddis house or a mud-dauber's set of pipes is a product of natural selection. Natural selection chose genes that fostered good building behaviour. Ancestral weaver birds varied in their building styles and skills; some of that variation was genetic, and was favoured or disfavoured by virtue of the success or failure of the resulting nests in protecting eggs and nestlings containing the genes concerned. In order for man-made buildings to qualify as extended phenotypes, it would be necessary for variation among buildings to be caused by variation in architects' genes. We can't absolutely rule that out but, to put it mildly, it doesn't strike me as a promising line of research. It wouldn't surprise me to find genetic variance in architectural talent. If one identical twin were good at three-dimensional visualization, I would expect that his twin would be too. But I'd be very surprised to find genes for gothic arches, postmodern finials or neoclassical architraves, whereas I would expect to find their equivalents in caddis larvae, mud-daubing wasps or dam-building beavers.

The extension to human architects was not the only 'too extended' that I had in mind in the title of my *Biology and Philosophy* paper. My main problem there was with a voguish (and rather tiresome) notion called 'niche construction'. A big example shows how this loose and vague idea confuses people. The free oxygen in our atmosphere is entirely put there by plants (including photosynthetic bacteria). Early in life's history there was no free oxygen. The green bacteria (and later plants) who put it there massively changed the niches of all subsequent life forms, including themselves. Most creatures today would die instantly without oxygen. That was *niche changing*, an incidental, not 'constructed', by-product of photosynthetic activity. Photosynthesis was naturally selected because of its immediate nutritional benefits for the green bacteria themselves. It was not naturally selected because of its effects on the atmosphere. Those green bacteria made oxygen not because they or their descendants or anyone else would benefit from breathing oxygen in the future. They made oxygen as a by-product, because they couldn't help it when photosynthesizing. After the oxygen was made, subsequent natural selection favoured those bacteria and other creatures capable of flourishing in oxygen. The niche was inadvertently changed, and everybody subsequently evolved to cope with what was at first a pollutant.

Natural selection implies a *discriminating* genetic advantage to the organism concerned, as opposed to general advantage to the world at large. When positive advantage accrues, meaning genetic advantage specifically for the individual doing it as opposed to the world at large, we have an extended phenotype. Otherwise we have no extended phenotype and no niche construction, merely niche changing.

A true extended phenotype, such as a bird's nest or a beaver dam or the subverted parental behaviour of a cuckoo's foster parent, has to be a Darwinian adaptation for the benefit of the genes that mediate it. 'Niche construction' is a phrase that can be meaningful if used with care. Since it is so often used without care and without full Darwinian understanding, it is a phrase that I would prefer to see not used. When it is used properly and with care, it becomes a special case of an extended phenotype, the special case where an animal changes its niche for the benefit of its own genes. A beaver dam is an example. There may not be many others.

The same confusion, between the extended phenotype and niche construction improperly used as a synonym for niche changing, was somewhat in evidence at the third of my 'aftermaths': a conference on extended phenotypes convened in 2008 in a large country house near Copenhagen. The organizer was David Hughes, a talented young Irish biologist now working in America, and he attracted a splendid cast list of distinguished scientists, including both critics and supporters of the extended phenotype. There's a good report on the conference in the journal *Science Daily*, with the title 'European evolutionary biologists rally behind Richard Dawkins' extended phenotype'.[1] The description 'European', by the way, was belied by the presence of American scientists including the distinguished geneticist Marc Feldman (one of the critics).

David Hughes is now the world's leading exponent in practice of the theoretical idea of the extended phenotype. He would be the ideal first director of the hypothetical 'Extended Phenotypics Institute' of the future, the fantasy pipe-dream I described as the climax to my *Biology and Philosophy* paper:

> After the formal unveiling by a Nobel Prizewinning scientist (Royalty wasn't considered good enough) the guests are shown wonderingly around the new building. There are three wings: the Zoological Artefact Museum (ZAM), the laboratory of Parasite Extended Genetics (PEG) and the Centre for Action at a Distance (CAD) . . . In all three wings, familiar phenomena are studied from an unfamiliar perspective: different angles on a Necker cube. [The scientists in all three wings

---

[1] http://www.sciencedaily.com/releases/2009/01/090119081333.htm.

pride themselves] on the disciplined rigour of their theory.[1] The motto carved over the main door of their Institute is a one-locus mutation of St Paul: 'But the greatest of these is clarity.'

It would now be necessary to add a Medical Wing to my fantasy Institute. The American biologist Paul Ewald is one of today's leaders, along with Randolph Nesse[2] and David Haig, of the burgeoning field of Darwinian medicine. I'm grateful to that inspired pioneer Robert Trivers for calling my attention to a fascinating paper by Paul and Holly Ewald on a Darwinian approach to cancer which makes use of the idea of the extended phenotype. It is well understood that the cells within a tumour are subject to natural selection within the tumour. But it's time-limited rather than open-ended natural selection: mutant cells that become 'better' (better at being cancerous, emphatically *not* better for the patient) outcompete less malignant cells within the tumour, becoming more numerous in the tumour. But that evolutionary process terminates with the death of the patient. And there exists a parallel, but

---

[1] In the context, this was a dig at the vague hand waving of 'niche construction theory'.

[2] Nesse's co-author, the great George C. Williams, is unfortunately no longer with us.

more long-term (because transgenerational) selection of genes in the rest of the body, to resist cancers, erect barriers against them, develop immunological tricks against them and so on. It's an asymmetric arms race, because the anti-cancer tricks have been honed against cancers of many past generations. The tricks of the tumours themselves have to be evolved afresh in each generation, for they begin their malign evolution anew in each body, starting as normal, healthy cells which then are naturally selected to evolve, step by step, the qualities needed to outcompete other cancer cells in the race to multiply.

The idea of an arms race between bodies and their cancers prompts interesting thoughts. Cancers are parasites, and particularly insidious ones because their cells are almost (but importantly not quite) identical to their hosts'. This makes them harder for the body (and for medical therapies) to discriminate against than 'foreign' parasites like tapeworms or bacteria. Over many generations, and many tussles against successive cancers, 'skills' for recognizing suspected cancer cells are honed. As in any such arms race, a balance must be struck between being too risk-averse (seeing danger where none exists) and too 'laid back' (failing to see danger where it really exists). This is analogous to the dilemma of a grazing antelope who sees a rustle in the long grass and has to decide whether it is a predator or just the wind. The jumpy antelope who reacts fearfully to every rustle ends each day undernourished because it kept interrupting its grazing to flee. The laid-back antelope who carries on grazing where others would flee is at risk of ending up in a leopard's stomach. Natural selection of antelope genes settles on a judicious balance between the risk-averse Scylla and the laid-back Charybdis. The immune system walks the same tightrope in detecting malignant cells. Too laid-back and the patient dies of cancer. Too 'jumpy', too risk-averse, and the immune system attacks harmless normal cells, wrongly 'suspecting' them of being cancerous. Well, can you think of a better explanation for auto-immune diseases like alopecia, psoriasis or eczema? Allergies too, of course, can be understood as risk-averse, 'trigger-happy' over-reaction of the immune system.

The original twist added by the Ewalds to this analysis is the importation of the idea of the extended phenotype. The tumour lives and evolves in a micro-environment provided by the surrounding cells of the body. The improved malignant tricks that tumour cells evolve in their within-body natural selection largely consist of manipulations of the micro-environment. For example, tumour cells, no less than other cells–indeed, probably more–need a good blood supply to feed and oxygenate them. Just as a beaver's genes work on beaver behaviour to construct the extended phenotype which dams a stream to make a lake, so the mutated and evolving genes in a tumour construct an extended phenotype which is an improved blood supply to the tumour. The cells of the enlarged or diverted blood vessels are not cancerous. They are manipulated by the cancer cells and, since this is a true Darwinian adaptation (for the benefit of the cancer, not the body), the changes in the blood supply constitute a true extended phenotype of mutated genes in the tumour. The Ewalds make full use of 'extended phenotype' terminology in their paper, and I am delighted that they regard the idea as helpful.

Constraints on perfection

In 1979, John Maynard Smith organized a conference at the Royal Society on 'The evolution of adaptation by natural selection'. John Krebs and I were each invited to give talks, and we decided to pool our efforts and write a joint one, on the subject of 'evolutionary arms races'. We already knew we could work well together because of our 1978 paper on 'Animal signals: information or manipulation?' (see page 464). I count John as an intellectual brother, although we see too little of each other nowadays. We have always laughed at the same absurdities without the need for explanation. When unpacking his effects as he moved back into the Oxford Zoology Department after a spell abroad, he came upon a useful item that made him think of me: 'Richard, if you should ever have need of a false beard . . .' Was he being prophetic? The day may yet come. As with my sister Sarah, I can count on John's past to include the same funny books and poems as my own: we effortlessly take each other's allusions. Though a little younger than me, he rightly won his FRS long before I did. Unlike me, he can cope with university politics and civil service administration, and combine them with doing excellent science. He became the knighted head of the British Food Standards Agency and is now a member of the House of Lords and the Head of Jesus, a beautiful old Oxford college.

The opening paragraph of our arms race paper, as delivered at the Royal Society conference in 1979, set the scene:

Foxes and rabbits race against each other in two senses. When an individual fox chases an individual rabbit the race occurs on the time scale of behaviour. It is an individual race, like that between a particular submarine and the ship it is trying to sink. But there is another kind of race, on a different time scale. Submarine designers learn from earlier failures. As technology progresses, later submarines are better equipped to detect and sink ships, and later-designed ships are better equipped to resist. This is an 'arms race', and it occurs over a historical time scale. Similarly, over the evolutionary time scale the fox lineage may evolve improved adaptations for catching rabbits, and the rabbit lineage improved adaptations for escaping.

We arranged our examples according to a four-way distinction between interversus intra-specific arms races (e.g. predator/prey versus male/male rivalry) and between symmetric versus asymmetric arms races (e.g. male/male rivalry versus parent/ offspring conflict). We considered how arms races end, whether in 'victory' for one side or in some kind of equilibrium. Inspired by an Aesop fable, we coined the 'Life/Dinner Principle' as a way for an arms race to end in 'victory': the rabbit runs faster than the fox because the rabbit is running for his life, while the fox is only running for his dinner. There is an asymmetry in the cost of failure on the two sides of the arms race. The asymmetry shows itself in economic terms. Both rabbit and fox would run like a Maserati if they could, but the machinery for running fast is costly. It has to be paid for out of other parts of the body's economy. The life/dinner asymmetry gives the rabbit an added inducement to divert precious resources into running speed.

A similar asymmetry arises in the 'Rare Enemy Effect'. Every one of the cuckoo's ancestors must have succeeded in fooling a foster parent, whereas the foster parent can look back on many ancestors that never encountered a cuckoo in their lives. The cost of failure is higher for the cuckoo than the host, so cuckoos, whose ancestors survived the more stringent side of the arms race, are better equipped to survive future encounters. The arms race idea has proved immensely fruitful and has pervaded many of my books. My friend the Cambridge zoologist N. B. Davies, who could fairly be seen as co-founder with John Krebs of the modern version of behavioural ecology, makes inspired use of the arms race idea in his classic field work on cuckoos.[1]

Perhaps the most overrated paper in my field, if not in all of biology, originated at the same Royal Society conference: S. J. Gould and R. C. Lewontin's 1979 'Critique of the adaptationist programme'. Lewontin and Gould were alpha males in the field, powerful ringleaders of the 1970s campaign (see page 132) against Edward O. Wilson (who, fortunately, was well

---

[1] Nick Davies is the leading contemporary authority on these remarkable birds. See, for example, his 2015 book, *Cuckoo: cheating by nature* (London, Bloomsbury).

able to look after himself). And the bullying tone continued at the 1979 Royal Society conference. Lewontin wasn't there, so Gould gave the lecture and was at his sneering best, playing for the horse laugh from the back row, and mysteriously ignoring the fact that his central thesis had been comprehensively undermined, earlier in the day, by the thoughtful and thorough presentation of Tim Clutton-Brock and Paul Harvey on 'Comparison and adaptation'. Perhaps Gould's failure to deal with Clutton-Brock and Harvey can be excused on the grounds that he had little time to alter his paper. But a brief nod in their direction, and a diminuendo on the sneers, would have been courteous.

The argument was about whether, when we look at some feature of an animal, it is right to assume that it has been shaped by natural selection–is it necessarily an 'adaptation'? Gould and Lewontin's attack on such alleged 'adaptationism' (an earlier coinage of Lewontin) was largely aimed at straw men, or at second-rate biology, a far cry from what we might call 'thoughtful adaptationism'. Clutton-Brock and Harvey undermined the Gould–Lewontin attack by demonstrating sophisticated quantitative techniques for testing adaptation hypotheses with true scientific rigour. These techniques, mostly statistical variants of the comparative method, have proceeded apace in subsequent years, in the hands both of Clutton-Brock and Harvey themselves, and of others including my sometime pupil Mark Ridley and later workers at Oxford fostered by Paul Harvey during his very successful years as our Professor of Zoology.

I'm sure I would be criticized as a rampant 'adaptationist', but my main contribution in print to the debate was actually called 'Constraints on perfection'–the chapter with that title in *The Extended Phenotype*. A thoughtful rather than straw version of adaptationism (not under that name) had been a major influence in Oxford zoology when I was an undergraduate. It was fostered by my own maestro, Niko Tinbergen, and by the school of E. B. Ford, founder of the subject of 'ecological genetics' and a devoted disciple of Sir Ronald Fisher, prodigious innovator in the field of statistics as well as that of population genetics. Ford was such a pernickety aesthete, it is hard to imagine him as a field worker, but he and his many talented colleagues, including Bernard Kettlewell, Arthur Cain and Philip Sheppard, did indeed go out into the woods and fields to measure the pressures of natural selection in nature. Their samplings of butterflies, moths and snails, along with the parallel school of American geneticists under Theodosius Dobzhansky (with whom Lewontin studied), found something quite unexpected. Selection pressures in the wild are hugely stronger than anyone had imagined. Seemingly trivial differences turned out to be massively reflected in differential mortality.

I have already mentioned Marek Kohn's *A Reason for Everything*, a spirited group portrait of the 'British School' of natural selectionists. Kohn rightly says that Ford left behind him 'an intensely selectionist atmosphere that enveloped Oxford zoology and a legend upon which he had worked as meticulously as on his Lepidoptera'. That legend included a cultivated misogyny. Miriam Rothschild (see pages 287–291) was one honoured exception, possibly because she was literally 'the Honourable'–the daughter of a Lord–and Ford was a snob. I only met him face to face once, although I attended all his lectures, and I saw him in the department often, using his outstretched hand to thread his way fastidiously through the plebeian throng at coffee time. He called it 'cocoa', refusing to admit the existence of Nescafé in much the same way as he preferred not to acknowledge dogs, calling them 'pussy'. (Kohn relates how Ford once startled a dog-owning lady by inquiring solicitously after her pussy.) The one time I met him socially, his shrewd, even cunning eyes led me to doubt the sincerity of his eccentric pose. On the other hand, the report, which I think originated with Philip Sheppard, that he had been seen in Wytham Wood at night, checking moth traps with a lantern which he swung to and fro while declaiming 'I am the Light of the World' suggests the opposite–if indeed he really thought he was unobserved.

Ford's *Ecological Genetics*, a beautifully written if rather egocentric treatise, leaves the reader in no doubt of the demonstrated power of natural selection. I imbibed the same spirit as an undergraduate through teachers such as Ford's junior colleague Robert Creed, John Currey, Niko Tinbergen (who, though not a geneticist, did field experiments on the survival value of animal behaviour which were staunchly adaptationist in style) and above all Arthur Cain, the most philosophically and historically sophisticated of the 'Oxford School'.

Arthur's adaptationism was more than staunch. If not quite over the top, it was close to the summit. It was also well thought out. Maynard Smith invited him to introduce the closing discussion of the 1979 Royal Society conference, and his animus towards Gould and Lewontin was palpable. He and I were sitting together in the front row before Gould began his talk, and Arthur was muttering to himself in a frenzy. He was especially incensed by an earlier published jibe by Lewontin against the Ford School as 'a British upper middle class activity', presumably an oblique reference to the genteel hobby of butterfly collecting; and he was rehearsing under his breath the reply that eventually appeared in his formal remarks: 'Presumably when prejudice is strong, facts can be dispensed with as well: my own background and upbringing could only be distinguished by the extreme purist from working class.' As we waited for Gould to begin, Arthur was bouncing up and down with nervous energy and he quoted to me Stanley Holloway's 'Let battle commence' (from the 'Sam: pick oop tha moosket' monologue).

In 1964, Arthur had written a paper called 'The perfection of animals', which included a trenchant attack on the idea of 'trivial', non-functional characteristics of animals. I drew upon it in introducing my 'Constraints on perfection' chapter:

Cain makes a similar point about so-called trivial characters, criticizing Darwin for being too ready, under the at first sight surprising influence of Richard Owen, to concede functionlessness: 'No one will suppose that the stripes on the whelp of a lion, or the spots on the young blackbird, are of any use to these animals . . .' Darwin's remark must sound foolhardy today even to the most extreme critic of adaptationism. Indeed, history seems to be on the side of the adaptationists, in the sense that in particular instances they have confounded the scoffers again and again. Cain's own celebrated work, with Sheppard and their school, on the selection pressures maintaining the banding polymorphism in the snail *Cepaea nemoralis* may have been partly provoked by the fact that 'it had been confidently asserted that it could not matter to a snail whether it had one band on its shell or two' (Cain, p. 48). 'But perhaps the most remarkable functional interpretation of a "trivial" character is given by Manton's work on the diplopod *Polyxenus,* in which she has shown that a character formerly described as an "ornament" (and what could sound more useless?) is almost literally the pivot of the animal's life' (Cain, p. 51).

Astonishingly, however, the most extreme adaptationist quotation I could find was not from Cain but from, of all people, Lewontin himself, writing in 1967 before his contrarian conversion set in: 'That is the one point which I think all evolutionists are agreed upon, that it is virtually impossible to do a better job than an organism is doing in its own environment.'

Starting with my Oxford bias towards adaptationism, my chapter went some way in what might seem to be the other direction, as I pinpointed some major *constraints* on perfection. Cain himself recognized that the animal we are looking at might be simply out of date, and he gave a working estimate of two million years as the upper bound for this. A more permanent constraint on perfection was suggested to me when I was an undergraduate by one of my tutors, John Currey (who also did research with Cain on snail population genetics). A branch of one of the cranial nerves, the recurrent laryngeal runs from the brain to the larynx. It doesn't go straight there, however. Instead, it dives down into the chest, loops around one of the main arteries leaving the heart, and proceeds back up the neck to the larynx. In a giraffe the detour is significant (British understatement) and it is presumably costly. The explanation lies in history, in the nerve's emergence in our fish ancestors before a discernible neck evolved. In those remote days, the most direct route of that nerve (its fishy equivalent) to what was then its target did indeed lie posterior to what was then the equivalent artery (supplying one of the gills). As I put it in *The Extended Phenotype*:

A major mutation might have re-routed the nerve completely, but only at a cost of great upheaval in early embryonic processes. Perhaps a prophetic, God-like designer back in the Devonian could have foreseen the giraffe and designed the original embryonic routing of the nerve differently, but natural selection has no foresight.

Years later, in a 2010 Channel Four television documentary called *Inside Nature's Giants*, I assisted in a revealing dissection of the recurrent laryngeal nerve of a giraffe which had died in a zoo. The scene had a dream-like quality which made it impossible to forget. The operating theatre was literally a theatre, the stage separated from the seated audience of veterinary students by a great wall of glass. The audience was in semi-darkness, and the fierce lights glaring down on the stage picked out the resemblance between the colour of the giraffe's patches and the orange overalls of the dissecting team with their uniform white wellington boots. The giraffe had one hind leg hoisted aloft by a derrick, which added to the surreal quality of the scene. From time to time I was invited by the television producer to advance to the glass wall and address the students with a microphone, explaining the evolutionary significance of the laryngeal nerve and its yards and yards of pointless diversion.[1]

Selection may be powerful, but it is impotent without genetic variation to select from. Pigs might fly, if only the necessary mutations to sprout wings (and change lots of other aerodynamically important things) were forthcoming. It's controversial how great a constraint this is, and it really belongs in the field of embryology.

---

[1] https://www.youtube.com/watch?v=cO1a1Ek-HD0.

I returned to the subject, in what I hope was a constructive manner, in *Climbing Mount Improbable*.

Another apparent constraint is imposed by costs of materials. In *The Extended Phenotype* I quoted from the 'Concorde' paper I had written with Jane Brockmann in 1980:

> An engineer, given carte blanche on his drawing board, could design an 'ideal' wing for a bird, but he would demand to know the constraints under which he must work. Is he constrained to use feathers and bones, or may he design the skeleton in titanium alloy? How much is he allowed to spend on the wings, and how much of the available economic investment must be diverted into, say, egg production?

It was an economic constraint of this kind that Jane and I invoked to explain the apparent Concordian behaviour of her digger wasps (pages 78–82).

The Darwinian engineer in the classroom

I have explained how my tutorials as an undergraduate at Oxford predisposed me towards the adaptationism which later came in for criticism; and how I, together with other Oxford colleagues, later became involved in a defence of a more cautious, more thoughtful adaptationism. When I became a tutor myself, I found that my adaptationist bias had pedagogical advantages. It furnishes a narrative flow to help in remembering factual details of biology.

As a lecturer and tutor, I continually sympathized with students facing the task of remembering huge numbers of facts, and I thought about how to make it easier. Medical students suffer the worst, and unfortunately my favourite teaching trick, which I am calling here 'the Darwinian engineer', probably would make little dent on the formidable array of sheer, unyielding facts that human anatomy presents. This makes me even more proud of my daughter Dr Juliet Dawkins's first-class degree, especially given that St Andrews is one of the few remaining medical schools that still teaches anatomy by hands-on dissection. The problem with anatomy, at least at the level of detail the best medical schools teach, is that so many of its facts are discrete snippets of information that resist being threaded together in a necklace of coherent narrative that might prompt memory. Certainly the broad highways of human anatomy make functional sense and can be taught accordingly, but the minute details of exactly which nerve goes over or under which artery—of literally vital importance for a surgeon—just have to be learned. If they make functional sense (and I expect they do) it is deeply buried, probably in internal intricacies of embryology, and hard to discern.

Zoology students have an easier life than medics, although it wasn't always so. Peter Medawar, in 1965, quoted one of eight examination questions set in 1860 for students of comparative anatomy at University College London:

> By what special structures are bats enabled to fly through the air? and how do the galeopitheci, the pteromys, the petaurus, and petauristae support themselves in that light element? Compare the structure of the wing of the bat with that of the bird, and with that of the extinct pterodactyl: and explain the structures by which the cobra expands its neck, and the saurian dragon flies through the atmosphere. By what structures do serpents spring from the ground, and fishes and cephalopods leap on deck from the waters? and how do flying-fishes support themselves in the air? Explain the origin, the nature, the mode of construction, and the uses of the fibrous parachutes of arachnidans and larvae, and the cocoons which envelop the young; and describe the skeletal elements which support, and the muscles which move the meoptera and the metap-

tera of insects. Describe the structure, the attachments, and the principal varieties of form of the legs of insects; and compare them with the hollow articulated limbs of nereides, and the tubular feet of lumbrici. How are the muscles disposed which move the solid setae of stylaria, the cutaneous investment of ascaris, the tubular peduncle of pentalasmis, the wheels of rotifera, the feet of asterias, the mantle of medusae, and the tubular tentacles of acinae? How do entozoa effect the migrations necessary to their development and metamorphoses? how do the fishes polypifera and porifera distribute their progeny over the ocean? and lastly, how do the microscopic indestructible protozoa spread from lake to lake over the globe? [1]

Medawar quoted this preposterous exam question in evidence against the widespread view that science, as it advances, becomes less and less easy to master because there is more and more to learn. His characteristically provocative reply was that we actually have to learn less than our Victorian predecessors, because multitudinous raw facts have become subsumed under

---

[1] Peter Medawar, 'Two conceptions of science' (1965), reprinted in *Pluto's Republic*.

relatively few general principles, the greatest of which was bequeathed by Darwin.

Medawar had a point; but, not for the first time, this laughing cavalier of the mind was exaggerating. He would have to admit that most papers in *Nature* and *Science* today can be read only by specialists in their respective fields. Nevertheless, weaving facts together into a functional story is a powerful aid to memory, and it's one that I started using early on in my lecturing career at Oxford and Berkeley, and especially as a tutor at Oxford. This is what I meant when I said that adaptationism can have pedagogical advantages. The particular way I use it as a teacher is to take a problem faced by an animal and pose it as an engineer might. Then I list various solutions that might occur to the engineer, naming the pros and cons of each. Finally I come on to the solution actually adopted by natural selection. This provides a narrative flow which grips and becomes an effortless guide to memory.

I put the technique through its paces in the second chapters of both *The Blind Watchmaker* (the example of bat sonar) and *Climbing Mount Improbable* (the example of spider webs) and I'll reprise the examples here to illustrate it. First, the bats. The problem facing a bat is how to find its way around at night. Given that birds have got daytime aerial hunting sewn up, bats were driven to hunt by night. And that posed a problem. It's dark. An engineer might think of various solutions, each with its own knock-on problems: emit your own light like some deep-sea fish; feel your way with long antennae like whip-scorpions; perfect an extreme sense of hearing like owls, so that the tiniest rustle betrays prey, an extreme sense of smell like moles or extreme sense of touch like star-nosed moles; or, finally, sonar: emit loud sounds and exploit the echoes. Of these engineering solutions, the one actually adopted by bats is sonar. In various ways bats time the echoes of their own ultrasonic shrieks, and calculate the position, and rate of change of relative position, of obstacles and prey.

But this raises further problems. Precise timing of the interval between a sound and its echo is improved if the sound is short. But the shorter and more staccato the sound, the harder it is to make it really loud; and it needs to be very loud because echoes are so faint. Could an engineer find a way to get the best of both worlds? One way is to make the sound not staccato. Have a longer cry but modulate its pitch: swoop down (or up) through an octave or so during the course of each shriek. Now the shriek is not short and it can therefore be loud. What is short is the time it spends at each pitch. When the echo bounces back, the brain 'knows' that high-pitched echoes are from early parts of the shriek, low-pitched echoes from late parts of the shriek. The Warden of my Oxford College at the time I was writing *The Blind Watchmaker*, the physicist Arthur Cooke, had worked on the top secret British radar project (then called RDF) in the Second World War, and he told me one evening at dinner that the same technique was used by radar engineers, under the name 'chirp radar'. Another engineering solution is to exploit the Doppler shift (the reason the ambulance siren drops in pitch as it rushes past you). Some bats make good use of this when tracking a moving target such as an insect prey.

Move on to the next engineering problem. Echoes, to repeat the point, are necessarily much fainter than the originating sounds: in danger of being too faint to hear. Possible engineering solutions: make the cries exceedingly loud, and/or make the ears exceedingly sensitive. But those two solutions tread on each other's toes. Extremely sensitive ears are in danger of being deafened by extremely loud cries. Early Second World War radar faced an analogous problem, and again Arthur Cooke told me at dinner that the engineers solved it by designing what they called 'send–receive' radar. And–would you believe?–the exactly equivalent solution is adopted by some bats. Temporarily switch your ears off just before you shriek, by tugging with a special-purpose muscle on the bones that transmit sound from the eardrum. Relax the muscle immediately after shrieking, so your ear is restored to peak sensitivity in time to hear the echo. This cycle: tug, shriek, relax muscle, listen for echoes, tug again . . . has to be repeated for every shriek; and amazingly the repeat rate can climb as high as 50 per second, faster than a machine gun, as the bat comes in for the kill, the final approach to an insect prey.

The pedagogic advantage of the 'Darwinian engineer' approach is that the facts become strung together in a memorable narrative rather than having to be learned piecemeal. Indeed, there's a good chance that students will *anticipate* facts even before they are told them, which is good training for how to dream up fruitful hypotheses that might be worth testing in research.

For example, bats often fly about in the company of hundreds of other bats. How might they solve the problem of their echoes being inadvertently jammed by all the other bat cries and echoes? Here's an idea that might occur to a student, thinking like a 'Darwinian engineer'. Imagine taking a movie film, cutting it up into its separate frames, shuffling the separate frames in a hat and splicing them together again at random. Now the story would no longer make sense: indeed, there would be no 'story', no consecutive narrative. In the same way, to an individual bat, all the other bats' echoes would sound like the equivalent of my movie film made of random frames: easily ignored because of their random unpredictability in relation to any 'story so far'. Only the individual bat's own echoes would form a coherent narrative, making sense when 'spliced' to their predecessors in the sequence of echoes. Experimental psychologists use the same kind of argument to solve the 'cocktail party problem': how do we manage to understand one conversation at a cocktail party when our ears are assailed by dozens of other conversations all around us?

I used the same 'Darwinian engineer' technique in chapter 2 of *Climbing Mount Improbable*, but this time using the example of spider webs instead of bat sonar. Once again, start with a problem: how might a spider extend the effective length of its prey-catching limbs? Again offer up various hypothetical solutions, culminating in the elegantly economical solution natural selection actually adopted: the silken web. And repeat the process for sub-problems and sub-sub-problems that subsequently open up. In a later chapter of the same book, called 'The fortyfold path to enlightenment', I followed the same formula when talking about the design of eyes. Here, I carried the 'design engineer' approach to what some might consider an absurd extreme, but I hope an instructive one. A lens is a simple device, but the computational problem it solves is actually surprisingly sophisticated. I chose to dramatize this by imagining a computer picking up rays of light and bending them through a precisely calculated angle in order to focus an image on a screen. That would be ridiculously complicated, yet the task is easily solved by a lens, a device so simple that–as I demonstrated in my Royal Institution Christmas Lectures–it can be approximated by a sagging, transparent plastic bag full of water, whose poorly focused image can be improved step by step up a smooth gradient on 'Mount Improbable'. It's a metaphor for how easy it can be in practice to evolve something apparently complicated in theory. Long touted since Darwin's own time as his nemesis, the eye is easily evolved and indeed has evolved several dozen times independently, all around the byways of the animal kingdom.

The value of the 'Darwinian engineer' approach for explaining things hit me much earlier, when I was inspired, separately, by two Cambridge eye physiologists, W. A. H. Rushton and H. B. Barlow. I had met Rushton when I was still a schoolboy, for he had two sons at Oundle, one of whom was my exact contemporary. We played the clarinet together in the school orchestra and Google tells me (no surprise) that he later made his career as an academic musicologist. Presumably because the eminent Professor Rushton had two sons at the school, he agreed to give a talk to the sixth form biology group.

Rushton made an interesting distinction between analogue and digital signalling systems. In an analogue telephone the continuously varying pressure wave of the spoken sound is transduced into a parallel voltage wave in a wire, which is transduced back again into sound in the earpiece at the other end. The problem is that, if the wire is long, the electrical signal attenuates and has to be boosted by an amplifier. Boosting inevitably introduces random noise. That doesn't matter if there are only a few boosting stations along the line. But given a sufficiently large number of boosting stations, the accumulated noise overwhelms the signal and the conversation becomes an unintelligible hiss. This is why nerves, at least long ones, cannot work like (analogue) telephone wires.

Nerves aren't wires carrying electric currents, they are even less hi-fi, more like fizzing trails of gunpowder acting as a fuse, with the added complication of the 'nodes of Ranvier', which can be seen as discrete boosting stations. The upshot is that a nerve has the noisy equivalent of hundreds of boosting stations strung out along its length. How might an engineer solve the noise problem? By abandoning all hope of conveying information via the height (voltage) of the wave. Instead, turn the wave into a spike, whose height is fixed or anyway irrelevant. Convey information not by the height of the spike but by the chattering pattern of a sequence of variable spikes. For example, signal a loud sound with a rapid burst of spikes in quick succession; a quiet sound by few spikes, well separated in time.

So, that's an interesting biological solution to an engineering problem. But, as with the bats and spiders, one solution leads on to the next problem, which needs another engineering solution. And this brings me to the second of my Cambridge influences, Horace Barlow. My first wife Marian and I met Horace (named after his grandfather Sir Horace Darwin, Charles's son) when we were at Berkeley, California, and attended his lectures as a visiting professor on sensory physiology. These lectures were notable for the fact that Horace usually turned up at least half an hour late. It was worth the wait. An immensely clever man, he was also a fount of idiosyncratic amusement. You could tell when a joke was on its way some seconds before it arrived, just by watching his face. The Barlow paper that inspired us dated from about ten years before we heard his lectures (it was the reason we went out of our way to attend them), and it completely changed my approach to teaching about sensory systems. Indeed, we both became obsessed with the Barlow paper, and for a period it dominated many of our science conversations with each other. The very name 'Horace Barlow' became a kind of shorthand between us to refer to a whole thread of thought that we were sharing at the time. My lectures on behavioural physiology to Berkeley students at the time became dominated by the 'Darwinian engineer' approach.

You remember I said just now that nerves signal loud sounds not by the height of the spike but by the frequency or timing of the spikes (and the same goes for high temperature, bright light etc.). That's true, but it raises a further engineering problem. If the frequency of nerve spikes is simply proportional to the intensity of the signal, the necessary information is indeed conveyed, but the process is wasteful–and wasteful in a profoundly interesting way. The wastage is remediable–by removing 'redundancy'. What is redundancy?

The state of the world at any one moment is pretty much the same as it was in the previous moment: the world doesn't change at random, capriciously. Like journalists reporting news, nerves reporting on the state of the world need to send a signal only when there is a *change.* Don't say: 'It's loud it's loud it's loud it's loud it's loud it's loud . . .' Instead, say: 'A loud sound has started. Assume no change until further notice.' This is where 'redundancy' comes in as a technical term in information theory. Once you know the current state of the world, further reports of the same state are *redundant.* Redundancy is the inverse of information. Information is a mathematically precise measure of 'surprise'. In the time domain, information means *changes* in the state of the world from one moment to the next, because only changes have surprise value. Redundancy in this context means 'sameness'. The receiver of multiple messages doesn't have to monitor all channels all the time: only those that signal a *change.* This could only fail to be helpful if the world changed randomly and capriciously all the time. Which, fortunately—well, obviously—it doesn't.

Redundancy filtering was Barlow's engineering solution to the problem of economical signalling in the time domain, and—sure enough—it is implemented by nervous systems in the form of *sensory adaptation.* Most sensory systems send a rapid burst of spikes every time they detect a change, after which the spike rate settles down to a low or even zero rate until there is another change.

There's an analogous engineering problem in the

spatial domain. If you think of an eye (or a digital camera) looking at a scene, most cells in the retina (or pixels in the camera) will be seeing the same thing as their neighbours in the retina (or camera). This is because the world's scenes are not capriciously random, pepper and salt, but are typically made up of large patches of uniform colour like the sky or a whitewashed wall. Away from an edge, every pixel sees the same as its neighbours, and to report it is a waste of pixels. The economical way to convey the information is for the sender to report on *edges* and for the receiver (the brain in this case) to 'fill in' the swathes of uniform colour between the edges.

Barlow pointed out that this engineering problem, too, has its neat, redundancy-reducing solution in biology. It's called *lateral inhibition*. Lateral inhibition is the equivalent of sensory adaptation, but in the spatial, not the temporal domain. Each cell in the array of 'pixels', in addition to sending nerve spikes to the brain, *inhibits* its immediate neighbours. Cells that are sitting in the middle of a patch of uniform colour are inhibited from all sides, and therefore send only few, if any, spikes to the brain. Cells that are sitting on the *edge* of a patch of colour receive inhibition from their neighbours on only one side. So the brain gets the majority of its spikes from edges: the redundancy problem is solved, or at least mitigated.

Barlow introduced his article–and it was this that especially grabbed Marian's and my imagination–with a mind-stretching thought experiment. Imagine that for every pattern the brain might ever want to recognize–every tree, every predator, every prey, every face, every letter of the alphabet, of the Greek alphabet–there was one nerve cell, hooked up to the retina in such a way that it fired when its 'own' shape fell on the retina. Each of these brain cells is wired up to a 'keyhole' combination of pixels so that it fires only when the correct 'keyhole' shape is seen. It also has to be wired up negatively to the 'anti-keyhole' (all the pixels other than the keyhole) otherwise it would fire when it saw a blank field of light covering the whole keyhole. That sounds fine, but on second thoughts it can't be true. Remember that all the shapes needing to be recognized by these overlapping keyholes may be presented in thousands of different orientations and from any distance. The number of overlapping keyholes (with the rest of the retina in each case being an anti-keyhole) would be so prodigiously large that their corresponding brain cells would have to be more numerous than all the atoms in the world. Fred Attneave, an American psychologist who independently thought up the same idea as Barlow, estimated that the volume of the brain would have to be measured in cubic light years!

The solution–redundancy reduction–extends beyond

sensory adaptation and lateral inhibition to a fascinating list of feature-detector neurones in the brain such as horizontal line detectors, vertical line detectors, 'bug detectors' and others, all of which can be represented as redundancy-reducing in the Barlow/Attneave sense. For example, a straight line can be represented as just its two ends, leaving the brain to 'fill in' the redundant intermediate points. As with the bats and the spider webs, the whole Barlow story can be told as an elegant and easily memorized sequence of problems, with engineering solutions giving rise to new problems, suggesting new engineering solutions and so on.

We should also expect that the 'detector' cells that evolve in the brain of an animal of a particular species will be tuned to detect not only features that are redundant in the sensory stream, but features that are functionally important for animals of that species–for example, the colour and shape of a sexual partner. These two combined would mean that a comprehensive list of the detector cells in an animal's brain should amount to a kind of indirect *description* of the important properties of the world in which the species lives.

And this idea, in turn, is related to another one, this time my own: 'the Genetic Book of the Dead'. The idea here is that the genes of an animal could in theory be read as a digital description of the environments in which its ancestors have survived.

'The Genetic Book of the Dead' and the species as 'averaging computer'

River Out of Eden begins by looking back at the reader's ancestors and reflecting–trivially when you think about it but still significantly–that not a single one of your ancestors died young or failed to achieve at least one heterosexual copulation. Every individual born inherits the genes of a literally unbroken line of successful ancestors. We inherit the genes that equipped this progenitorial elite, as I called them, to be an elite. The exact means by which an individual becomes a successful ancestor varies from species to species but, however they do it, we are all descended from individuals who were good at it. 'Good at it' means good at flying in the case of birds, bats and pterosaurs, good at digging in the case of moles, aardvarks and wombats, good at hunting in the case of lions, hawks and pike, good at fighting in the case of male deer, elephant seals and parasitic fig wasps.

There is a sense, therefore, in which the DNA of a species could, in principle, be read out as a kind of description of the way of life at which that species excels. I have mentioned this idea of 'the Genetic Book of the

Dead' in several of my books, but I argued it most fully in the chapter of that name in *Unweaving the Rainbow*. Here's one of the ways I introduced it:

> A species is an averaging computer. It builds up, over the generations, a statistical description of the worlds in which the ancestors of today's species members lived and reproduced. That description is written in the language of DNA. It lies not in the DNA of any one individual but collectively in the DNA–the selfish cooperators–of the whole breeding population. Perhaps 'readout' captures it better than 'description'. If you find an animal's body, a new species previously unknown to science, a knowledgeable zoologist allowed to examine and dissect its every detail should be able to 'read' its body and tell you what kind of environment its ancestors inhabited: desert, rain forest, arctic tundra, temperate woodland or coral reef. The zoologist should be able to tell you, by reading its teeth and its guts, what it fed on. Flat, millstone teeth and long intestines with complicated blind alleys indicate that it was a herbivore; sharp, shearing teeth and short, uncomplicated guts indicate a carnivore. The animal's feet, and its eyes and other sense organs spell out the way it

moved and how it found its food. Its stripes or flashes, its horns, antlers or crests, provide a read-out, for the knowledgeable, of its social and sex life.

I called the species an 'averaging computer', but why is it the *species* that is the averaging computer, not the individual organism? Because, at least in sexually reproducing animals, any one individual genome is but an ephemeral sample of the gene pool that has been sieved and winnowed down the generations, averaging the conditions and adversities which individuals of ancestral generations braved and survived. The species gene pool is a kind of negative image of the average environment of individuals of the species. If we think of natural selection as a sculptor, chiselling rough raw materials towards ever-increasing perfection, the entity that is chiselled is the species gene pool. Each individual's genome is a sample of that gene pool, and the survival (or failure) of the individual depends (among other things) upon the set of genes that it was lucky enough (or unlucky enough) to draw from the pool. I first tried to convey the idea of genes' success being dependent on their genetic companions in *The Selfish Gene* in 1976, with my metaphor of reshuffled rowing crews, where the oarsmen stand for genes and the successively re-crewed boats stand for organisms. Like many metaphors, this one should not be pushed too far, but it does convey the important idea that the best genes, in the long run, will tend to survive in the gene pool, even though many copies of them perish because they are dragged down by inferior fellow crew members in particular bodies. It is the gene pool that improves in the long run as natural selection chips its way down the generations. It is but a short step from here to the image of the Genetic Book of the Dead. It's important to understand that the environment is not directly imprinted on the genes–that would be Lamarckism. Rather, the genes vary at random and the ones that fit the environment survive to populate the gene pool of the future.

I think it was while giving tutorials that it first occurred to me that a sufficiently knowledgeable zoologist should in principle be able to read out, from the anatomy, physiology and DNA of a species, how and where it lived, who its enemies were, the weather it had to contend with and so on. I was teaching the principles of taxonomy, the science of animal classification. Animals that are unrelated but have similar ways of life tend to resemble each other in superficial features, which are in danger of distracting us from the features they share with their true taxonomic relatives. Dolphins superficially resemble marlins because both swim fast near the surface of the sea, but these superficial resemblances are outnumbered by the features by which dolphins resemble land mammals, and those by which marlins resemble other fish. Numerical methods exist for estimating these competing resemblances, independently of whether they are 'ancestral' or 'recent'.

Such 'numerical taxonomy' methods are less fashionable nowadays than they were when I learned them as an undergraduate from Arthur Cain, but they are good for illustrating the point. You measure everything you can find about a whole lot of species, feed all the measurements into a computer, and ask the computer to come up with a figure for the *distance* between each species and each other species. Distance here doesn't, of course, mean spatial distance. It means how much they resemble each other: their distance from each other in a multidimensional, mathematical 'resemblance space'. What you hope to find is that, although dolphins and marlins are pulled a little bit 'closer' to each other than they 'should' be, because of their similar ways of life, these similarities (streamlining and so on) are swamped by the much more numerous differences stemming from the fact that one is a mammal and the other a fish: they've had a very long time to diverge from each other since the Devonian period. The numerical calculations 'filter off' the superficial (minority) resemblances by swamping them, leaving us with the 'fundamental' (majority) resemblances that indicate pedigree relationships.

It occurred to me, while thinking aloud together with pupils in tutorials, that those numerical methods might in principle be stood on their head. Instead of filtering off the 'superficial' functional characteristics (like the streamlined shape of dolphins and marlins), leaving the 'true' taxonomic characters, we could do the opposite: go out of our way to filter off the taxonomic characters that stem from relatedness, and concentrate on the minority of functional resemblances. How might this be done? Imagine that we construct a set of pairs of animals. The first of each pair thrives in water, the second on dry land. Yet, taxonomically speaking, each animal is more closely related to its pair than to any of the others on 'its side' of the pairings: {otter, badger} {beaver, gopher} {yapok, opossum} {water shrew, land shrew}{water vole, vole} {pond snail, land snail} {water spider, land spider} {marine iguana, land iguana}. Suppose we make hundreds of measurements on all these animals (and lots more similar pairs)–anatomical measurements, physiological measurements, biochemical measurements, genetic sequences—and then take them all and throw them into a computer, telling the computer which member of each pair is aquatic, which terrestrial. Now ask the computer (this is not as easy as it sounds, but methods exist for doing it) questions like this: 'What do the aquatic animals in each case have in common, as opposed to their terrestrial counterparts?' We might get a bit subtler than this. Instead of asking our animals to tick a box, either aquatic or terrestrial, we might place them along a gradient of aquaticness and look for quantitative correlations along the gradient. We might even make so bold as to ask: 'What measurements of an animal would I have to *multiply* by what *factor*, in order to morph it from terrestrial to aquatic?'

We could then do the same thing for pairs of arboreal versus ground-dwelling species: {squirrel, rat} {tree frog, frog} {tree kangaroo, wallaby}; then the same for pairs of underground versus above-ground species: {mole, shrew} {mole cricket, cricket} {mole rat, rat} and so on. In the case of aquatic versus terrestrial, we might expect webbed feet to pop up as one answer, and that's pretty obvious. But I would hope the computer would find less obvious answers, buried deep inside the animals. Something about blood chemistry, for example. And, to bring us back to the Genetic Book of the Dead, we could do the same exercise with genes. Are there genes that tie aquatic animals to other aquatic animals in spite of their not being particularly closely related? We normally expect genetic comparisons to tell us which animals are closely related. With respect to most of their genes, marine and land iguanas, being close cousins, will certainly come out closely resembling each other. But I would also hope to do the opposite: find a few genes that marine iguanas have in common with other marine creatures, and don't share with land iguanas or other dry-land animals–perhaps a gene concerned with excretion of salt.

It was considerations such as this, talked through and argued about with pupil after pupil in tutorials down the years, that led me to coin phrases like 'Genetic Book of the Dead', and to suggest that a sufficiently knowledgeable zoologist, when presented with an unknown animal, should eventually, with the help of a computer, be able to reconstruct the way of life of that animal–more strictly, of its ancestors. In particular, the genes that helped the animal's ancestors to survive are in principle decipherable as a coded description of its ancestral world: ancestral predators, ancestral climate, ancestral parasites, ancestral social system.

And in those tutorials where my pupils and I threw these ideas around, I was mindful of my own tutor Arthur Cain, and his dictum that 'the animal is what it is because it needs to be'. On one occasion as a graduate student I found myself in the Royal Oak pub in Oxford (known as the doctors' pub because the old Radcliffe Infirmary was opposite) having a solitary supper of, I am chagrined to say, bacon and eggs. Arthur coincidentally happened to be doing the same thing in the same pub, so we joined each other (like, I am also chagrined to recall, the two 'travelling men' who founded the Gideons). We talked about taxonomy and adaptation, and Arthur at one point illustrated his theme by suggesting that a squirrel might be described as a rat which had moved a certain distance away from a rat-like ancestor, along the 'arboreality dimension'. That image stayed with me and informed the chapter of *Unweaving the Rainbow* called 'The Genetic Book of the Dead' and also the idea of 'the Museum of All Possible Animals' which dominated two chapters of *Climbing Mount Improbable* (see below). But the 'Museum' was more directly inspired by my attempts at computer modelling, which began while I was writing *The Blind Watchmaker*.

Evolution in pixels

Chapter 3 of *The Blind Watchmaker*, 'Accumulating small change', occupied as much time and effort as the other ten chapters put together. This was because of the weeks and months I spent writing the suite of computer programs, called Blind Watchmaker, designed to breed 'computer biomorphs' on the screen by artificial selection. The word 'biomorph' was borrowed from my friend Desmond Morris, whose surrealist paintings depict quasibiological forms which, by his own entirely believable account, 'evolve' from canvas to canvas. Desmond's painting *The Expectant Valley* had been used for the cover of *The Selfish Gene*. I bought the original at one of Desmond's exhibitions, because the price (£750) was exactly equal to the advance given me by Oxford University Press, and the omen pricked my fancy. When, a decade later, I spoke to Desmond about *The Blind Watchmaker* he was so taken with the title that he set to work, there and then, on a painting with the same title. And this new painting–though it had more to do with the title than the contents of the book– later graced the covers of both the Longman and the Penguin editions of *The Blind Watchmaker*.

I wrote my computer biomorph program in Pascal, a now largely superseded language, which itself was a direct descendant of the (even more thoroughly superseded) Algol 60 language that I had learned as a graduate student. I had continual recourse to the Apple Macintosh 'Toolbox', the repertoire of hard-wired machine code programs that gives the Mac its characteristic (and notoriously imitated) 'look and feel'; and the half-dozen technical manuals of the Mac toolbox became my much thumbed, increasingly grubby and messily annotated bible.

I was also continually running to the ever-patient Alan Grafen for help and advice: not that he was a more experienced Mac programmer than I was—rather the reverse—but he has undenied advantages in the IQ department. As P. G. Wodehouse might have put it: 'North of the collar stud, Alan stands alone.' Or, as Marian said of him: 'He has the most annoying habit of being right.' During the course of my programming marathon Alan once rather endearingly said he felt sorry for me, because I was mired in a peculiarly difficult piece of coding but I had got into it too deep to back out. That sounds Concordian and, to an extent, it was: backing out would have meant throwing away all the work I had put in so far. But there was more to it than that. I was driven to persist—and for this I dare to take credit, even to feel a little proud—by a biological intuition, almost like an instinctive nose for what I, as a biologist, could scent must work. I was propelled forwards by a conviction that something truly exciting must eventually emerge from my biomorph-generating algorithm, if only I persisted and got myself out of the mire of complexity.

The key to it was the fractal nature of the embedded 'embryology' of my biomorphs, the recursive tree-growing procedure whose quantitative details were controlled by a set of nine (more in later versions of the program) numbers, which I called genes. Obviously, if you change the numerical values of the genes you'll change the morphology of the biomorph. Less obviously, the change is often in a biologically interesting direction. I imported Darwinism (though not sex) by (asexually) 'breeding' daughter biomorphs from parent biomorphs using artificial selection. The computer offered up a choice of daughter biomorphs with slightly mutated genes, and the human chooser picked the one that was to give birth to the next generation–and so on for an indefinite number of generations. The numerical values of the genes were concealed: just like a breeder of cattle or roses, the biomorph breeder saw only the consequences of genetic change, the morphology on the computer screen.

In my dreams, then, I foresaw that something interesting and unexpected would emerge. But I never dared hope that my biomorphs would evolve their way from botany to entomology!

When I wrote the program, I never thought that it would evolve anything more than a variety of tree-like shapes. I had hoped for weeping willows, cedars of Lebanon, Lombardy poplars, seaweeds, perhaps deer antlers. Nothing in my biologist's intuition, nothing in my 20 years' experience of programming computers, and nothing in my wildest dreams, prepared me for what actually emerged on the screen. I can't remember exactly when in the sequence it first began to dawn on me that an evolved resemblance to something like an insect was possible. With a wild surmise, I began to breed, generation after generation, from whichever child looked most like an insect. My incredulity grew in parallel with the evolving resemblance . . . I still cannot conceal from you my feeling of exultation as I first watched these creatures emerging before my eyes. I distinctly heard the triumphal opening chords of *Also sprach Zarathustra* (the '*2001* theme') in my mind. I couldn't eat, and that night 'my' insects swarmed behind my eyelids as I tried to sleep.

There are computer games on the market in which the player has the illusion that he is wandering about in an underground labyrinth, which has a definite if complex geography and in which he encounters dragons, minotaurs or other mythic adversaries. In these games the monsters are rather few in number. They are all designed by a human programmer, and so is the geography of the labyrinth. In the evolution game, whether the computer version or the real thing, the player (or observer) obtains the same feeling of wandering metaphorically through a labyrinth of branching passages, but the number of possible pathways is all but infinite, and the monsters that one encounters are undesigned and unpredictable. On my wanderings through the backwaters of Biomorph Land, I have encountered fairy shrimps, Aztec temples, Gothic church windows, aboriginal drawings of kangaroos, and, on one memorable but unrecapturable occasion, a passable caricature of the Wykeham Professor of Logic.

The latter paragraph touches on one of the main biological lessons that I took away with me from this programming exer-cise. The inner eye of my imagination saw 'biomorph land', a multidimensional landscape of morphology, a nine-dimensional hypercube in which all possible biomorphs lurked, every one connected to every other one by a navigable trajectory of step-by-step, gradual evolution. In theory, though less tidily because the number of genes is not fixed, we can imagine all possible *real* animals as sitting in

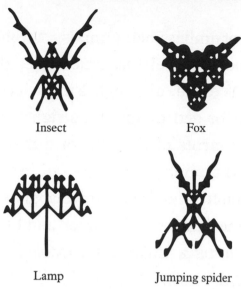

Insect   Fox

Lamp   Jumping spider

A selection of biomorphs bred by the Blind Watchmaker program

an *n*-dimensional hypercube, and I called this 'genetic space' in chapter 3 of *The Blind Watchmaker*. Most of the inhabitants of this monstrous (I use the word advisedly) hypercube not only have never existed but never could have survived if they had done: 'However many ways there may be of being alive, it is certain that there are vastly more ways of being dead' (a sentence that, I am pleased to see, has made it into the *Oxford Dictionary of Quotations*). Actual animals are islands in this hyperspace, vastly spaced out from one another as if in some Hyperpolynesia, surrounded by a fringing reef of closely related animals and separated from other islands by largely impassable wastes of impossible animals. Actual evolution is represented by timelines, trajectories through the hypercube. You see, although I'm no good at writing equations or getting my sums right, I do perhaps have the rudiments of the soul of a mathematician. Or so I would aspire.

Dan Dennett later fruitfully elaborated the idea under the name 'Library of Mendel' and I too carried it further in *Climbing Mount Improbable* in my fanciful Museum of All Possible Animals:

> Imagine a museum with galleries stretching towards the horizon in every direction . . . preserved in the museum is every kind of animal form that has ever existed, and every kind that could be imagined. Each animal is housed next door to those that it most resembles. Each dimension in the museum–that is each direction along which a gallery extends–corresponds to one dimension in which the animals vary . . . the galleries criss-cross one another in multidimensional space, not just the ordinary three-dimensional space that we, with our limited minds, are capable of visualizing.

In *Climbing Mount Improbable*, I introduced this 'Museum' using the rather special example of mollusc shells. It had been understood for a while that a shell is a (logarithmically) expanding tube, growing at the margin. If we ignore the cross-sectional shape of the tube (for example, by assuming it's a circle), the form of every shell is determined by only three numbers which, in *Climbing Mount Improbable*, I dubbed *flare*, *verm* and *spire*. *Flare* determines the rate of expansion of the tube as it grows; *spire* determines the departure from the plane. *Spire* is zero in a typical ammonite (all in one plane) but has a high value in, say, *Turritella*. *Flare* is high in a cockle (indeed, the 'tube' expands so rapidly it runs out before it reaches the point of looking like a tube at all), low in *Turritella*. *Verm* takes longer to explain in words, but high *verm* is epitomized by *Spirula* in the picture. As the American palaeontologist David Raup realized, if there are only three numbers governing variation in shape of a set of animals, all those animals can be accommodated in a simple mathematical space–a three-dimensional cube. We don't need a hypercube: an actual cube will do. And by the same token I realized that I could write a snail version of my biomorph program, with only three genes instead of nine. Instead of choosing which of a set of tree-like biomorphs to breed from, I could present snailomorphs or–let's not mix our languages–conchomorphs. By choosing the favoured breeder, generation after generation, it should be possible to evolve from any shell to any

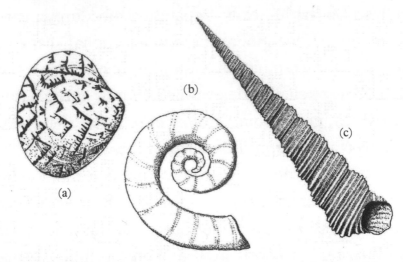

Shells to illustrate flare, verm and spire: (a) high flare: Liconcha castrensis, *a bivalve mollusc; (b) high verm:* Spirula; *(c) high spire:* Turritella terebra.

other. Evolution would be a step-by-step trajectory through the cube of all possible shells.

To write the program, I had only to substitute a new three-gene snail embryology module in place of the original nine-gene biomorphs embryology. All the rest was the same. And it did, indeed, prove very easy to breed any shell starting from any other shell, simply by choosing, in every generation, the shell that most resembled the target. 3-D printers hadn't been invented in those days. If I'd had one, I would have 'printed' the entire cube. As it was, I had to be content with printing the six edges of the cube on flat squares of paper, which I glued to the outside of a cardboard box. There's a photo in the picture section of Lalla holding the 'snail box'.

Presumably real-life evolution is free to wander any-where in the cube—the virtual Museum of All Possible Shells. However, as Raup had earlier noted, there are some sizeable 'no go' areas (volumes, rather) in which, al-though the mathematics would permit it, no shells have ever actually survived. This is because these forms would be functionally inviable. Mutants that strayed into these 'Here Be Dragons' zones simply died. Below are four mathematically possible denizens of an untenanted region of the cube. They don't exist as shells although, interest-ingly, they do exist as horns of antelopes and other bovids.

But it isn't strictly true that the Museum of All Pos-sible Shells is a three-dimensional cube. It's this only if we ignore the cross-sectional shape of the growing tube and assume that it is, for example, a perfect circle. I tried making it a variable ellipse instead of a circle, by adding a fourth gene to the original three. But real

life isn't as geometrically perfect as that. For many shells, the cross-sectional shape is not easy to specify mathematically (though it is of course in principle possible), so I resorted to entering it into the program free-hand. Apart from this modification to the embryology module, the program remained the same, with only three genes, and I was able to breed an encouragingly realistic menagerie of shells on my computer screen (see below).

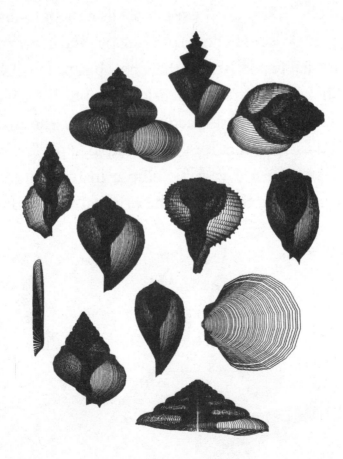

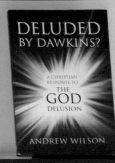

DELUDED BY DAWKINS?

A CHRISTIAN RESPONSE TO THE GOD DELUSION

ANDREW WILSON

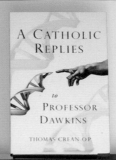

A Catholic REPLIES

to PROFESSOR DAWKINS

THOMAS CREAN O.P.

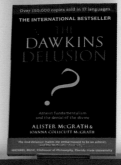

Over 150,000 copies sold in 17 languages

THE INTERNATIONAL BESTSELLER

THE DAWKINS DELUSION?

Atheist fundamentalism and the denial of the divine

ALISTER McGRATH & JOANNA COLLICUTT McGRATH

The God Delusion makes me embarrassed to be an atheist, and McGrath touches the nerve.
MICHAEL RUSE, Professor of Philosophy, Florida State University

THE DAWKINS LETTERS

CHALLENGING ATHEIST MYTHS · DAVID ROBERTSON

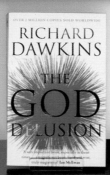

OVER 2 MILLION COPIES SOLD WORLDWIDE

RICHARD DAWKINS

THE GOD DELUSION

A very important book, especially in these times...a magnificent book, really magnificent' Ian McEwan

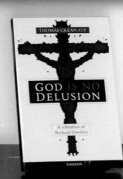

THOMAS CREAN, O.P.

GOD IS NO DELUSION

A refutation of Richard Dawkins

IGNATIUS

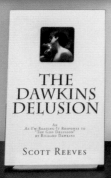

THE DAWKINS DELUSION

An 'As-I'm-Reading-It' Response to 'The God Delusion' by Richard Dawkins

SCOTT REEVES

CHALLENGING RICHARD DAWKINS

WHY RICHARD DAWKINS IS WRONG ABOUT GOD

KATHLEEN JONES

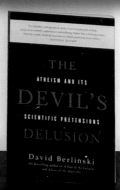

THE DEVIL'S DELUSION

ATHEISM AND ITS SCIENTIFIC PRETENSIONS

David Berlinski

THE GOD SOLUTION

Are you ready?

Ian Stott

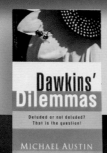

Dawkins' Dilemmas

Deluded or not deluded? That is the question!

MICHAEL AUSTIN

MIKE KING

THE GOD DELUSION REVISITED

'But was there ever dog that praised his fleas?' A small selection of the more than twenty religious books provoked by *The God Delusion*, along with the 'dog' itself.

The Simonyi Lectures. Charles Simonyi is the far-sighted benefactor of Oxford University's Chair of Public Understanding of Science. A man of many interests and enthusiasms, he lives with his contemporary art collection in a fabulous house in Seattle (*below right*), and in 2009 went on a trip into space. He is seen here (*below left*) between his fellow astronauts.

As the first Professor of Public Understanding of Science, I endowed the Simonyi Lectures, and was fortunate to attract a galaxy of stars to present them. *Opposite, clockwise from bottom left*: Jared Diamond, Daniel Dennett, Richard Gregory and Steven Pinker. *This page, from top*: Martin Rees, Richard Leakey, Carolyn Porco, Harry Kroto and Paul Nurse.

A JOURNEY THROUGH THE JOYS OF SCIENCE

BREAK THE SCIENCE BARRIER

RICHARD DAWKINS

Television. *Above left*: *Break the Science Barrier* was the fir TV documentary I presented for Channel 4. More recently I have worked with Russell Barnes (*above*, rear centre) an his crew, including Tim Crag cameraman (*right*), and Adam Prescod, sound man (front) o productions including *Faith Schools Menace*, filmed in Bel (*left*) and *The Genius of Charl Darwin* (*below*) – in which I had a potential Attenborough moment with a gorilla: if only hadn't been in a zoo.

ssell and I also worked together on *The Genius of Charles Darwin*, filming in a Nairobi slum *bove*). For *Root of All Evil?* we visited Lourdes (*below left*) and Jerusalem, where I donned the igatory hat for visiting the Wailing Wall (*below right*).

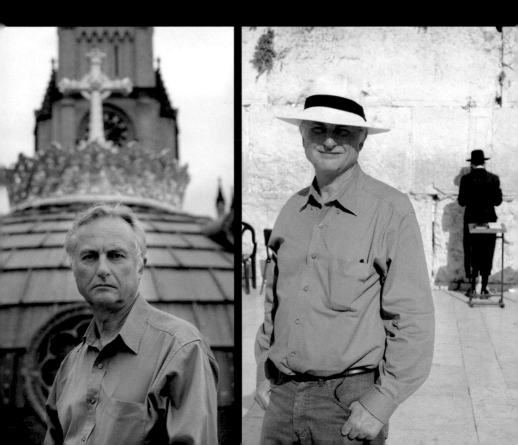

Images of evolution. *Right*: Debunking the myth 'As if a hurricane, blowing through an aircraft junkyard, could assemble a Boeing 747' – nice shot, but it ended up on the cutting-room floor. *Below*, Lalla holding my cube of all possible biomorph shells.

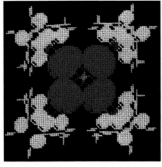

At the time I met Lalla, I was working flat out on computer biomorphs, monochrome and colour (*above, right*), and she was inspired to create some embroidered chair covers (*above*), each stitch representing one pixel. The one on the right isn't a biomorph, though you would be forgiven for thinking so: it's actually a skeleton of a glass sponge.

All about memes. *Left*: with Dan Dennett and Susan Blackmore at one of Sue's 'Memelab' gatherings in Devon. At one of these gatherings I propagated the 'Chinese Junk' meme (*centre*). *Below*: Boyhood experience with the clarinet equipped me to play the EWI at the end of the Saatchi & Saatchi memetic extravaganza at Cannes.

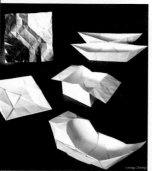

MUTATION IN THE MIND

The feast. Guests foregathering for my seventieth birthday dinner in New College Hall.

In addition to the original tree embryology and the snail embryology, were there yet more embryology modules that could be imported into my evolution program? I had long been fascinated by D'Arcy Thompson's 'transformations'. That great Scottish zoologist (see pages 123–4) had been one of those who inspired Raup and later me in our shell work. But he was best known for his demonstration that a biological form could be transformed into a related form by a mathematical transformation. You can visualize it by drawing an animal form, say the crab *Geryon*, on

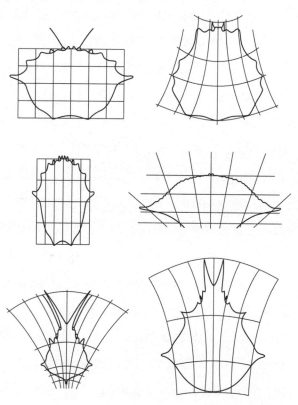

a sheet of stretched rubber. Then you find that you can transform the shape into a variety of related crabs by stretching the rubber in mathematically specified ways. Here is D'Arcy Thompson's representation of this process. *Geryon* is drawn on squared graph paper ('rubber') at top left. The (alas only approximate) form of five other crabs is obtained by distorting the graph coordinates (stretching the 'rubber') in five different mathematically elegant ways.

I had long been fascinated to dream of 'What might D'Arcy Thompson have done with a computer?' Indeed, I once set it as an exam question in the final honour school of zoology at Oxford. I don't think anybody answered it–perhaps, sadly, because none of their lectures had equipped them to do so, and (I suppose it's understandable) nervous exam candidates like to play it safe. Now I wanted to try to answer my own question by modifying my biomorph program. The genes, instead of controlling the development of a tree, would mathematically control the stretching of virtual 'rubber' in the computer. As with the conchomorphs, it would be necessary to rewrite only the kernel embryology routine of the original biomorphs program. All the rest could remain the same. It should be possible to 'evolve' from *Geryon* to, say, *Corystes* through step-by-step selection. Following D'Arcy Thompson himself, I was prepared to overlook the fact that these crabs are all modern species, none descended from any other. I was captivated by the idea that related animals can be seen as stretched, twisted, distorted versions of each other, distorted versions of their neighbours in the great Mathematical Museum of All Animals.

The mathematical and computer skills required were beyond me even if I'd had the time to exercise them, so I joined a consortium at Oxford to bid for a grant to hire two programmers. One was to work on my 'D'Arcy Thompson' project and the other on an unconnected project concerned with agriculture. The programmer who came to work on 'my' project was Will Atkinson, and he proved to be everything I could have wished for.

The 'genes' in Will's 'D'Arcy' program did a variety of things. Some changed the 'stretched rubber' from rectangle to trapezium, the magnitude of the distortion determined by the numerical value of the gene. Some changed one or both 'axes' to logarithmic, or did various other mathematical transformations. The biological shape drawn on the rubber changed progressively as the observer chose favoured 'progeny' for 'breeding', just as in my original biomorphs program.

Elegantly written as Will's program was, the forms that it 'evolved' seemed to get less and less 'biological' as the generations went by. The evolving animals increasingly looked like degenerate versions of their ancestors, rather than newly viable transformations; not like real evolutionary descendants, as my original evolving biomorphs did. Will and I worked out the reason for this, and it is an informative one. The 'D'Arcymorphs' don't have an embryology. What evolves from one generation to the next is not the animal forms themselves but the 'rubber' on which they are drawn.

And D'Arcy Thompson's original transformations, after all, were never really evolutionary, since the animals he drew were both adult and modern. Adult animals don't change into other adult animals. Embryonic processes evolve from the embryonic processes of ancestors. Julian Huxley (my sometime predecessor as tutor in zoology at New College) modified D'Arcy Thompson's method to transform embryos into adults, and this, as Peter Medawar pointed out, is a more biologically realistic usage. The reason my original biomorphs were 'fertile', even 'creative', in generating biological form is that they had an embryology: a recursive, branching tree, which had a kind of built-in propensity to keep evolving in freshly interesting directions. The 'conchomorphs', too, had their own (very different but still biologically interesting) embryology, capable of generating a rich variety of biologically realistic forms. Isn't real-life embryology 'creative' like that? Do embryologies even evolve to become better at generating evolution? Could there be a kind of higher-level selection of embryologies, choosing those that are evolutionarily fecund? This was the germ of my idea of the 'evolution of evolvability', and I'll return to it in a moment.

My original biomorphs in *The Blind Watchmaker*, with their nine genes, meandered their evolutionary way within the confines of a nine-dimensional hyper-cube. Evolutionary trends consisted of an inch-by-inch walk through this particular hypercube, this nine-dimensional Museum of All Biomorphs. I was inter-ested in possible ways of escaping from the hypercube altogether, out into a larger hypercube. One way of doing this was by substituting a completely different embryology, for example replacing tree embryology with snail embryology, and I pursued it. But before I did that I was interested in exploring the consequences of increasing the number of genes affecting my exist-ing embryology, the embryology of the original tree biomorphs. This would be equivalent to expanding, to more than nine, the number of dimensions of math-ematical space available for evolution, and I hoped it would give me insight into real biological evolution. I did it in two stages. The second stage introduced genes for colour, and made its debut in *Climbing Mount Improbable*. The first stage–still monochrome–appeared in an appendix added to the 1991 reprint of *The Blind Watchmaker*. There I upped the number of genes from nine to sixteen. The branching tree embryology remained at the core, and the new genes (once again, genes were just numbers) implemented various ways of drawing this basic biomorph. 'Segmentation' genes drew a series of biomorphs, in line astern, mimicking the segments of an earthworm or a centipede. One gene determined how many segments were drawn, another controlled the distance between segments, another implemented 'gradients' of progressive change as you went from front to back. Segmented biomorphs (see opposite) resembled arthropods even more than my

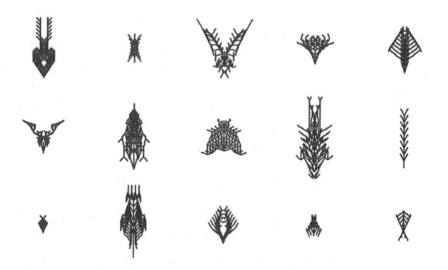

'Zarathustra' insects. They look 'biological', don't they, even if you can't pin them down to particular, real-life species? Another suite of genes 'mirrored' biomorphs in various planes of symmetry.

The sixteen-dimensional hypercube, with its new symmetry genes and segmentation genes, permitted the evolution of a much wider repertoire of biomorphs than were permitted in the original nine-dimensional space. It was even possible to breed a rather imperfect alphabet, with which I ineptly attempted to sign my name (see below). It would have been completely impossible to breed an alphabet with the original nine genes, and the imperfections of the letters discoverable in the sixteen-dimensional hypercube suggest that further genes would be needed to increase the flexibility of biomorph evolution.

ΓICHΑΓ▷ ▷ΑⱲⰍINϟ

Such thoughts led me back to biology and to propose the idea of the evolution of evolvability.

The evolution of evolvability

The year after *The Blind Watchmaker* was published, I was invited by Christopher Langton, visionary inventor of the science of artificial life, to the inaugural conference of his new discipline at the Los Alamos National Laboratory in New Mexico. It was sobering to see where the original atomic bomb was developed and recall, in the midst of the long peace, the dark oracular words of Robert Oppenheimer after the first atom bomb test in the desert:

> We knew the world would not be the same. Few people laughed, few people cried, most people were silent. I remembered the line from the Hindu scripture, the *Bhagavad-Gita* . . . . 'Now I am become Death, the destroyer of worlds.' I suppose we all thought that, one way or another.

The people who gathered for the first Artificial Life conference were very different from Oppenheimer's colleagues, but I could imagine that the atmosphere was a little bit similar: pioneers coming together to work on a completely new and strange enterprise, albeit ours was constructive and theirs as destructive as can readily be imagined. In addition to Chris Langton himself, I was pleased to meet various luminaries of the nearby Santa Fé Institute, including Stuart Kauffman, Doyne Farmer and Norman Packard. The latter two had been comrades-in-arms in an adventurous–indeed perilous–attempt to break the bank at Las Vegas, using

principles of Newtonian physics with miniature computers concealed in their shoes and operated by their
toes. The whole story is entertainingly told by Thomas
Bass, in yet another of those books whose title I decline
to mention because it was gratuitously changed as it
crossed the Atlantic.

Something of the same risky spirit, together with
the dreamlike atmosphere of the New Mexico desert,
seemed to be embodied in a charming young woman
whom I also met at the conference, and who drove me
to her home in the desert outside Santa Fé. She tried
to persuade me to take Ecstasy. I hadn't heard of it
before (this was 1987), and I now think I was right to
decline her offer although it felt cowardly at the time.
But something about her soft beauty, her strange adobe
house, the 'New Age' music she played for me, the
ghostly silence of the desert and the crisp clarity of the
air which, as in a dream, shrank the distance to the
mountains, gave me a high with no need for the drug.
Somehow that little interlude in her company, especially the hundred-mile view of the mountains almost
psychedelically magnified on the south-western horizon, sums up for me the atmosphere at that remarkable
conference.

I entitled my talk 'The evolution of evolvability'
and, so far as I am aware, my lecture, followed by my

paper in the published proceedings of the conference, represents the debut of that now much used phrase. I used a Mac to demonstrate the extra freedom to evolve in the expanded 'biomorph space' granted by the increase from nine to sixteen genes, and I then went on to expound the biological moral.

It's too easy for an arch-adaptationist like me to think that natural selection can achieve anything, without limit. But selection can work only on the mutations that embryology throws up (this was one of the 'constraints on perfection' that I had listed in *The Extended Phenotype*, five years earlier). Evolutionary change is a crawl through the multidimensional corridors of the Museum of All Possible Animals. But some of the corridors, if not totally blocked, are harder to negotiate than others, and evolution, like water trickling down a hill, will seek the path of least resistance. The point about the evolution of evolvability is this. Maybe some previously blocked, or quantitatively impeded, corridors in the museum can be suddenly unblocked by the evolutionary invention of an innovation in embryology. The first segmented individual, back in Precambrian antiquity, may or may not have been better at surviving than its unsegmented parents. But the embryological revolution that birthed it triggered a new burst of evolution, as if floodgates had been suddenly opened.

Could there then be a kind of higher-level natural selection, choosing whole lineages by virtue of the evolutionary 'fertility' of their embryologies? To me, as a
committed Darwinian adaptationist back in the 1980s,
this was verging on a heretical idea, but it was one that
excited me.

The first segmented animal must have had unsegmented parents. And it must have had at least two segments. The essence of segments is that they are like
one another in complex respects. A centipede is a train
with a long series of identical leg-bearing trucks in the
middle, a sensory engine at the front and a genital ca-
boose at the back. The segments of the human spine are
not identical, but all have the same pattern of a vertebra, dorsal and ventral nerves, muscle blocks, repeated
blood vessels and so on. Snakes have hundreds of vertebrae and some species have hugely more vertebrae
than others, most of them identical to their neighbours
in the 'train'. Since all snake species are cousins of one
another, individual snakes must from time to time be
born with more (or fewer) vertebrae than their parents,
and always a whole number more (or fewer). You can't
have half a segment. You can go from 150 segments
to 151, or to 155, but not to 150.5 or 149.5 segments.
Segments are all or none. We now understand pretty
well how this comes about—by what are called homeotic

mutations. Amazingly–a stunning discovery that long post-dates my undergraduate days reading zoology–it is the same homeotic mutations that mediate segmentation in both vertebrates and arthropods, and genes can even be transplanted from mice to fruit flies and have something tantalizingly like the same effect.

The year before my 'Evolution of evolvability' lecture, in *The Blind Watchmaker* I had written of 'Stretched DC-8' as opposed to 'Boeing 747' macromutations. The distinguished astronomer Sir Fred Hoyle (not the first or the last physicist to blunder ineptly[1] into biology) expressed his scepticism of Darwinism with the image of a hurricane blowing through a junkyard and having the luck to assemble a Boeing 747. He was talking about the origin of life (abiogenesis), but his metaphor has become a favourite with creationists casting doubt on evolution itself. The point they miss, of course, is the power of *cumulative* natural selection, the slow climb up the gentle slopes of Mount Improbable. There's a photo of me in the picture section standing in an aircraft graveyard, keeping a weather eye out

---

[1] Even arrogantly, as when he tried to show that the famous fossil bird, *Archaeopteryx*, was a fake, arguing that no physicist would accept such poor evidence as biologists do. He was a genuinely distinguished physicist, whose elucidation of how the chemical elements are forged in the interiors of stars should have won him a Nobel Prize–indeed, *did* win a Nobel Prize for a colleague involved in the same enterprise.

for hurricanes that might spontaneously put together a Boeing 747.

I invoked another airliner, the Stretched DC-8, in a contrasting metaphor. This was a version of the DC-8 airliner, lengthened 11 metres by the insertion of two extra plug-ins, 6 extra metres in the forward fuselage and 5 metres in the aft fuselage. It was a DC-8 with two homeotic mutations. We can think of each row of seats in the additional portions of the fuselage, with its associated tray tables, lights, ventilators, call buttons, music ports etc. as a segment, a duplicate of the pre-mutation segments. My biological point was that while there is a fundamental objection to a radically new, complex animal, or complex organ, being produced in a single mutational leap (Hoyle's 747), there is no principled objection to the duplication of whole segments, no matter how complex each segment might be (my DC-8). You can't invent a vertebra from scratch. But you can make a second vertebra, given that a first already exists, in a single mutation. The embryological machinery that can make one segment can make two segments, or ten. And we now even know about the homeotic mechanism that does it.

Embryological mechanisms can also easily stretch every one of a series of segments. I'd still call the result a 'Stretched DC-8' even though that isn't the way the airliner 'mutated'. This is because it's not a major jump in complexity, as a hypothetical '747 mutation' would be. A giraffe has the same number of neck vertebrae as any ordinary mammal: seven. The giraffe's neck achieves its great length by stretching all seven cervical vertebrae. I strongly suspect that it happened gradually, but there would be no objection, of the principled, insuperable '747' type, to the proposition that the neck shot out in a single macro-mutation, affecting all seven vertebrae simultaneously. The existing embryological machinery to make neck vertebrae, with all their associated complexity of nerves, blood vessels and muscles, was all present and correct. All that was needed was a quantitative tweak in some growth field, to stretch all seven simultaneously and massively. And the same would have been true if–as in snakes–the elongation had been achieved by duplicating the vertebrae rather than by stretching each one.

The authoritarian regime in George Orwell's *1984* prescribed a daily 'Two Minutes Hate' against a renegade party member called Goldstein (shades of Trotsky, or the 'fallen angel' myth of Satan). Substitute 'scorn' for 'hate' and you have some idea of the prevailing reaction, in the Oxford Zoology Department of my undergraduate days, to the German American geneticist Richard Goldschmidt, largely under the influence of E. B. Ford. Goldschmidt's 'hopeful monster' idea of the evolutionary importance of macro-mutations is indeed misguided in the contexts in which he proposed it (for example, in the very 'Oxford' bailiwick of butterfly mimicry), but since he never strayed from conscionable 'Stretched DC-8' territory towards 'Boeing 747' macro-mutational fantasy, Goldschmidt was not in principle beyond the pale. And it would be hard to fault the title 'hopeful monster' for the first segmented animal–not that anybody has ever seen a fossil of that long-dead Model-T of morphological mass production.

Macro-mutations (mutations of large effect) do occur. There is no principled objection to a macro-mutation being incorporated into a gene pool as the norm, although it seldom happens. My *principled* objection is to the idea of a macro-mutation putting together a brand new, complex, functioning organ or system, with many parts whose simultaneous combination would be too much of a coincidence: something like an eye, with its retina, lens, focusing muscles, aperture-controlling machinery and so on. There is no principled objection to the idea that the 'four-eyed fish', *Anableps*, could have acquired its two extra eyes in a single macro-mutation. Indeed, that is probably how it did happen, a lovely example of Stretched DC-8 evolution, by homeotic mutation. The embryonic machinery of the pre-mutated ancestor already 'knew' how to make an eye. But any one of those eyes, or indeed any vertebrate eye, could not have been made from scratch in a single mutational step: such '747 evolution' would be inadmissibly miraculous. The machinery of a vertebrate eye had originally to be built up gradually, step by step.

Hereabouts, by the way, lies the answer to the silly claim, originating with Stephen Gould and often reiterated, that Darwin, as a 'gradualist', would have been opposed to so-called 'punctuated' evolution. Darwin was a 'gradualist' only in the sense that he would have had no truck with 747 macro-mutations. Although Darwin obviously didn't use airliner terminology, the nature of his objections was such as to preclude only macro-mutations of the 747 kind, not the Stretched DC-8 variety.

The evolution of language could be an interesting test case for discussion. Could the ability to speak have arisen in a single macro-mutation? As I mentioned on page 290, the main qualitative feature that separates human language from all other animal communication is syntax: hierarchical embedment of relative clauses, prepositional clauses etc. The software trick that makes this possible, at least in computer languages and presumably in human language too, is the recursive subroutine. A subroutine is a piece of code that, when called, remembers whence it was called and returns there when it is finished. A recursive subroutine has the additional ability to call itself and then return to an outer (more global) version of itself. I went into this in detail in *Appetite for Wonder*, so will here content myself with the summary diagram below. The sentence was composed by a computer program that I wrote, capable of generating an infinite number of perfectly grammatical (if lacking in semantic content) sentences, recognizable as syntactically correct by any native speaker of English. I have parsed this particular sentence using brackets and a typeface which shrinks with the depth of embedment. Notice how the subsidiary clauses embed themselves within the main sentence, rather than being tacked on at the end.

It takes almost no effort to write a program capable of generating any number of grammatically correct (though semantically empty) sentences of this kind. But only if your computer language allows recursive

The adjective noun
(of the adjective noun
(which adverbly adverbly verbed
(in noun (of the noun (which verbed)))))
adverbly verbed.

subroutines. It could not, for example, have been written in the original IBM Fortran language, or any of its contemporary rivals. I wrote it in the only slightly younger language, Algol 60, and it could easily be written in any of the more modern programming languages developed after the 'macro-mutation' of recursive subroutines was introduced.

It looks as though the human brain must possess something equivalent to recursive subroutines, and it's not totally implausible that such a faculty might have come about in a single mutation, which we should probably call a macro-mutation. There's even some suggestive evidence that a particular gene called Fox P2 might be involved, since those rare individuals with a mutated version of this gene can't talk properly. More tellingly, this is one of the minority of regions of the genome where humans are unique among the Great Apes. However, the evidence on Fox P2 is unclear and controversial and I won't discuss it further. The reason I am prepared to contemplate macro-mutation in this case is a logical one. Just as you can't have half a segment, there are no intermediates between a recursive and a non-recursive subroutine. Computer languages either allow recursion or they don't. There's no such thing as half-recursion. It's an all or nothing software trick. And once that trick has been implemented, hierarchically embedded syntax immediately becomes possible and capable of generating indefinitely extended sentences. The macro-mutation seems complex and '747-ish' but it really isn't. It's a simple addition– a 'stretched DC-8 mutation'–to the software, which abruptly generates huge, runaway complexity as an emergent property. 'Emergent': important word, that.

If a mutant human was born, suddenly capable of true hierarchical syntax, you might well ask who she could talk to. Wouldn't she have been awfully lonely? If the hypothetical 'recursion gene' was dominant, this would mean that our first mutant individual would express it and so would 50 per cent of her offspring. Was there a First Linguistic Family? Is it significant that Fox P2 actually does happen to be a genetic dominant? On the other hand, it's hard to imagine how, even if a parent and half her children did share the software apparatus for syntax, they could immediately start using it to communicate.

I'll briefly mention again the possibility, which I discussed in *An Appetite for Wonder*, that this recursive software might have been used for some pre-linguistic function such as planning a hunt for an antelope or a battle against a neighbouring tribe. Each phase of a cheetah's hunt has a series of appetitive routines, calling subordinate routines, each one terminated by a 'stopping rule' which signals a return to the point in the superior program from which the subroutine was called. Could this subroutine-based software have paved the way for linguistic syntax, just waiting for the last macro-mutation to fall into place, the mutation that allowed a subroutine to call itself–recursion?

Noam Chomsky is the genius mainly responsible for our understanding of hierarchically nested grammar, as well as other linguistic principles. He believes that human children, unlike the young of any other species, are born with a genetically implanted language-learning apparatus in the brain. The child learns the particular language of her tribe or nation, of course, but it is easy for her to do so because she is simply fleshing out what her brain already 'knows' about language, using her inherited language machine. Hereditarian tendencies in intellectuals nowadays (though not always in the past) tend to be associated with the political right, and Chomsky, to put it mildly, hails from the opposite pole of the political spectrum. This disjunct has sometimes struck observers as paradoxical. But Chomsky's hereditarian position in this one instance makes sense and, more to the point, interesting sense. The origin of language may represent a rare example of the 'hopeful monster' theory of evolution.

Less dramatic than hopeful monsters such as might have initiated segmentation or, arguably, language, there could have been lots of embryological innovations which, while not conferring dramatic survival advantages on their first individual possessors, opened floodgates to future evolution. And so we return to the evolution of evolvability. In coining this phrase at the Los Alamos conference, I meant to include a kind of higher-order natural selection which we notice with hindsight. A new innovation, whether or not it directly improves the survival of the individual in the short term, leads to multiple evolutionary branchings such that its descendants inherit the earth. Segmentation was my primary example, and language might be an especially dramatic one, but there are others. The early adaptations that enabled fish to leave the water and invade the land didn't just help those pioneers to a new source of food, or a new way to escape marine predators. They pioneered new living environments, not just for individual survival in the short term but for clades blossoming through future ages. Just as Darwinian selection favours adaptations that help individuals to survive, so there can be a higher-order, non-Darwinian selection (or Darwinian only in a vague and arguably confusing sense) among *lineages* for the quality of evolvability. Such was the point I made in my 'evolution of evolvability' lecture at the Los Alamos conference, and I illustrated it with my computer biomorphs and the new vistas of evolution that were opened up to them when I rewrote the program with new genes for segmentation, and symmetry in various planes.

In the question session after my lecture (sympathetically chaired by the distinguished theoretical biologist Stuart Kauffman) somebody jokingly asked whether my biomorph program, in addition to breeding an alphabet, could breed money. In a flash, I was able to bring up on the screen a passable dollar sign (see the 's' in my signature on page 375), and thus my talk came to its end in good-humoured laughter.

Kaleidoscopic embryos

Although my Los Alamos talk was called 'The evolution of evolvability', I didn't at that stage carry the theme as far as I might have. The chapter of *Climbing Mount Improbable* called 'Kaleidoscopic embryos' went further, and in a direction that I find rather satisfying. I've already mentioned the 'mirror genes' that I introduced in one of the later versions of my biomorph program. The genes that control animal symmetry in various planes can also be thought of as inserting 'mirrors' in the embryo, like the mirrors in a kaleidoscope. Most, but not all, animals have this kind of mirror running down the midline, which makes them symmetrical in the left/right direction. A mutation in the third leg of an insect might theoretically affect the right side only, but actually it is mirrored on the left side too. Technically this mirroring is a constraint, for it restricts freedom to evolve: without it, perfect symmetry could still be achieved–'contrived' might be more apt–by separate mutations on the two sides, along with a whole lot of exotic asymmetries. But if we assume (as is plausible for reasons I discussed in *Climbing Mount Improbable*) that there is some more global benefit to left/right symmetry itself, evolutionary improvement is speeded up if mutations are automatically mirrored on both sides. Therefore, rather than being seen as a constraint (which it strictly is), the imposition of symmetry (a midline 'mirror' in the embryonic kaleidoscope) can be seen as the opposite–as an evolutionary advance in evolvability.

The same is true of other planes of symmetry, although these are less common in real biology. On the left in the illustration on the following page is a computer biomorph with four-way symmetry (two 'kaleidoscopic mirrors' at right angles). In the middle is the skeleton of a radiolarian (an exquisite microscopic single-celled creature) and on the right is a stalked jellyfish (obviously not to the same scale). All of them

have 'two mirrors' at right angles, buried deep in their embryology. In the case of the biomorph, I know this is true because I wrote the embryology's software. In the case of the two real animals, I don't know for sure but I would bet my shirt that the four-way symmetry is a default constraint in the embryology. My conjecture is that whatever was the innovation in the fundamental embryology that set up this kaleidoscopic constraint, it had an advantage; and I would want to call that innovation an evolutionary advance in evolvability.

Echinoderms (starfish, sea urchins, brittle stars etc.) mostly have five-way symmetry. Once again, it seems to me almost obvious that the relevant symmetry rule is lodged deep within the embryology, such that a mutation in a detail at the tip of one arm of a starfish, say, is mirrored in all five arms (this generalization is not negated by the fact that occasionally starfish with more than five arms turn up). And once again, given that the symmetry is for some reason a good thing for a starfish to have, 'mirroring' the mutations is a short cut (when compared with piecemeal change in each arm separately) to achieving change without departing from five-way symmetry. It therefore deserves to be considered under the heading of 'evolution of evolvability'. And it is significant that all my efforts to breed five-way symmetrical biomorphs on the computer screen failed. It's almost obvious. Five-way symmetry could be achieved only by a radical rewrite of the embryology routine–which again makes the point that we really are talking about the evolution of evolvability. The 'echinoderm' biomorphs that I managed to breed on the screen are all 'cheats' (see illustration opposite). They look superficially like a sand dollar, a sea lily, a sea urchin, a brittle star and two starfish respectively, but not one of them is five-way symmetrical.

At the time of the Los Alamos conference there were no colour Macs. When I finally got one, the obvious next step in expanding the genome of my

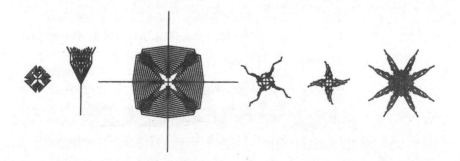

biomorphs was to add a new suite of genes for colour. At the same time I added genes to modify the lines with which the basic trees of the embryology algorithm were drawn. Simple lines were still allowed, but I introduced a new gene to change their thickness, and other new genes to change them from simple lines to rectangles or ovals; to control whether these shapes were filled or open; and to control the colour of the lines and of the fill. These extra genes opened up new floodgates of evolution, tempting the selecting human to breed biomorphs that looked more and more like exotic flower designs, table mats and butterflies. I had a fancy to take the computer out into the garden and offer real bees and butterflies the chance to choose the 'flowers' and 'butterflies' on the screen. I hoped that real insects would breed simulacra of real species of flowers, starting from non-flower-like beginnings. Unfortunately, it turned out–as I should have anticipated–that the bright daylight that brings the insects out to forage is the very same bright daylight that makes the screen difficult to see. As so often happens with seemingly bright ideas, I shelved the project and never came back to it. Perhaps night-flying moths? Would a modified version of a touch-sensitive screen like an iPad's respond directly to the buffeting of a moth?

I was creating colour biomorphs at about the time I met Lalla. Embroidery is one of her many talents—at that time she hadn't moved on to mosaics, painting ceramics or (her current art forms) weaving and drawing with a sewing machine—and she was inspired by the coloured four-way symmetrical biomorphs to embroider cushions and chair covers in which the stitches of the embroidery corresponded exactly to pixels on the computer screen (see picture section). They are still much admired twenty years later.

All my biomorph-style programs employed artificial selection, not natural selection. I could only dream about the much more difficult problem of how to simulate *natural* selection in an interesting way. The fact that it's difficult is itself instructive. One could imagine building into my biomorph program a selection criterion such as 'spikiness' or 'roundness'. And indeed I did exactly that as an experiment. This bypassed the human eye as selection agent, and it did work. But it wasn't biologically very interesting. In order to simulate survival in a 'world', it would be necessary to construct that world, with its own 'physics', its own (ideally three-dimensional) geography, its own rules for how biomorphs should interact with other objects and other biomorphs in that world, rules for how not to occupy the same physical space as other objects, and so on. In the years since *The Blind Watchmaker* was published, clever programmers have developed such artificial worlds with their own 'physics', for example Steve Grand with his Creatures, Torsten Reil with his Natural Motion, and the various fantasy environments of the Second Life type. Out of my league, and anyway I've shaken off the addiction to programming.

Arthromorphs

The evolution of evolvability is all about floodgates opening to new creative improvement. The Los Alamos conference where I introduced the idea became a kind of metaphor for the idea itself, because that conference really did release something a bit like a creative surge in my own mind (and probably in other participants' minds as well). For me that surge was to reach its climax in *Climbing Mount Improbable*, which I regard as my most under-rated book (it is the least read of them all, although it is probably the most innovative after *The Extended Phenotype*).

And now here's another gate that conference opened. It was there that I met Ted Kaehler. One of Apple's star programmers, Ted has the sort of creatively original mind we have come to associate with that artistically innovative corporation. He was there partly to assist with computer demonstrations (including mine), but his expertise and interests extend way beyond such technicalities, and I had many discussions with him on evolutionary ideas. I saw more of him later, when he was working with Alan Kay's Apple-sponsored education project in Los Angeles, the project whose high-pressure think-tank I was briefly privileged to join when I stayed with the lovely Gwen Roberts and did most of my work on the colour biomorphs (see page 406). Ted and I brainstormed with increasing enthusiasm—it's a wonderful feeling, as I described in the wasp chapters, when joint thinking goes fast and well. We obsessed about the evolution of evolvability, especially segmentation; and together we hatched a plan to write a new biomorph-style artificial selection program concentrating on segmented arthropod-like artificial creatures, and incorporating other overtly biological principles of embryology. We called our new artificial creatures 'arthromorphs'.

The original Blind Watchmaker biomorphs had nine genes. The 'Los Alamos' version had sixteen. The colour version had thirty-six. Each enlargement of the genome opened what I have been calling floodgates, unleashing an expansion of evolutionary 'creativity', albeit constrained in 'constructive' ways, for example by segmentation or 'kaleidoscopic mirrors'. But each of these enhancements relied on a major intervention by the programmer. I had to go back to the drawing board and write a whole lot of new code. And, in a way, that is a suitable metaphor for the evolution of evolvability, for I do think that in real biology the radical watershed events we are talking about–like the origin of segmentation, the origin of multicellularity, the origin of sex, or the origin of five-way symmetry in echinoderms–are rare and rather catastrophic upheavals, a little bit analogous to a major rewrite of a computer program. Indeed, the analogy even extends to 'debugging', for we can be sure that when a revolutionary mutation is incorporated into the gene pool by selection it will have knock-on effects, which need to be ironed out in its wake: ironed out by subsequent selection in favour of a retinue of minor mutations that smooth away the adverse side-effects of a generally beneficial major mutation.

But real biology knows an intermediate tier of mutation, less revolutionary than the origin of multicellularity, sex, segmentation or novel 'mirrors' of symmetry, but more radical than the ordinary point mutations whereby a Watson–Crick nucleotide changes to another of the gang of four–C, T, G or A. This intermediate category includes duplications (or the inverse, deletions) of whole stretches of chromosome. Gene duplication is the main way in which genomes get bigger. In *The Ancestor's Tale* (specifically in The Lamprey's Tale) I described the process for the particular case of haemoglobin. To recap briefly: we have five different 'globin' chains, coded by different genes in different parts of the genome. And the point is that all five are descended from a single ancestral globin coded by a single ancestral gene. The ancestral gene (which is still the only one possessed by our remote and primitive cousins the lampreys) was successively duplicated in evolution to make the multiple 'globin genes' we have today. Normally when we speak of evolutionary divergence we mean the splitting of an ancestral species into two. Two populations of walking, breathing animals split and go their separate ways. Here, on the other hand, while we are still talking about evolutionary divergences, we are talking about splits that each occurred *within* a single individual, so that the two descendant molecular lineages persisted, *side by side*, within the bodies of future individuals through all future generations.

Incidentally, I am often asked whether our improved understanding of genomics has changed what I would say if I were to rewrite *The Selfish Gene.* The answer is no—a reluctant no in some ways, for scientists take pride in changing their minds when new evidence comes in. My 'gene's-eye view' of 1976 is, if anything, strengthened by new considerations such as the gene duplication discussed in The Lamprey's Tale. This is because we now see evolutionary divergence at the gene level *within* individuals, which downplays the importance of the individual (as opposed to the gene) as the level at which selection acts.

Ted and I, when drawing up the specification for our Arthromorphs program, didn't attempt to simulate haemoglobin-style gene duplication per se. However, our new program did incorporate a form of gene duplication (and deletion), which proved to be highly instructive. Whereas all my previous biomorph programs had a fixed repertoire of genes (nine, sixteen or thirty-six for the three versions), arthromorphs had a variable number of genes, the number of genes itself being subject to mutation. Can you see that we were moving in the direction of letting evolution do its own rewriting of the software, whereas previously, for each macro-mutated advance in the evolvability of biomorphs, I had had to sit down and write a whole lot more code?

Segmentation was deeply woven into the fabric of arthromorph embryology, although it was genetically allowable to have only a single segment. Left/right mirroring was a default constraint: all arthromorphs were left/right symmetrical. Each segment consisted of an oval body part (its shape and size under genetic control) with the capacity to grow a pair of symmetrical limbs, each limb with the capacity to bifurcate in a claw. So far, so arthropodan. The number of joints in each limb was under genetic control, as were the size of each and

the angle of each joint; and the same went for the size
and angle of the terminal claws.

Where it starts to get embryologically more interesting is that groups of neighbouring segments (in series)
share fields of influence. For example, the first three
(say) segments might be nearly the same as each other
but more different from the next two segments, and
different again from the next four segments–a structure redolent of head, thorax, abdomen (see Arthromorph 1 below). Each of these (of course it didn't have
to be three: that number was itself subject to genetic
variation) groups of segments we called a *tagma* (plural
tagmata), which is the correct name in arthropod biology. But the segments within a tagma didn't have to be
literally identical. Each segment was influenced by its
own tagma-specific genes, which were free to mutate
independently of the other segments. The comparative

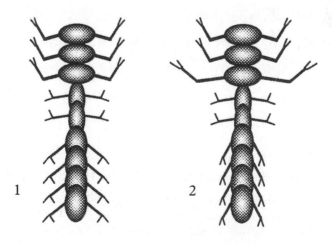

uniformity within a tagma was achieved by multiplying the genetic quantities of each segment by a number (a 'gene') specific to that tagma. Arthromorph 2 pictured here is similar to Arthromorph 1, except that segment 3, though recognizably a member of tagma 1, has longer legs than the other two segments of tagma 1. Tagma 3 as a whole has different legs, too.

At a higher level, there were other genes which multiplied the values of all the genes of the whole organism, across all tagmata. Finally, we added 'gradient' genes, which multiplied the other genetic effects by an increasing (or decreasing) number as you proceeded posteriorly along the organism (or along a tagma). Increases (and decreases) in numbers of tagmata, and of segments within each tagma, were achieved by gene duplication (or deletion).

Such was arthromorph embryology, and you'll notice that it was more complicated than biomorph embryology in biologically interesting ways. It pushed me to the limit of my programming skill, so that I had to lean on Ted's superior experience. I did the coding myself (in Pascal, which was not Ted's preferred language and has now, as I've already mentioned, been largely superseded), but Ted guided me with emailed suggestions written in a kind of pseudo computer language pretty much akin to a formal subset of English.

At times, I suspect he must have got a little impatient with my slowness–not up to the standards of a professional Apple software engineer–but he was always very kind and we got it finished in the end. Once the difficult embryology routine was written, it was a simple matter to embed it in a version of the original biomorph program to handle the choosing for 'breeding' on the screen. Below is a 'zoo' (perhaps flea circus is a better term) of a selection of the countless arthromorphs that

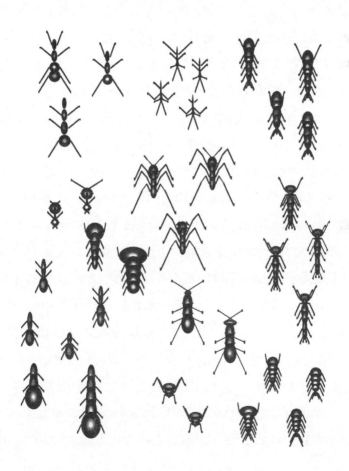

I was able to breed by artificial selection, once the program was finished.

Christopher Langton's series of Artificial Life conferences continued, as did a related series called Digital Biota. Chris himself was present at the second of these, held at Magdalene College, Cambridge in 1996, and I was invited to give the keynote address, on the title 'The view from real life': obviously an attempt to ground the geeks in real biology as they explored the enchantments of their virtual worlds. The conference, for me, was chiefly memorable because of a wonderful impromptu speech by Douglas Adams (reprinted in *The Salmon of Doubt*) and because it was there that I met Steve Grand, author of *Creation: Life and How to Make It*, a tour de force to match the virtuosity of his artificial life program, Creatures. It was also there that I was introduced to the amazing possibilities of virtual worlds in which 'avatars', owned by players from all over the world, could roam amid fantastic castles and palaces, casinos and streets, all built and even policed as a communal project. Although I am intrigued by the feats of programming that go into such virtual worlds, I find something a little creepy about the extremes reached by people who, through their avatars, inhabit them. The phrase 'get a life' has become a cliché, but you can't help feeling it ought to be hung up in prominent town squares in Second Life, where the inhabitants have even been known to 'marry' people whom they have never met in the flesh, and subsequently get 'divorced' over 'unfaithfulness' in cyberspace. Ah well, perhaps it is the way of the future and one day I shall virtually eat my own real words.[1]

The cooperative gene

My emphasis on the species gene pool (rather than the individual genome) as the text of a Genetic Book of the Dead also serves to shore up another central plank of my world-view: the cooperative gene. This is the obvious but important idea that, from a functional point of view, the gene is impotent outside the context of the other genes in the gene pool–which is to say the other genes with which it has to share a large number of individual bodies, distributed in space and time. There is a whole chapter, 'The selfish cooperator' devoted to it in *Unweaving the Rainbow,* but it is an idea that was foreshadowed (despite the book's title) in *The Selfish Gene*:

---

[1] After this chapter was written, I was contacted by Alan Canon, a virtuoso programmer from Kentucky, who volunteered to resuscitate the Arthromorphs program, and the other programs of my 'Watchmaker Suite', so that they will run on modern computers. You can download the latest version of Watchmaker Suite from https://sourceforge.net/projects/watchmakersuite.

A gene that cooperates well with most of the other genes that it is likely to meet in successive bodies, i.e. the genes in the whole of the rest of the gene pool, will tend to have an advantage.

For example, a number of attributes are desirable in an efficient carnivore's body, among them sharp cutting teeth, the right kind of intestine for digesting meat, and many other things. An efficient herbivore, on the other hand, needs flat grinding teeth, and a much longer intestine with a different kind of digestive chemistry. In a herbivore gene pool, any new gene that conferred on its possessors sharp meat-eating teeth would not be very successful. This is not because meat-eating is universally a bad idea, but because you cannot efficiently eat meat unless you also have the right sort of intestine, and all the other attributes of a meat-eating way of life. Genes for sharp, meat-eating teeth are not inherently bad genes. They are only bad genes in a gene-pool that is dominated by genes for herbivorous qualities.

This is a subtle, complicated idea. It is complicated because the 'environment' of a gene consists largely of other genes, each of which is itself being selected for its ability to cooperate with *its* environment of other genes.

I would happily bring out a new book called *The Cooperative Gene,* but the book itself would be identical, word for word, with *The Selfish Gene.*[1] There is no paradox here. The selfish genes that survive do so in their environment. That environment of course includes the external environment that we can see: the climate, the predators and parasites, the food supply and so on. But an even more important part of the environment of any gene is the other genes in the gene pool of the species–which means the set of genes with which it is statistically likely to share a body. A gene in isolation can have no phenotypic effects, and the phenotypic effects that it does have depend on the other genes that are present in the body in the short term, the gene pool in the long term. Natural selection, at each locus independently, favours whichever allele *cooperates* with the other genes with whom it shares a succession of bodies: and that means it cooperates with the alleles at those other loci, which cooperate in their turn. Cooperation is the name of the game. The effect is that cartels of mutually cooperating genes build up in gene pools. If a member of one cartel were plucked out

---

[1] As it happens, my colleague and former pupil Mark Ridley, whose views on evolution are closely similar to mine, has published a book called *The Cooperative Gene.* That, at least, is the title of the American edition. The original British edition is called *Mendel's Demon.*

and plonked down in another cartel, the result would not be a success. My understanding of this important point was heavily influenced by research of the E. B. Ford school at Oxford. Ford and his colleagues showed by hybridization experiments that complex characteristics of moths break down when genes are exposed to a foreign 'genetic climate', the foreign genetic climate of another species. This work made a big impression on me in my undergraduate tutorials with Ford's junior colleague Robert Creed. Here's how I put it in *Unweaving the Rainbow*. I apologize for such long quotations, but I have no better words to express a point which has been many times misunderstood.

It is tempting to speak of the 'whole cheetah' or the 'whole antelope' as being selected, 'as a unit'. Tempting, but superficial. Also lazy. It requires some extra thinking work to see what is really going on. Genes that program the development of carnivorous guts flourish in a genetic climate that is already dominated by genes programming carnivorous brains. And vice versa. Genes that program defensive camouflage flourish in a genetic climate that is already dominated by genes programming herbivorous teeth. And vice versa. There are lots and lots of ways of making a living. To mention only a few mammal examples, there is the cheetah way, the impala way, the mole way, the baboon way, the koala way. There is no need to say that one way is better than any other. All of them work. What is bad is to be caught with half your adaptations aimed at one way of life, half aimed at another.

This kind of argument is best expressed at the level of the separate genes. At each genetic locus, the gene most likely to be favoured is the one that is compatible with the genetic climate afforded by the others, the one that survives in that climate through repeated generations. Since this applies to each one of the genes that constitute the climate–since every gene is potentially part of the climate of every other–the result is that a species gene pool tends to coalesce into a gang of mutually compatible partners.

And that is the very important sense in which genes are simultaneously 'selfish' and 'cooperative': a cornerstone of what, in my more presumptuous moments, I might call my 'world-view'. It is a much more coherent and penetrating way to think about the evolution of cooperation than limp-wristed hand-waving about selection of the organism 'as a unit'.

Universal Darwinism

In 1982, the centenary of Darwin's death was marked by commemorations around the world, perhaps most prominently at Cambridge, where the young Charles read for his undergraduate degree in theology, and where he 'walked with Henslow' and collected beetles. I felt honoured to be invited to speak, and I chose as my title 'Universal Darwinism'. My idea was that natural selection is more than just the driving force of evolution in the life forms that we know, on this planet; there is, as far as we know, no other force capable of filling the role of ultimate responsibility for adaptive evolution. 'Adaptive' is a necessary word in that sentence. Random genetic drift is responsible for much, if not most, of evolutionary change at the molecular level. But it cannot be responsible for functional, adaptive evolution. Only natural selection, as far as we know and as far as anybody has so far imagined, produces organs that work as if an engineer had designed them: wings that fly, eyes that see, ears that hear, stings that paralyse. Or so I asserted. My stated implication was that, if we ever discover life elsewhere in the universe, it will turn out to be Darwinian life: it will have evolved along some local equivalent of Darwinian principles.

My argument was not logically unassailable, but I think it's still strong. It became, indeed, my answer to the question 'What do you believe but cannot prove?' in John Brockman's annual 'Edge' series. It amounted to: 'Nobody has ever thought of a viable alternative to natural selection.' I have to admit that, when you put it like that, it is vulnerable to disproof the moment somebody actually comes up with an alternative. But scientists are allowed to have hunches, and the burden of my contribution was the strong hunch that no such disproof will ever–could ever–transpire. I made what I think is an unassailable case *in principle*, not just in fact, against all known alternatives to natural selection, most notably the Lamarckian theory invoking 'use and disuse' and the inheritance of acquired characteristics. Hitherto, biologists had gone along with Ernst Mayr, centenarian founding father of the neo-Darwinian synthesis, who thought that Lamarck's hypothesis was in principle a good one, slain only by what T. H. Huxley would have called an ugly fact: acquired characteristics are *not* in fact inherited. An implication of Mayr's view was that, if there were a planet where acquired characteristics were inherited, evolution on that planet could be Lamarckian and would work just fine. This was what I went out of my way to deny–I think convincingly, and nobody has ever published a refutation.

It's a fact that muscles that are much used for a particular purpose grow bigger, and become better for that purpose. Lift weights and your muscles enlarge. Walk around with bare feet and the skin of your soles gets tougher. If you run marathons you become better able to run marathons: your heart, lungs, leg muscles and many other things become better for the purpose. So, on our hypothetical planet with Lamarckian evolution, stronger muscles, leathery feet and trained lungs would be passed on to the next generation. Lamarck thought it was by this principle that improvements evolve. The usual objection is that acquired characteristics are not, as a matter of ugly fact, inherited. My objections were different–principled rather than factual, and threefold.

First, even if acquired characteristics were inherited, the principle of use and disuse is too crude and unfocused to mediate all but a few examples of adaptive evolution. The lens of an eye is not washed clear by photons streaming through it. Muscular enlargement represents a rather rare example of an improvement that can come about by use and disuse. Only natural selection has chisels fine and sharp enough, and sufficiently accurately aimed, to sculpt the multitude of subtle and often tiny improvements of evolution. The principle of use and disuse is too crude and inept. On the other hand, any genetically mediated improvement, no matter how subtle and no matter how deeply buried inside the cellular chemistry of the organism, is grist to natural selection's fine-grinding mill.

Second, only a minority of acquired characteristics are improvements. Yes, muscles grow when you use them, but most parts of the body wear out with repeated use and become smaller, less perfect, often pitted and scarred. It has become almost a cliché that religious circumcision over many generations has signally failed to effect any evolutionary reduction in the foreskin. Lamarckian evolution would have to employ some sort of 'selection' mechanism to sort out the few improvements (like leathery feet) from the many dis-improvements (wearing out of hip joints etc.)–and that sounds a lot like Darwinian selection!

Our bodies are not walking inventories of ancestral scars and broken limbs, notwithstanding popular belief. My mother had a beloved dog called Bunch, who had a habit of limping on three of his four legs (luxating patella is a common problem in small dogs). A neighbour had an older dog, Ben, who had lost one hind leg in an accident and limped, perforce, on the only three legs left to him. She tried to persuade my mother that Ben must be Bunch's father!

Indulge me in a shameless moment of sentiment. Looking, this week, through a tattered folder of poems that my parents collected for each other over many years, painstakingly copied out by hand, I found the following in my mother's handwriting, obviously written immediately after the death of Bunch. It is unfinished, as can be seen from the crossings out and corrections, but I think it is beautiful enough to be reproduced here. And if you can't be sentimental in an autobiography, when *can* you?

> Dear little ghost of happiness
> Who lopes beside me down the years,
> No dog of flesh and blood will take
> Your place, nor ever stem the tears
> That come unbidden from the heart.
> For you were very part of me–
> And all the fields and all the ways
> Down woodland ride or open hill
> Are empty places now for me.
>
> *You are not there–you are not there*

Dear Bunch. It's all nonsense to say you can't miss a dog as much as a person. Or, to put it another way, for purposes of mourning, a dog can *become* a 'person'.

My third objection to any kind of evolution based on the inheritance of acquired characteristics might not necessarily apply universally to all life forms everywhere. It is, however, an objection in the case of any life form whose embryology is 'epigenetic' (as Earthly embryology is) rather than 'preformationistic'. And it remains arguable (for another day) that preformation-istic embryology is in principle unworkable. What do those technical terms 'epigenetic' and 'preformationist' mean? They go way back in the history of embryology. I would now call them 'origami' embryology and '3-D printer' embryology respectively. Origami embryol-ogy, as I wrote in *The Ancestor's Tale* and *The Greatest Show on Earth*, creates a body by following a recipe or program of instructions to grow tissues and fold them, invaginate them, refold them, turn them inside out. It's of the nature of origami (epigenetic) embryology that it is irreversible. You can't take a body and reverse engi-neer the instructions that made it, any more than you can take an origami bird or boat and reverse engineer the folding sequence that gave rise to it; or take a gour-met dish and reconstruct the words of its recipe.

Preformationist (or 'blueprint') embryology is quite different. It is reversible and it doesn't exist in the biol-ogy of our planet. This is why it is wrong to describe DNA as a 'blueprint'. Blueprint embryology, if it existed, would be reversible. You can reconstruct the blueprint of a house by measuring its rooms and scal-ing down. You can't reconstruct the DNA of an animal, no matter how detailed and meticulous your measure-ments of its body.

Preformationist or 'blueprint' embryology is well

portrayed by a 3-D printer. The 3-D printer is a natural extension of the ordinary paper printer. It builds up the object it is 'printing' layer by layer. I first saw one of these astonishing machines as the guest of Elon Musk in his SpaceX rocket factory. This particular 3-D printer was–as an exhibition tour de force–printing chessmen. Unlike a milling machine, which plays the role of a computer-controlled sculptor and *subtractively* carves objects out of a block of metal, a 3-D printer builds the object up additively, layer upon layer. You can present it with serial scans of an existing 3-D object in layers, and it will then build up the layers again in the new, copied object. Life as we know it develops epigenetically,[1] not preformationistically.

One could (just, perhaps, only just, only perhaps) imagine a preformationistic life form, somewhere in the universe, whose embryology works like a 3-D printer, scanning the parent's body and then building the child up layer by layer. And such a life form could, in theory, pass on acquired characteristics to the next

---

[1] Don't by the way, be confused by the fact that the word 'epigenetics' has recently been hijacked as a label for a fashionable and over-hyped idea that changes in gene *expression* (which of course happen all the time during the course of normal embryonic development, otherwise all the cells of the body would be the same) can be passed on to future generations. Such transgenerational effects may occasionally happen and it's a quite interesting, if rather rare, phenomenon. But it's a shame that, in the popular press, the word 'epigenetics' is becoming misused as though cross-generational transmission was a part of the very definition of epigenetics, rather than a rare and interesting anomaly.

generation. Whatever scanning mechanisms copied the present body could copy its acquired alterations (presumably including the scars of injury, mutilation, and wear and tear). But everything we know about the DNA/protein-based life of this planet is inimical to the idea of scanning the parental body and rendering the scanned information into the genes for transmission to the next generation. That isn't, nor could it be, how DNA works. You can't reconstruct an animal's genome from its body. And the only way we know to construct a body from its genes is to grow an embryo in a womb or egg. Moreover–to repeat the circumcision point–scanning the body would pick up all the injuries as well as the 'use and disuse' improvements.

So, I concluded, Ernst Mayr was wrong to say: 'Accepting his premises, Lamarck's theory was as legitimate a theory of adaptation as that of Darwin. Unfortunately, these premises turned out to be invalid.' No, they didn't 'turn out' to be invalid; they couldn't do the job even if they were valid, as a matter of principle. And I hope I gave Francis Crick cause to revise these words: 'No one has given *general* theoretical objections why such a mechanism must be less efficient than natural selection.'

At the end of my talk, Stephen Gould stood up and eloquently demonstrated, not for the first time, the way in which massive erudition can sometimes overload the mind to the extent of obscuring the point that matters. He articulately and fluently pointed out that, in the late nineteenth and early twentieth centuries, various alternatives to natural selection were fashionable: mutationism and saltationism, for example. This is historically true but crashingly beside the point. Like Lamarckism (as I also had argued in my Cambridge lecture), neither mutationism nor saltationism nor any other nineteenth-century 'ism' is in principle capable of mediating *adaptive* evolution.

Take 'mutationism' for example. William Bateson (1861–1926) was one of many geneticists (he coined the word) who thought that Mendelian genetics somehow superseded natural selection and that mutation, without selection, was a sufficient explanation of evolution. I gave two quotations from him in *The Blind Watchmaker:*

'We go to Darwin for his incomparable collection of facts [but] . . . for us he speaks no more with philosophical authority. We read his scheme of Evolution as we would those of Lucretius or Lamarck.'

And again,

'The transformation of masses of populations by imperceptible steps guided by selection is, as most of us now see, so inapplicable to the fact that we can only marvel both at the want of penetration displayed by the advocates of such a proposition, and at the forensic skill by which it was made to appear acceptable even for a time.'

What utter nonsense. Gould was certainly right, as a matter of historical fact, that other theories of evolution than those of Darwin and Lamarck were going around in the nineteenth and early twentieth centuries, and Bateson was one of the perpetrators. My point was not to deny the history but to show that those other theories, along with Lamarckism, were in principle wrong. Always had to be wrong. And that should all have been clear from an armchair even before evidence disproved them. Darwinian natural selection is not only supported by evidence. Where *adaptive*, functionally *improving* evolution is concerned, natural selection is certainly the only theory we know that is in principle capable of doing the job, and the generalization will–or such is my hunch–stretch to theories we don't know.

I didn't argue for universal Darwinism very well at the Cambridge conference, badly underestimating the time it would take to develop the theme. And in those days I was less skilled at concealing my discomfort when I made a mistake of that kind. This wasn't the only occasion on which I ran out of time, and I recall the familiar sensation of red overheating and literally sweating with anxiety and panic. During the coffee break after what I perceived as my failure, I sat on, disconsolately motionless, in the emptying lecture hall. A sweet friend who saw my distress came up behind me and silently kissed the top of my head with her hands gently resting on my shoulders. The warmth of feminine *tendresse* is one of the good reasons for staying alive. When I came to reprise the story of universal Darwinism in the last chapter of *The Blind Watchmaker*, I told it better.

Memes

In the last chapter of (the original edition of) *The Selfish Gene* I advocated a version of universal Darwinism, in the course of downplaying the gene, which had been the starring hero of the rest of the book. Any self-replicating coded information, I argued, could step into evolution's play as understudy to DNA. And maybe on some distant planet it has. I should have added–but didn't make it sufficiently clear until *The Extended Phenotype*–that the understudy would need an additional qualification: the power to influence the probability of its own replication. Indeed, before I had the title of *The Extended Phenotype*, I remember that

Geoffrey Parker, pioneering evolutionary theorist of Liverpool University, asked me what my next book would be about and I said 'power'. Geoff got the point immediately, and I can't think of many other people who would have got it from that single word.

In 1976, when introducing the idea of universal Darwinism in *The Selfish Gene*, what other examples of potentially powerful replicators–hypothetical alternatives to DNA–could I turn to? Computer viruses would have done it, but they had only just been invented in some squalid little mind, and even if I'd thought of them I wouldn't have wanted to advertise the idea. I mentioned the possibility of strange replicators on alien planets and continued:

> But do we have to go to distant worlds to find other kinds of replicator and other, consequent, kinds of evolution? I think that a new kind of replicator has recently emerged on this very planet. It is staring us in the face. It is still in its infancy, still drifting clumsily about in its primeval soup, but already it is achieving evolutionary change at a rate that leaves the old gene panting far behind.

It is true that cultural evolution is orders of magnitude faster than genetic evolution. But I would have been jumping the gun if I had implied that natural selection of memes should take all credit for cultural evolution. It might, but that would have been a bolder claim than I set out to make. The evolution of language, for example, clearly owes more to drift (memetic drift) than to anything resembling selection. I went on to coin the word itself:

> The new soup is the soup of human culture. We need a name for the new replicator, a noun that conveys the idea of a unit of cultural transmission, or a unit of *imitation*. 'Mimeme' comes from a suitable Greek root, but I want a monosyllable that sounds a bit like 'gene'. I hope my classicist friends will forgive me if I abbreviate mimeme to *meme*. If it is any consolation, it could alternatively be thought of as being related to 'memory', or to the French word *même*. It should be pronounced to rhyme with 'cream'.

Just as genes are selected for their mutual compatibility so, in principle, might memes be. The large literature on memetics has adopted the word 'memeplex' as a contraction of 'meme complex'. In *The Selfish Gene* I reiterated the idea of cooperating gene complexes (I used the phrase 'evolutionarily stable set of genes') and then tentatively drew the memetic parallel, as follows:

> Mutually suitable[1] teeth, claws, guts, and sense organs evolved in carnivore gene pools, while a different stable set of characteristics emerged from herbivore gene pools. Does anything analogous occur in meme pools? Has the god meme, say, become associated with any other particular memes, and does this association assist the survival of each of the participating memes? Perhaps we could regard an organized church, with its architecture, rituals, laws, music, art and written tradition, as a co-adapted stable set of mutually assisting memes.

And here's another interesting possibility: if we acknowledge that memes, as well as genes, might be naturally selected, then mutually compatible complexes of memes and genes together might be favoured, each in their respective domains of selection. Thus, if the Genetic Book of the Dead is a description of ancestral environments, why would those ancestral environments

---

[1] This was probably a misprint for 'stable', introduced by the human equivalent of today's hilarious 'autocorrecting' software. If so, it happens to be a felicitous one, for either of the two words is appropriate. A rare example, perhaps, of an advantageous memetic mutation.

of genes not include ancestral memes? Wouldn't ancestral social practices, ancestral religions, ancestral marriage customs and habits of warfare have constituted an important part of the worlds in which ancestral genes survived? And vice versa.

In addition to regional differences in climate, exposure to the sun, exposure to cows' milk etc., there are important differences between populations in culture, religion, traditions, marriage customs and so on, which could have exerted different selective effects on genes. That's not at all implausible. The populations concerned have been geographically more or less separated for long enough. So the Genetic Book of the Dead might include a description of ancestral cultures. Another way to put that is that genes and memes cooperate with each other in mutually compatible cartels. This is what E. O. Wilson meant when he spoke, long ago, of 'gene–culture coevolution'. Is there a 'Memetic Book of the Dead'? And does it include descriptions of ancestral genes as well as memes? I leave that as a field for the reader to plough, adding as a seed the suggestion that cultural, linguistic or religious barriers to gene flow and meme flow could play the same role as geographic barriers in fostering evolutionary divergence. Interestingly, such cultural barriers could work even where geographic distances between populations are too small to do so. If the enmities between neighbouring valleys in the New Guinea highlands have isolated their warring populations sufficiently to foster the evolution of a thousand mutually unintelligible languages, what does that do to gene flow between them? If the Genetic Book of the Dead is a description of ancestral worlds including their cultures, why should there not be a Memetic Book of the Dead whose descriptions include ancestral genes? And, *mutatis mutandis*, why should the Genetic Book of the Dead not include a description of ancestral memes?

I've been a little detached from it, but there has grown up a large literature on memetics, including many books with 'meme' in the title. Significant advances in meme theory have been made by, among others, Susan Blackmore (in *The Meme Machine*), Robert Aunger (in *The Electric Meme*) and Daniel Dennett (in several books including *Consciousness Explained*, *Darwin's Dangerous Idea*, *Breaking the Spell* and *Intuition Pumps*). Both Dennett and Blackmore see memetics as playing a crucial role in human evolution, including the evolution of mind. Sue Blackmore has convened a series of 'Memelab' workshops in the eccentrically beautiful Devon house she shares with her husband, the television science presenter Adam Hart-Davis (who, incidentally, was involved, when at Oxford University Press, with the publication of *The Selfish Gene*). Each one organized as a complete weekend house party, with the participants sleep-ing in the house and eating meals together, these workshops capture, for me, some of the wonderfully companionable thinking-aloud feeling that I have always found so congenial. If the weather is kind the workshop ends in a windy climb up a Dartmoor tor. On one memorable occasion, Dan Dennett was able to be present and, as he usually does, he raised the game of the rest of us.

Chinese junk and Chinese Whispers

I wrote the foreword to Sue Blackmore's *The Meme Machine*, and used the occasion to offer an answer to one of the main criticisms of meme theory. The criticism was that memes, unlike genes, don't have high-fidelity replication. As the generations go by, the critic complains, the information will degenerate, a condition fatal to evolution. In DNA, the sequence ATGCGATTC will be precisely copied (or, if miscopied, will contain a definite, discretely identifiable error). But when a meme, such as a nursery rhyme, is copied–say from father to child–the replication is imprecise. The child's voice is higher, her vowels are not pronounced in exactly the same way, and so on. Therefore memes are not like genes and can't be the basis for evolution because the replication fidelity is not high enough.

The criticism is superficially plausible but demonstrably wrong. I answered it with a series of thought experiments, versions of the childhood game of 'Chinese Whispers' ('Telephone' to American children). Imagine twenty children in a line. I whisper a sentence to the first child. It might be 'Down in a deep dark dell sat an old cow munching a beanstalk.' The first child whispers what she has heard to the second child, and so on down the line. The twentieth child then sings out the 'evolved' version of the sentence. It might have become distorted, provoking hilarity. But, if the phrase is short, and especially if it means something in the children's own language, there's a good chance that it will survive, intact, to the end of the line. Now, if we take a particular instance of the game where the phrase emerges correctly at the far end of the line, I make the following point. It doesn't matter that each child's vocalization of the phrase is not a precise copy of the preceding child's. One child might have an Irish accent, the next a Scottish accent, the next a Yorkshire accent and so on. Because the phrase is meaningful to them in their own shared language, each child will 'normalize' it. Scottish vowels are different from Yorkshire vowels, but the difference doesn't detract from the content. An Australian child hears the words, recognizes them as drawn from the shared lexicon of the English language, and correctly renders them in such a way that the next child, with an American accent, can understand the phrase and pass it on.

There might be a 'mutation' somewhere along the way. For example, say child number 14 changes 'dark' to 'dank', and the mutant form, 'dank', is then copied to the end of the line. That would be interesting in itself. But let's take the case where no such mutation occurs. Now suppose that an experimenter with a tape recorder eavesdrops on each whispering child down the line. She separates the nineteen recordings as separate little tapes and scrambles them in a hat. Independent observers are then given the tapes and asked to rank them in order of resemblance to the original message as spoken to the first child. You know what the result would be. Assuming no mutations of the dark/dank kind, there will be no tendency for tapes that come from early in the sequence to be better than those that come late. There is no tendency for degeneration as the copying process proceeds. That alone should be enough to confound the criticism I mentioned.

But let's take the thought experiment one stage further. Suppose the message is in a language the children do not know. Suppose it is *Arma virumque cano, Troiae qui primus ab oris.* Again, we know what will happen, without needing to do the experiment. The children, knowing no Latin, can only imitate phonetically. The final sounds emerging from the twentieth child will have lost almost all resemblance to the original sound spoken to the first child. Moreover, if we do the experiment of shuffling the tapes in a hat, once more we know what the result will be. Independent observers will certainly be able to rank the tapes in order of resemblance to the original message: there will be a steady deterioration in resemblance to the original, from the first tape to the last.

You could do parallel experiments not with words but with skills: for example, a skill in carpentry (simulating the transmission from master carpenter to apprentice and so on through twenty 'generations'). Here, the equivalent 'normalization', playing a role equivalent to the shared language, will, I suspect, be an appreciation of what the skill is designed to achieve. If, for instance, the master is teaching the apprentice how to hammer in a nail, the apprentice will probably not imitate the precise number and strength of hammer blows. Rather, he will imitate the *goal* the master is trying to achieve, namely 'head of nail flush with wood' and he'll go on hammering until the goal is achieved. The goal is what he will imitate, and that is what will be passed on to the next 'apprentice'.

In my Blackmore foreword, I used the example of another manual skill: folding paper to make an origami Chinese junk. Its plausibility as a meme is illustrated by the fact that, when I introduced it to my boarding school, it spread like a measles epidemic. Even more interesting, I learned the skill from my father who had, in his turn, learned it when it spread as an epidemic through the very same school a quarter-century earlier.

When imitating an origami skill, each child imitates, not the exact hand movements of the preceding child but a 'normalized' version consisting of a perception of what the preceding child is *trying* to do. For example, the 'apprentice' child will infer that the 'teacher' child is trying to fold exactly into the middle. If the 'teacher' is clumsy and his fold actually overshoots the midline slightly, the 'apprentice' will ignore the error and try to fold exactly into the midline. And now, the equivalent of the 'tapes in the hat' is to ask independent observers to rank the nineteen Chinese junks. Assuming no major mutations (which again would be interesting in their own right), there will be no tendency for junks later in the sequence to show any 'degeneration' compared to junks early in the sequence. There'll be good ones and bad ones dotted around the line because, I strongly suspect, skilled children will not attempt to imitate obvious incompetence like folding a little way past the midline, but will 'normalize'.

There may be good objections to the meme/gene analogy but 'degeneration' because of low-fidelity replication is not one of them.

If we wanted to perform an experiment in memetics, what might we do? Perhaps take a word with a conventional pronunciation, invent a 'mutant' mispronunciation and broadcast it, daily, to tens of thousands of people; then, later, investigate whether the mutant pronunciation takes over the meme pool and becomes the norm. An expensive protocol, unlikely to attract funds from a grant-giving agency. Fortunately, however, by sheer happenstance, the expensive part of the experiment is sometimes done for us. Trains of the London underground system have an on-board tannoy, which daily broadcasts, to tens of thousands of passengers, the names of the stations. The pre-mutant pronunciation of Marylebone station is (something like) 'marry-le-bön'. The mutant pronunciation on the Bakerloo Line, spoken by a young woman's recorded voice, is 'marley-bone'. All that remains for the experiment is to ask a random sample of commuters on the Baker-loo Line how they pronounce the name, and repeat the sample at yearly intervals; then investigate the dispersive spread of the meme by sampling the British population as a whole. My guess is that the mutant form has already started to propagate quite widely. I facetiously suggest that the last bastion to fall will be the Marylebone Cricket Club, the famous MCC.

Models of the world

Under Horace Barlow's influence, I came to see an animal's sensory systems, especially the set of tuned recognition neurones in the brain, as a kind of model of the world in which the animal lives. In the same vein, I presented the genes of the animal as a digital description of past worlds–a kind of statistical average of the living conditions of the animal's ancestors, environments in which the ancestors had survived. I saw the gene pool of a species as an averaging computer, averaging the properties of ancestral worlds. Similarly, brains, as they learn, average the statistical properties of the world that the individual animal has experienced during its own lifetime. As the sculptor which is natural selection chips away at a gene pool, turning it into a descriptive model of averaged ancestral worlds, so individual experience sculpts brain models of the current world. In both cases the models are updated by data from the world, but on the timescale of generations for the gene pool model, the timescale of individual development for the brain model. I've long liked this poem by Julian Huxley (with whom, for some reason, I identified as an undergraduate), and I quoted it in *A Devil's Chaplain*:

> The world of things entered your infant mind
> To populate that crystal cabinet.
> Within its walls the strangest partners met,
> And things turned thoughts did propagate their
>   kind.
> For, once within, corporeal fact could find
> A spirit. Fact and you in mutual debt
> Built there your little microcosm—which yet
> Had hugest tasks to its small self assigned.
> Dead men can live there, and converse with
>   stars:
> Equator speaks with pole, and night with day;
> Spirit dissolves the world's material bars—
> A million isolations burn away.
> The Universe can live and work and plan,
> At last made God within the mind of man.

I would now add another verse, in imitation of the Huxley style:[1]

---

[1] I encountered Julian Huxley only once, when he was old and I was young. The Department of Zoology at Oxford had commissioned a joint

Ancestral worlds invade your species' genes,
Encoding long forgotten deaths and lives.
Digital texts enshrining what survives,
Distilled from genomes now in smithereens.
What went before? What happened? Who can
say?
Yet all is written in your DNA.

Reverting to Julian Huxley's poem and the brain models constructed on the timescale of individual development, many of my public lectures during the 1990s were devoted to the theme. I was especially inspired by the virtual reality software to which I was introduced during my Christmas Lectures year, and which I demonstrated to the children at the Royal Institution. I brought it all together in the chapter of *Unweaving the Rainbow* called 'Reweaving the world'.

When we think we are looking 'out' at the real world, there is a strong sense in which we are looking at a simulation, constructed in the brain but constrained by infor-

---

portrait of its three elder statesmen, Alister Hardy, John Baker and E. B. Ford. Sir Julian was invited to unveil it. As he read out each page of his speech, he put it to the bottom of the pile in his hand. After he read the last sheet and put it to the bottom, he simply started again, reading the top sheet. To the delight of the mischievous students present, he read right through the whole speech twice and was about to embark on a third reading when his wife bustled forward, seized him by the arm and hustled him off the stage.

mation flowing in from the real world. It is as though the brain contains cupboards full of models waiting to be pulled out at the behest of information flowing in from the sense organs. In a sort of 'exception that proves the rule' sense, visual illusions persuade us of this, as Richard Gregory has shown us in his books (and in his Simonyi Lecture in Oxford). The Necker Cube is a famous illusion, which I also used in *The Extended Phenotype* (see page 448) as an analogy for my two ways of looking at natural selection: the gene's-eye view and the 'vehicle' view. It is a two-dimensional pattern which is equally compatible with two alternative three-dimensional models in the cupboard. The brain could have been designed to plump for one of the two models and stick to it. Actually what it does is take first one model out of the cupboard, 'see' that for a few seconds, then put it back and take the other model out. So what we see is first one cube, then the other, then the first again, and so on.

Other famous illusions such as the Devil's Tuning Fork (see below), the Impossible Triangle (which I demonstrated during my Royal Institution Christmas Lectures[1]) and the Hollow Mask Illusion all make the same point more dramatically.

---

[1] Lecture 5, 20 minutes in: http://richannel.org/christmas-lectures/1991/richard-dawkins#/christmas-lectures-1991-richard-dawkins-the-genesis-of-purpose.

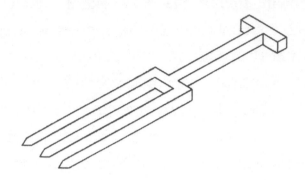

We move through a constructed world, a world of virtual reality. If we are sane, undrugged and awake, the constructed virtual reality through which we walk is constrained by sense data in ways that conduce to our survival: it is the real world, not a world of dreams or hallucinations, in which we must survive. Computer software allows us to walk through imaginary worlds, fantasy worlds, Greek temples or Fairylands, science-fiction landscapes of alien planets. When we turn our head, accelerometers in the helmet register the movement, and the images the computer presents to our eyes shift in register. We seem to be turning inside the Greek temple and now see a statue that had been 'behind us'. And when we dream at night, the brain's own virtual reality software emancipates itself from reality and we walk through splendid mansions of the mind, or run from constructed monsters in panic-ridden nightmares.

In *Unweaving the Rainbow* and in my lectures of the 1990s, I fantasized about a surgeon of the future, in a separate room from her patient, walking through his intestine, which is realistically simulated using data from an endoscope inside him. When she turns her head to the side, the tip of the endoscope swings in sympathy. She forges on through the virtual (yet constrained by endoscopic reality) intestine until she locates the tumour ahead of her. She wields the virtual chain saw in her toolkit, and the appropriate microsurgical blade at the tip of the endoscope mirrors her large arm movements in miniature and delicately cuts out the tumour. A parallel fantasy had a plumber of the future doing the same kind of thing, walking–or even swimming–down a virtual drain while his movements are mirrored in the real drain by a small robot deployed to clear a blockage. The key is constraint: the virtual worlds in which we move are not purely fantastic but guided along tramlines constrained to be usefully close to reality.

Our human mental cupboards are especially well furnished with face models, and we eagerly retrieve them given even the smallest encouragement through our optic nerves. This accounts for the numerous stories of Jesus or the Virgin Mary appearing in a slice of toast or a damp wall. The Hollow Mask Illusion (which also featured in my Christmas Lecture[1]) is the most spectacular manifestation of our face-happy model-retrieving facility. It's also significant that there is a named brain deficiency, prosopagnosia, sufferers from which can see normally except that they can't recognize faces, even faces of people they know well and love.

I reprised the theme in *The God Delusion*, showing how wrong we are to be impressed by visions and apparitions, ghosts and djinns, angels and Blessed Virgins. Our brains are masters of the art of virtual reality. To knock up a vision of a haloed, glowing, robed figure is child's play, and so is a still small voice in a storm. Many people are sincerely convinced that they have a personal experience of God: he speaks to them, appears to them in dreams and waking reveries. They need to be less impressed. Study Richard Gregory and the psychologists. Recognize the power of illusion. Understand how easily illusion morphs into delusion. God Delusion, for example.

The argument from personal incredulity

In *The Blind Watchmaker* I coined the phrase 'argument from personal incredulity' as a summation of the creationist's principal 'argument'. A less sarcastic

---

[1] Lecture 5, 18 minutes in: http://richannel.org/christmas-lectures/1991/richard-dawkins#/christmas-lectures-1991-richard-dawkins--the-genesis-of-purpose.

rendering would be 'argument from statistical improbability'–or 'argument from complexity', for statistical improbability is the relevant measure of complexity and the relevant provoker of incredulity. Here's how the argument always goes. A complex biological structure is extolled, having many parts arranged in a precise way. Any random rearrangement of the parts wouldn't work. The number of possible rearrangements is calculated and found, of course, to be astronomically large. Therefore the complicated arrangement can't have come about by chance. Therefore–here's where the argument shoots itself in the foot–God must have done it.

Darwin himself devoted part of a chapter to what he called 'organs of extreme perfection and complication'. He began with a famous sentence, much quoted by creationists:

> To suppose that the eye with all its inimitable contrivances for adjusting the focus to different distances, for admitting different amounts of light, and for the correction of spherical and chromatic aberration, could have been formed by natural selection, seems, I freely confess, absurd in the highest degree.

You can tell, can't you, from the tone of Darwin's sentence, that he wasn't going to leave it there. Doesn't his tone send a clear signal that a 'but' or a 'yet' is about to follow? He might even have been leading his readers on, beckoning them towards him so that the punch, when it came, had the greater impact: 'Yet reason tells me that . . .' Google finds a mere 39,300 occurrences of the latter clause, compared to 130,000 for the immediately preceding phrase, 'absurd in the highest degree'. As Darwin himself said elsewhere, 'Great is the power of steady misrepresentation . . .'

What's wrong with the argument from statistical improbability, of course, is that natural selection is not a theory of chance. Natural selection is the non-random filtering of random variation, and the reason it works is that the improvement is cumulative and gradual. In *The Blind Watchmaker* I illustrated this with the metaphor of a combination lock, for example guarding the door of a bank vault. In the BBC *Horizon* film with the same name, I actually hammed up an attempt to open a real bank vault with a random number. The whole point of a combination lock is that you need stupendous luck to break in by random dial-twiddling. But, if the lock had a fault such that the vault progressively creaked open another chink every time the number on the dial moved in the right direction, any fool could break in. That's the equivalent of gradualistic natural selection.

My later metaphor of Mount Improbable did the

same explanatory duty. As briefly mentioned above, when I worked with Russell Barnes on the television film *Root of All Evil?* we simulated Mount Improbable in the 'Garden of the Gods' at Colorado Springs. I was filmed standing atop a sheer precipice to represent the 'creationist' or 'huge stroke of luck' side of the mountain, where achieving the improbable in a single step is equivalent to leaping from the foot to the summit in a single bound. Then the camera was re-sited and I had to tramp stolidly up the gentle, gradual slope on the 'evolutionary' side of the mountain: no problem evolving organs of limitless complexity, given only enough time and a gentle gradient of improvement with no sudden leaps. Television being what it is, of course, the gradient and the precipice were actually on different mountains (the 'Inspector Morse Effect', whereby the melancholy inspector is filmed entering one Oxford college and debouching into the quadrangle of another).

Of all the many reasons people give for theistic belief, the argument from statistical improbability is the one I meet by far the most often. Frequently, as I said, it goes with a naive mathematical calculation of the stupendous odds against something as complex as an eye or a haemoglobin molecule coming into existence 'by chance'. It also extends to the idea of the Big Bang as the origin of all things. Here are a couple of examples from a Jehovah's Witness pamphlet, and they are entirely typical of the genre:

> Imagine that someone told you that there was an explosion at a printing plant and that the ink spattered onto the walls and ceilings and formed an unabridged dictionary. Would you believe it? How much more unbelievable is it that everything in the orderly universe came about as a result of a random big bang?

> If you were walking through the forest and discovered a beautiful log cabin, would you think: 'How fascinating! The trees must have fallen in just the right way to make this house.' Of course not! It's just not reasonable. So why should we believe that everything in the universe just happened to come about?

I have to confess that this kind of thing frustrates me and sometimes provokes me to an impatience which I (only slightly) regret. This is for three reasons. First, if it were really true that the odds stacked against a naturalistic explanation of apparent design really were monumental on this scale, outnumbering the atoms in the universe, only an equally monumental fool would fall for it. I hate to descend to an argument from authority, but is it too much to ask that this might perhaps arouse a little inkling of a misgiving in the mind of the creationist? Isn't it worth at least a fleeting thought that those who trot out these gigantic improbabilities might, just conceivably, have missed the point? Scientists get things wrong sometimes. They seldom get things wrong by 80 orders of magnitude.

My second reason for irritation is that the 'argument against blind chance' misses so much of real value, mainly the spectacular power and elegance of science, epitomized by Darwin's theory. Supremely powerful yet supremely simple, it really is one of the most beautiful ideas ever to occur to a human mind, and the uninitiated are missing it. Worse, if they thrust their misunderstanding on children they are denying children that beauty, the beauty of intellectual consummation.

And, third, the argument from statistical improbability (or complexity) is irritating because the astronomical odds against complexity arising by blind chance are simply a restatement of the *problem* that must be solved by any theory of existence, whether it is the Big Bang, or evolution, or the God theory. Anybody can see that the answer to the riddle of existence cannot be blind chance, or sudden springing into existence from nothing. This is especially true in the case of life, because the illusion of design there is so stunningly persuasive. The problem is to find the *alternative* to blind chance. The improbability of life is precisely the problem we need to solve. The God theory conspicuously doesn't solve it: merely restates it. Natural selection, being gradual and cumulative, solves the problem and is probably the only process that could solve it. It is clearly futile to try to solve the problem of life's complexity by postulating another complex entity called God. The same applies, albeit less obviously, to the problem of cosmological origins. The more the creationist piles on the statistical improbability, the more he shoots himself in the foot.

The God delusion

That point, about the statistical improbability of apparent design, pervades *The Blind Watchmaker* and *Climbing Mount Improbable*, and I nominated it explicitly as the central argument of *The God Delusion* (it is, of course, not original). The publication of the latter book led to a large number of alleged replies to the point about God being complex and therefore no solution to the riddle of complexity. The replies are all the same and all equally weak. They can be summed up in one sentence: 'God is not complex but simple.' How do we know? Because theologians say so, and they're the authorities on God, are they not? Easy. Win the argument by fiat! But you cannot have it both ways. Either God is simple, in which case he doesn't have the knowledge and design skills to provide the explanation of complexity that we seek. Or he is complex, in which case he needs explaining in his own right, no less than the complexity that he is being invoked to explain. The simpler you make your god, the less qualified he is to explain the complexity of the world. And the more complex you make him, the more does he require an explanation in his own right.

Peter Atkins dramatized the point in his beautifully written *Creation Revisited*, where he postulated a 'lazy god' and then, step by step, whittled down what the lazy god would need to do in order to make the universe we see. He concluded that the lazy god would need to do so little he might as well not bother to exist at all. As for granting God the complex *supplementary* skills he is supposed to deploy–listening to the thoughts of seven billion people simultaneously (not to mention conversing with dead ones too), answering their prayers, forgiving their sins, meting out posthumous punishment or reward, saving some cancer patients but not others– that only adds to the problem, and in a big way.

Darwinian evolution uniquely solves the problem of life's statistical improbability, because it works cumulatively and gradually. It really does broker a legitimate traverse from primordial simplicity to eventual complexity–and it is the only known theory capable of doing so. Human engineers can make complex things by design, but the whole point is that human engineers need to be explained too, and evolution by natural selection explains them at the same time as it explains the rest of life.

Of course, there's much more in *The God Delusion* than the central argument about statistical improbability. There are passages on the evolutionary origins of religion, on the roots of morality, on the literary value of religious scriptures, on religious child abuse and many other things. I like to think it's a humorous and humane book, far from the angry and strident polemic that is sometimes alleged. Some of the humour is satire, even ridicule, and it's true that the targets of such humour often have a hard time distinguishing good-natured ridicule from hate speech. One of the things I learned from Peter Medawar is that precisely aimed satirical ridicule is not the same thing as vulgar abuse (see also page 432 below). Nevertheless, religiously motivated critics often seem incapable of seeing the difference. One even suspected me of Tourette's Syndrome, though it is hard to believe he had read the book–he probably just fell in love with his own simile!

Given the level of vitriol visited on this book, it is quite surprising that, in all my hundreds of public appearances, including many in the so-called 'Bible Belt' of the United States, I have hardly ever experienced any heckling of any sort to my face, indeed, scarcely ever been subjected to any critical questioning. This is actually quite a disappointment, for I have found myself enjoying the rare exceptions–in particular the occasion when I was invited to lecture at Randolph Macon Women's College in Virginia (they now take men too). Randolph Macon is a decent liberal arts college, with high standards. But Liberty 'University', founded by the infamous Jerry Falwell, is in the same town, and a substantial busload from Liberty came across and occupied the front row of the lecture theatre at Randolph Macon. They monopolized the question and answer session, lining up in a body behind the microphones in the two aisles. Their questions were courteous to a fault, but all were frankly motivated by fundamentalist Christianity, profession of which is an entrance requirement imposed by that 'university'. I had, of course, no difficulty in disposing of each one in turn, to cheers from the Randolph Macon women. One questioner began by telling us that Liberty University possesses a dinosaur fossil labelled as three thousand years old. He asked me to explain how they might go about demonstrating the true age of such a fossil.[1] I explained that fossils are dated by several different radioactive clocks, running at very different speeds, and all independently agree that dinosaurs are no less than 65 million years old. I added:

> If it's really true that the museum of Liberty University has a dinosaur fossil which is labelled as being 3000 years old, then that is an educational disgrace. It is debauching the whole idea of a university, and I would strongly encourage any members of Liberty University who may be here to leave and go to a proper university.

That got the biggest cheer of the evening–because Randolph Macon is a proper university. Another of the questions that evening, 'What if you're wrong?' (Google it), together with my answer, has gone viral.

The only hostile heckling I have experienced was in Oklahoma where, in an enormous sports stadium, one man stood up in the middle of my speech and started yelling 'You have insulted my Saviour!' He was hustled out by uniformed bouncers, not by my wish. That same event, at the University of Oklahoma, was the only occasion where an attempt was made to prevent

---

[1] https://www.youtube.com/watch?v=qR_z85O0P2M

me speaking by legal means. State Representative Todd Thomsen introduced a bill into the state legislature of which the following is an extract (when you see a page littered with lots of Whereases you know to brace yourself for trouble).

WHEREAS, the University of Oklahoma, as a part of the Darwin 2009 Project, has invited as a public speaker on campus, Richard Dawkins of Oxford University, whose published opinions, as represented in his 2006 book 'The God Delusion', and public statements on the theory of evolution demonstrate an intolerance for cultural diversity and diversity of thinking and are views that are not shared and are not representative of the thinking of a majority of the citizens of Oklahoma; and

WHEREAS, the invitation for Richard Dawkins to speak on the campus of the University of Oklahoma on Friday, March 6, 2009, will only serve to present a biased philosophy on the theory of evolution to the exclusion of all other divergent considerations rather than teaching a scientific concept.

NOW, THEREFORE, BE IT RESOLVED BY THE HOUSE OF REPRESENTATIVES OF THE 1ST SESSION OF THE 52ND OKLAHOMA LEGISLATURE:

THAT the Oklahoma House of Representatives strongly opposes the invitation to speak on the campus of the University of Oklahoma to Richard Dawkins of Oxford University, whose published statements on the theory of evolution and opinion about those who do not believe in the theory are contrary and offensive to the views and opinions of most citizens of Oklahoma.

Rep. Thomsen further alleged that I was paid $30,000 for the lecture, and tried to get officials of the university penalized for squandering public money in this way. He ended up with egg on his face because I neither received, nor asked for, a penny. Moreover, his bill failed to pass. It is truly astonishing that his main objection to my lecturing on evolution was that I held 'views that are not shared and are not representative of the thinking of a majority of the citizens of Oklahoma'. What does Rep. Thomsen think a university is *for*?

Here's an example of the kind of thing in *The God Delusion* to which, I suspect, critics take exception as savage or strident, aggressive or offensive, but which I see as good-natured satire: a touch of the stiletto perhaps, but miles away from the bludgeon or from vulgar abuse. After pointing out that Roman Catholicism, though avowedly monotheistic, has leanings towards polytheism, with the Virgin Mary a goddess in all but name and the saints attracting personal supplication as demigods each in his own specialist field of expertise, I continued:

Pope John Paul II created more saints than all his predecessors of the past several centuries put together, and he had a special affinity with the Virgin Mary. His polytheistic hankerings were dramatically demonstrated in 1981 when he suffered an assassination attempt in Rome, and attributed his survival to intervention by Our Lady of Fatima: 'A maternal hand guided the bullet.' One cannot help wondering why she didn't guide it to miss him altogether. Others might think the team of surgeons who operated on him for six hours deserved at least a share of the credit; but perhaps their hands, too, were maternally guided. The relevant point is that it wasn't just Our Lady who, in the Pope's opinion, guided the bullet, but specifically Our Lady *of Fatima*. Presumably Our Lady of Lourdes, Our Lady of Guadalupe, Our Lady of Medjugorje, Our Lady of Akita, Our Lady of Zeitoun, Our Lady of Garabandal and Our Lady of Knock were busy on other errands at the time.

Wounding sarcasm perhaps; but 'strident'? I don't think so; and certainly not symptomatic of 'Tourette's Syndrome'. I think it's legitimate satire and would like to think quite funny, but it gave grave offence not only to Catholics but even to the rightly admired Melvyn Bragg, a non-religious cultural commentator and enabler who is well on his way to national treasurehood. Such censure arises only, I suspect, because we have taken on board a convention that religion is off limits to criticism, even to the gentle mockery indulged in the above passage. The point was well put by Douglas Adams in his impromptu speech in Cambridge (see page 566) some years before *The God Delusion*:

> Religion . . . has certain ideas at the heart of it which we call sacred or holy or whatever. What it means is, 'Here is an idea or a notion that you're not allowed to say anything bad about; you're just not. Why not?—because you're not!' If somebody votes for a party that you don't agree with, you're free to argue about it as much as you like; everybody will have an argument but nobody feels aggrieved by it. If somebody thinks taxes should go up or down you are free to have an argument about it. But on the other hand if somebody says 'I mustn't move a light switch on a Saturday,' you say, 'I *respect* that.'

Why should it be that it's perfectly legitimate to support the Labour party or the Conservative party, Republicans or Democrats, this model of economics versus that, Macintosh instead of Windows–but to have an opinion about how the Universe began, about who created the Universe . . . no, that's holy? . . . We are used to not challenging religious ideas but it's very interesting how much of a furore Richard creates when he does it! Everybody gets absolutely frantic about it because you're not allowed to say these things. Yet when you look at it rationally there is no reason why those ideas shouldn't be as open to debate as any other, except that we have agreed somehow between us that they shouldn't be.

I emphasized this double standard again in the preface to the paperback edition of *The God Delusion* (built around the commonly heard weasel phrase, 'I'm an atheist *but* . . .'; I've already mentioned Salman Rushdie's more recent noticing of the 'but brigade'). I compared the relatively understated language of my book to the savagery we take for granted in theatre criticism, in political commentary, even in restaurant reviews: '. . . .he most disgusting thing I've put in my mouth since I ate earthworms at school' '. . . quite the worst restaurant in London, maybe the world . . .'

That notorious passage about the eight Catholic Virgin goddesses occurs in chapter 2 of *The God Delusion*. It is the long opening sentence of that same chapter that has undoubtedly given the most offence–and even led to charges of 'anti-Semitism', as I have noted in an earlier chapter.

> The God of the Old Testament is arguably the most unpleasant character in all fiction: jealous and proud of it; a petty, unjust, unforgiving control-freak; a vindictive, bloodthirsty ethnic cleanser; a misogynistic, homophobic, racist, infanticidal, genocidal, filicidal, pestilential, megalomaniacal, sadomasochistic, capriciously malevolent bully.

And yet, whether apologists like it or not, every single word of that list is eminently defensible. Examples abound in the Bible. I considered listing them here but soon realized that the illustrative quotations would fill a book. And my goodness, there's an idea: a book! I know of nobody better qualified to write such a book than my friend Dan Barker. I put it to him and he jumped at it.

Dan used to be a preacher. As I wrote in my foreword to his 2008 book, *Godless: How an Evangelical Preacher Became One of America's Leading Atheists*,

> The young Dan Barker was not just a preacher, he was the kind of preacher that 'you would not want to sit next to on a bus'. He was the kind of preacher who would march up to perfect strangers in the street and ask them if they were saved; the kind of doorstepper on whom you might be tempted to set the dogs.

Dan knows his Bible as intimately as Charles Darwin knew his beetles and his barnacles, and I am delighted to say he took my suggestion and is now writing a book devoted to illustrating, chapter and merciless verse, every word of that opening sentence of my chapter 2, in order.

Of course, Christian apologists reply, we all know about the awkward and embarrassing passages in the Old Testament. But whatabout[1] the New Testament? Yes, you can find some gently humane wisdom in the teachings of Jesus. The Sermon on the Mount is so good, one wishes more Christians would follow it. But the core

---

[1] 'Whataboutery' is a new abstract noun now in the process of entering our language (it has a Wikipedia entry, but has yet to make it into the *Oxford English Dictionary*). It is most often used to downplay a negative point by diverting attention to something else.

myth of the New Testament (for which, to be fair, St Paul not Jesus is to blame) shares the obnoxiousness of the Genesis myth of Abraham's near-sacrifice of Isaac,[1] from which it may be derived. I made this point in *The God Delusion* and later reprised it in a P. G. Wodehouse parody that I wrote for a Christmas anthology in 2009. I was unfortunately obliged, for copyright reasons, to change the names of Jeeves, Bertie and the Reverend Aubrey Upjohn, Bertie's headmaster at the school where he had once won a prize for Scripture Knowledge.

'All that stuff about dying for our sins, redemption and atonement, Jarvis. All that "and with his stripes we are healed"' carry-on. Being, in a modest way, no stranger to stripes administered by old Upcock, I put it to him straight. "When I've performed some misdemeanour"–or malfeasance, Jarvis?'

'Either might be preferred, sir, depending on the gravity of the offence.'

'So, as I was saying, when I was caught perpetrating some malfeasance or misdemeanour, I expected the swift retribution to land fairly and squarely on the Woofter trouser seat, not some other poor sap's innocent derrière, if you get my meaning?'

---

[1] Ishmael in the Islamic version of the myth.

'Certainly, sir. The principle of the scapegoat has always been of dubious ethical and jurisprudential validity. Modern penal theory casts doubt on the very idea of retribution, even where it is the malefactor himself who is punished. It is correspondingly harder to justify vicarious punishment of an innocent substitute. I am pleased to hear that you received proper chastisement, sir.'

'Quite, Jarvis.'

'I am so sorry sir, I did not intend . . .'

'Enough, Jarvis. This is not dudgeon. Umbrage has not been taken. We Woofters know when to move swiftly on. There's more. I hadn't finished my train of thought. Where was I?'

'Your disquisition had just touched upon the injustice of vicarious punishment, sir.'

'Yes, Jarvis, you put it very well. Injustice is right. Injustice hits the coconut with a crack that resounds around the shires. And it gets worse. Now, follow me like a puma here. Jesus was God, am I right?'

'According to the Trinitarian doctrine promulgated by the early Church Fathers, sir, Jesus was the second person of the Triune God.'

'Just as I thought. So God—the same God who made the world and was kitted out with enough nous to dive in and leave Einstein gasping at the shallow end, God the all-powerful and all-knowing creator of everything that opens and shuts, this paragon above the collarbone, this fount of wisdom and power—couldn't think of a better way to forgive our sins than to turn himself over to the gendarmerie and have himself served up on toast. Jarvis, answer me this. If God wanted to forgive us, why didn't he just forgive us? Why the torture? Whence the whips and scorpions, the nails and the agony? Why not just forgive us? Try that on your Victrola, Jarvis.'

'Really sir, you surpass yourself. That is most eloquently put. And if I might take the liberty, you could even have gone further. According to many highly esteemed passages of traditional theological writing, the primary sin for which Jesus was atoning was the Original Sin of Adam.'

'Dash it, Jarvis, you're right. I remember making the point with some vim and *élan*. In fact, I rather think that may have been what tipped the scales in my favour and handed me the jackpot in that scripture knowledge fixture. But do go on, Jarvis, you interest me strangely. What was Adam's sin? Something pretty fruity, I imagine. Something calculated to shake hell's foundations?'

'Tradition has it that he was apprehended eating an apple, sir.'

'Scrumping? That was it? That was the sin that Jesus had to redeem—or atone according to choice? I've heard of an eye for an eye and a tooth for a tooth, but a crucifixion for a scrumping? Jarvis you've been at the cooking sherry. You are not serious, of course?'

'Genesis does not specify the precise species of the purloined comestible, sir, but tradition has long held it to have been an apple. The point is academic, however, since modern science tells us that Adam did not in fact exist, and therefore was presumably in no position to sin.'

'Jarvis, this takes the chocolate digestive, not to say the mottled oyster. It was bad enough that Jesus was tortured to atone for the sins of lots of other fellows. It got worse when you told me it was only one other fellow. It got worse still when that one fellow's sin turned out to be nothing worse than half-inching a D'Arcy Spice. And now you tell me the blighter never existed in the first place. Jarvis, I am not known for my size in hats, but even I can see that this is completely doolally.'

'I would not have ventured to use the epithet myself, sir, but there is much in what you say. Perhaps in mitigation I should mention that modern theologians regard the story of Adam, and his sin, as symbolic rather than literal.'

'Symbolic, Jarvis? Symbolic? But the whips weren't symbolic. The nails in the cross weren't symbolic. If, Jarvis, when I was bending over that chair in the Rev Aubrey's study, I had protested that my misdemeanour, or malfeasance if you prefer, had been merely symbolic, what do you think he would have said?'

'I can readily imagine that a pedagogue of his experience would have treated such a defensive plea with a generous measure of scepticism, sir.'

'Indeed you are right, Jarvis, Upcock was a tough bimbo. I can still feel the twinges in damp weather. But perhaps I didn't quite skewer the point, or nub, in re the symbolism?'

'Well, sir, some might consider you a trifle hasty in your judgment. A theologian would probably aver that Adam's symbolic sin was not so very negligible, since what it symbolised was all the sins of mankind, including those yet to be committed.'

'Jarvis, this is pure apple sauce. "Yet to be committed?" Let me ask you to cast your mind back yet again to that doom-laden scene in the beak's study. Suppose I had said, from my vantage point doubled up over the armchair, "Headmaster, when you have administered the statutory six of the juiciest, may I respectfully request another six in consideration of all the other misdemeanours, or peccadilloes, which I may or may not decide to commit at any time into the indefinite future. Oh, and make that all future misdemeanours committed not just by me but by any of my pals." Jarvis, it doesn't add up. It doesn't float the boat or ring the bell.'

'I hope you will not take it as a liberty, sir, if I say that I am inclined to agree with you. And now, if you will excuse me, sir, I would like to resume decorating the room with holly and mistletoe, in preparation for the annual yuletide festivities."[1]

There are good verses as well as nasty ones in both the Old and the New Testaments. But there has to be some criterion for choosing which verses are good and which bad. To avoid circularity, that criterion must come from outside the scriptures. It is hard to piece together where our dominant criteria for morality come from, but they are clearly exhibited in what I called the 'shifting moral *Zeitgeist*'. We today are twenty-first-century moralists, unmistakably labelled with

---

[1] This is an extract from 'The Great Bus Mystery' in Ariane Sherine, ed., *The Atheist's Guide to Christmas* (London, HarperCollins, 2009).

twenty-first-century values. Even the most advanced and progressive thinkers of the nineteenth century, men like T. H. Huxley, Charles Darwin and Abraham Lincoln, would appal us with their racism and sexism if they were to enter a modern dinner party or web chatroom. Huxley and Lincoln both took the inferiority of black men for granted, and many of the US Founding Fathers owned slaves. Most of the world's democracies introduced female suffrage as late as the 1920s, France in 1944, Italy in 1946, Greece in 1952 and Switzerland not until a staggering 1971. Justifications for opposing women's suffrage unbelievably included: 'It's not necessary because women just vote with their husbands anyway.' The moral *Zeitgeist* moves inexorably in one direction, with the result that even the most progressive thinkers of the nineteenth century tend to lag behind the least progressive thinkers of the twenty-first. It is by the standards of a civilized twenty-first-century conversation that we cherry-pick the Bible and decide that this verse is bad but that one good. And since we evidently have preferred and agreed standards for cherry-picking, why bother to go to the Bible at all, if we seek moral guidance? Why not go straight to our moral *Zeitgeist* and cut out the scriptural middleman?

There are, on the other hand, good reasons to go back to scripture as literature, because, as I also said in *The God Delusion*, our whole culture is so bound up with it that you can't take your allusions or understand your history if you are biblically illiterate. Indeed, I filled two pages with close-packed biblical quotations, phrases familiar to everyone but whose scriptural origins are known to few. I am strongly in favour of teaching children *about* religion, even as I passionately oppose indoctrinating children in the *particular* religious tradition into which they happen to have been born. I have repeatedly called attention to the strange fact that, although we would cringe if we met a phrase like 'Existentialist child' or 'Marxist child' or 'Postmodernist child' or 'Keynesian child' or 'Monetarist child', our whole society, secular as well as religious, blithely fails to cringe on hearing 'Catholic child' or 'Muslim child'. We need to raise consciousness of the unacceptability of such phrases, exactly as feminists succeeded in raising consciousness about phrases like 'one man one vote'. Please, *never* speak of a Catholic child or a Protestant child or a Muslim child. Speak, instead, of 'a child of Catholic parents' or 'a child of Muslim parents'. Alarmist demographic calculations concluding, for example, that 'France will have a Muslim majority by the year so and so' are entirely based on the gratuitous presumption that children automatically inherit the religion of their parents. That is an assumption that must be fought, not unthinkingly taken for granted.

It's a recurrent question, one that has often been put to me since publication of *The God Delusion*, whether we should be conciliatory and 'accommodationist' when arguing with religious people, or whether to be totally frank. I mentioned this earlier in connection with public questions put to me by Lawrence Krauss and Neil deGrasse Tyson. I suspect that each of the two approaches works well, but with different audiences. I once heard a well-received lecture entitled 'Don't be a dick' in which the speaker asked the audience for a show of hands: 'If somebody called you an idiot, would you be more or less likely to be persuaded to their point of view?' Needless to say, the vote was overwhelmingly negative. But the speaker should have asked a different question. 'If you were a third party, sitting on the fence, listening to an argument between two people, and one of them gave good reasons to think the other was an idiot, would that bias you in favour of one or the other?' I hope I never stoop to gratuitous personal insults, but I do think humorous or satirical ridicule can be an effective weapon. It must hit its target accurately. The American satirical cartoon *South Park* once included me in a lampoon. It's an instructive illustration because half of it was an accurately targeted satirical '*touché* moment' (a future century in which the atheist 'movement' had split schismatically into warring factions) and the other half was not aimed at any target at all and could not be called satire in any sense (a cartoon of me buggering a bald transsexual).

If there are passages of *The God Delusion* which can be read by the sensitive as strongly critical if not actually 'strident', the book both ends and begins gently. The last section, entitled 'The mother of all burkas', is an extended metaphor. The life-impoverishing slit in the burka stands for the narrowness of a pre-scientific world-view, and I go on to illustrate various ways in which the slit can be widened, with consequent enhancement of life and its joys. Science widens it, for example, by showing what a tiny fraction of the electromagnetic spectrum is visible to our senses.

The beginning of the book is a generous reminiscence about a chaplain at my old school who, as a boy, was lying with his face in the grass and was inspired, by a moment of revelation, to embrace the religion which was to become his life's path. 'Suddenly the micro-forest of the turf seemed to swell and become one with the universe and with the rapt mind of the boy contemplating it.' I respected his epiphany enough to say that, 'in another time and place, that boy could have been me under the stars, dazzled by Orion, Cassiopeia and Ursa Major, tearful with the unheard music of the Milky Way, heady with the night scents of frangipani and trumpet flowers in an African garden'.

The reference to Ursa Major was consciously prompted by the memory of a poem, written by my mother as a girl, which concluded with the following lines:

> The Great Bear stands upon his head,
> His paws among the apple boughs
> That, dark against a darker sky,
> Wave in the wind and tap their twigs
> With little sounds forlorn and sad
> Within the night's dark emptiness.

My opening page ended with a warmly indulgent reminiscence of how we used to distract our chaplain in divinity lessons by asking him to recall his wartime service in the RAF; and I quoted, in his honour, John Betjeman's gently affectionate poem, 'Our padre'.

> Our padre is an old sky pilot,
> Severely now they've clipped his wings,
> But still the flagstaff in the Rect'ry garden
> Points to Higher Things.

After my book was published, I was delighted when an old boy of the same school sent a verse in to my website, RichardDawkins.net:

> I knew your flying chaplain,
> As my Housemaster I oughta.
> While you embraced his liberal views
> I just embraced his daughter.

Whatever the faults of a British private school education, Oundle must have something going for it if it produces alumni who can turn out that sort of thing.

Full circle

I'll end where I began, on my seventieth birthday, among a hundred guests at the dinner that Lalla put on for me in New College Hall. After the choir had sung nostalgic songs, after speeches by Lalla herself, by Alan Grafen, my star pupil and later mentor, and by Sir John Boyd, former Ambassador to Japan and then Master of Churchill College, Cambridge, I made a speech of my own. It culminated in a little poem (verse, rather; I don't think I would dignify it as a poem) filled with parodic allusions–to A. E. Housman (favourite of my youth, and also of Bill Hamilton who indeed reminded me of the melancholy protagonist of *A Shropshire Lad*), to the Book of Psalms, to George and Ira Gershwin, to our national game of cricket, to Shakespeare, G. K. Chesterton, Andrew Marvell, Dylan Thomas and Keats.

Now of my three score years and ten
Seventy won't come again:
And take from seventy springs the lot . . .
Subtraction tells you what I've got.
But only if you're so alarmist
As to believe the ancient psalmist.
For what is said in holy writ
I'm one who doesn't care a bit.
Away with actuarial mystics!
I'll throw my lot with hard statistics.
The bible may be old and quaint . . .
Necess'rily so . . . it ain't
(I'll go along with George and Ira).
Across the Reaper's bows I'll fire a
Warning shot. I'm not about
To let life's Umpire give me out,
'Leg before', or 'caught and bowled',
At least until I'm really old
And reach that bourn—the one we learn,
From which no travellers return:
That decent inn—no Marriott—
Presaged by time's winged chariot.
Still time to gentle that good night.
Time to set the world alight.
Time, yet new rainbows to unweave,
Ere going on Eternity Leave.

Acknowledgements

For advice, help and support of various kinds, I should like to thank Lalla Ward, Rand Russell, Marian Stamp Dawkins, Sally Gaminara, Hilary Redmon, Gillian Somerscales, Sheila Lee, John Brockman, Alan Grafen, Lars Edvard Iverson, David Raeburn, Michael Rodgers, Juliet Dawkins, Jane Brockmann, Lawrence Krauss, Jeremy Taylor, Russell Barnes, Jennifer Thorp, Bart Voorzanger, Miranda Hale, Steven Pinker, Lisa Bruna, Alice Dyson, Lucy Wainwright, Carolyn Porco, Robyn Blumner, Victor Flynn, Alan Canon, Ted Kaehler, Eddie Tabash, Lary Shaffer, Richard Brown.

Picture acknowledgements

E very effort has been made to trace copyright hold-ers, but any who have been overlooked are invited to get in touch with the publishers.

Credits read from top left clockwise on double-page spreads.

Section one

Sir Peter Medawar: © Godfrey Argent Studio; Nikolaas Tinbergen: courtesy Lary Shaffer; Douglas Adams: © LFI/Photoshot; Carl Sagan, *c.* 1984: NASA/Cosmos; David Attenborough and RD: © Alastair Thain; John Maynard Smith: courtesy the University of Sussex; Bill Hamilton: photo courtesy Marian Dawkins

Cutting from the *Gainesville Sun*, 11 May 1979: courtesy Jane Brockmann; view of the Smithsonian Tropical Research Center, Panama, 1977: © STRI; Michael Robinson; Fritz Vollrath: photos supplied by the author; *Sphex ichneumoneus*: courtesy Jane Brockmann

Schlosshotel, Kronberg: © imageBROKER/Alamy; Karl Popper, 1989: IMAGNO/Votava/TopFoto; first meeting of the Human Behavior and Evolution Society, Evanston, Illinois, August 1989: courtesy Professor Edward O. Wilson; RD at Melbu, 1989: photo by Tone Brevik courtesy Nordland Akademi, Melbu; Betty Pettersen, 1992: courtesy Nordland Akademi, Melbu; general view of Melbu: photo Odd Johan Forsnes courtesy Nordland Akademi, Melbu; Jim Lovell and Alexei Leonov; Starmus conference, June 2011: both © Max Alexander; spaceman drawing: STARMUS courtesy Garik Israelien

RD and Neil DeGrasse Tyson, Howard University, September 2010: Bruce F Press, Bruce F Press Photography; RD and Lawrence Krauss: photo supplied by the author

Section two
RD and Lalla, 1992: © Norman McBeath; RD and the Archbishop of Canterbury, Rowan Williams, Oxford, February 2012: Andrew Winning/Reuters/Corbis; Robert Winston and RD, Cheltenham Literature Festival, October 2006: © Retna/Photoshot; RD and Joan Bakewell, Hay Festival, May 2014: © Keith

Morris News/Alamy; RD at a book-signing: Mark Coggins

Alan Grafen and Bill Hamilton at the Great Annual Punt Race, mid-1970s: both courtesy Marian Dawkins; RD, Francis Crick, Lalla, Richard Gregory, Oxford, early 1990s: photo by Odile Crick supplied by the author; RD having received an honorary degree from Richard Attenborough, University of Sussex, July 2005: photo courtesy the author; Mark Ridley, c. 1978: courtesy Marian Dawkins; Great Annual Punt Race, c. 1976: photo Richard Brown

Royal Institution Christmas Lectures, London 1991 and Japan 1992, left-hand page: all stills from *Growing Up in the Universe*; right-hand page: all © The Yomiuri Shimbun

RD in front of the Triton: courtesy Edith Widder; Raja Ampat, Indonesia: ©Images & Stories/Alamy; RD in a canoe: photo Ian Kellet supplied by the author; RD on Heron Island: photo supplied by the author; Heron Island, Great Barrier Reef © Hilke Maunder/ Alamy; Edith Widder in Triton; RD, Mark Taylor and Tsunemi Kubodera in Triton; giant squid: all courtesy Edith Widder

Paternal clan chart (slightly amended): © Oxford Ancestors; RD and James Dawkins: photo supplied by the author

Section three

Commander Gennady Padkalka (top), Charles Simonyi (centre) and Flight Engineer Michael R. Barratt on Expedition 19 launch day in Kazakhstan, March 2009: NASA/Bill Ingalls; Charles Simonyi at home in Seattle, *c.* 1997: © Adam Weiss/Corbis; Martin Rees, Astronomer Royal, May 2009: © Jeff Morgan 12/Alamy; Richard Leakey in Kenya, January 1994: David O'Neill/ Associated Newspapers/Rex; Paul Nurse, October 2004: © J. M. Garcia/epa/Corbis; Harry Kroto, 2004: Nick Cunard/Rex; Carolyn Porco presenting the first images transmitted from the spacecraft Cassini at a news conference in Pasadena, July 2004: Reuters/ Robert Galbraith; Steven Pinker, 1997; Jared Diamond, Aspen, Colorado, February 2010: © Lynn Goldsmith/ Corbis; Daniel Dennett, Hay Festival, May 2013: © D. Legakis/Alamy; Richard Gregory: Martin Haswell

Enemies of Reason team; Tim Cragg and Adam Prescod filming *Genius of Charles Darwin*: both courtesy ClearStory; RD at the Wailing Wall, Jerusalem; RD, Lourdes: both Tim Cragg, *Root of All Evil*; RD and gorilla: Tim Cragg, *Genius of Charles Darwin*; RD and Russell Barnes, *Faith Schools Menace*: courtesy ClearStory

RD in a junkyard: photo courtesy author; Daniel Dennett, Sue Blackmore and RD at Memelab, Devon,

2012; making Chinese junks at Memelab: both photos by Adam Hart-Davis, courtesy Sue Blackmore; chair covers and Lalla with cube: all supplied by the author; RD at home, 1991: Hyde/Rex

70th birthday dinner, New College, 2011: photo Sarah Kettlewell

About the Author

R ichard Dawkins was first catapulted to fame with his iconic work *The Selfish Gene*, which he followed with a string of bestselling books. Part one of his autobiography, *An Appetite for Wonder*, was published in 2013.

Dawkins is a Fellow of both the Royal Society and the Royal Society of Literature. He is the recipient of numerous honours and awards, including the Royal Society of Literature Award (1987), the Michael Faraday Award of the Royal Society (1990), the International Cosmos Prize for Achievement in Human Science (1997), the Kistler Prize (2001), the Shakespeare Prize (2005), the Lewis Thomas Prize for Writing about Science (2006), the Galaxy British Book Awards Author of the Year Award (2007), the Deschner Prize (2007) and the Nierenberg Prize for Science in the Public Interest (2009). He retired from his position as Charles Simonyi Professor for the Public Understanding of Science at Oxford University in 2008 and remains a Fellow of New College.

In 2012, scientists studying fish in Sri Lanka created *Dawkinsia* as a new genus name, in recognition of his contribution to the public understanding of evolutionary science. In the same year, Richard Dawkins appeared in the BBC Four television series *Beautiful Minds*, revealing how he came to write *The Selfish Gene* and speaking about some of the events covered in this autobiography.

In 2013, Dawkins was voted the world's top thinker in *Prospect* magazine's poll of over 10,000 readers from over 100 countries.

HARPER LUXE

THE NEW LUXURY IN READING

We hope you enjoyed reading
our new, comfortable print size and found it
an experience you would like to repeat.

Well – you're in luck!

HarperLuxe offers the finest in fiction and
nonfiction books in this same larger print size and
paperback format. Light and easy to read, HarperLuxe
paperbacks are for book lovers who want to see
what they are reading without the strain.

For a full listing of titles and
new releases to come, please visit our website:

www.HarperLuxe.com

HARPER LUXE